Mobile Communication Systems

J.D. Parsons, DSc(Eng), FEng, FIEE
David Jardine Professor of Electrical Engineering
University of Liverpool

and

J.G. Gardiner, PhD, CEng, FIEE
Professor of Electronic Engineering
University of Bradford

Blackie
Glasgow and London

Halsted Press,
a division of
John Wiley & Sons, Inc., New York

Blackie and Son Limited,
Bishopbriggs, Glasgow G64 2NZ
7 Leicester Place, London WC2H 7BP

Distributed in the USA by
John Wiley and Sons, Inc., New York.
Orders from USA only should be sent to
John Wiley & Sons, 1 Wiley Drive
Somerset, New Jersey 08873

First published 1989

British Library Cataloguing in Publication Data

Parsons, J.D.
Mobile communication systems
1. Mobile radio systems
I. Title II. Gardiner, J.G.
621.3841'65

ISBN 0-216-92261-5

Library of Congress Cataloging-in-Publication Data

Parsons, J.D. (John David)
Mobile communication systems/J.D. Parsons, J.G. Gardiner.
p. cm.—(Blackie series in information technology)
Includes bibliographies and index.
ISBN 0-470-212136-6: $44.95 (U.S.)
1. Mobile communication systems. I. Gardiner, J.G. II. Title.
III. Series.
TK6570.M6P37 1989
621.3845—dc19

88-32711
CIP

Photosetting by Thomson Press (India) Ltd., New Delhi
Printed in Great Britain by Bell & Bain (Glasgow) Ltd.

Preface

During the past decade there has been a dramatic change in the nature of mobile communications technology and its impact on the general communications environment. In the 1970s, mobile radio was a minority activity in communications, based on relatively unsophisticated technology. The 1980s, however, have seen the emergence of analogue cellular systems and the definition of future digital systems, and the predicted demand for these services is such that investigations into the use of higher frequency bands have already begun. It is predicted that, by the late 1990s, the 'personal communications' world will have resulted in the majority of adults in Europe and North America being dependent on radio-connected terminals of various kinds for more than 50% of their total telecommunications needs. The technology which will form the basis of this revolution has now been defined, at least in outline, and the fixed and mobile equipment that will be used in systems of the future will bear little resemblance to that available even ten years ago.

It is impossible within the confines of a single, relatively short book to cover all the subject areas needed for a study of this exciting and expanding field of technology. We have, perforce, been selective and have chosen those topics which we believe to be of primary importance at the present time. Even within these topics, however, we have been further constrained to concentrate only on aspects directly relevant to mobile communications, an example of this being Chapter 4 which is concerned with modulation. We have concentrated exclusively on land mobile radio as used in the civilian scenario, and have not covered satellite mobile systems or military mobile radio. Although some topics, such as trunking, are important, we have been able to find room for only a brief treatment of the principles; readers requiring further information should consult the references at the end of the relevant chapter.

In the period up to the mid 1970s, both fixed and mobile equipment was fairly primitive and systems engineering was largely concerned with using this equipment to maximize the use of the available radio channels. In modern systems, the radio channel is still a major factor and, following Chapter 1 which gives some background for readers with little prior knowledge, we deal with the properties and characteristics of the channel with regard to both signals and noise. We have deliberately chosen to present the information about multipath and short-term effects in Chapter 2, ahead of that on median signal strength prediction and long-term effects in Chapter 3; this seems logical to us, although there may be those who would have preferred to see the order of presentation reversed. Modulation is covered in Chapter 4, before the final chapter on the channel, which deals with noise. Again, opinions may differ on the order of presentation, but we felt that some background on modulation was needed before addressing the problem of how impulsive noise affects communication systems in Chapter 5. Diversity reception, a subject that has been researched for many years, is of increasing importance as mobile systems move into the digital era. Chapter 6 is a summary of the principles and practice which, it is hoped, will pave the way for those who wish to delve deeper into the literature.

A great deal of modern systems engineering now focuses on digital signal processing, channel and speech coding, spectrally efficient modulation methods and, perhaps most importantly, integration of the radio segment into complete systems using fixed-line networks with computer management, and with the public switch telephone network (PSTN). A detailed treatment of these topics is well beyond the scope of this book, but cellular systems embrace them all to a greater or lesser extent. In any case, cellular systems are of fundamental importance in their own right and three chapters at the end of the book are devoted to them: Chapter 7 deals with the fundamentals, Chapter 8 discusses analogue systems, using TACS as the primary example and comparing it briefly with other systems; Chapter 9 introduces the digital cellular systems of the future and shows how the techniques mentioned above influence decisions about channel bandwidth, multiple access techniques and system capacity.

Throughout the book we have, in the interests of brevity and readability, declined to reproduce lengthy mathematical derivations which can, in any case, be found elsewhere. Nevertheless, we have tried wherever possible to use the mathematical notation which is in common usage, to make it easier for readers to understand relevant published papers with a minimum of difficulty.

No book of this kind results from the efforts of only the authors whose names appear on the title page. We have been privileged, over the years, to supervise the research work of many students who have tackled difficult problems, come forward with original ideas, stimulated our thoughts and generally contributed much to our knowledge. In the context of this book, particular mention must be made of Dr Anwar Bajwa, a former student and academic colleague of J.D. Parsons, and now a valued friend; he contributed much during the early stages of writing, and his influence and advice are gratefully acknowledged. Finally, grateful thanks are due to our wives, Mary and Sheila, who have encouraged us not only in the writing of this book, but throughout our careers; they have made many scarifices which are much appreciated.

JDP
JGG

Contents

1 Introduction to mobile communications

1.1 Background

Mobile radio plays a very important role within the commercial, business and public safety sectors in almost all industrialized parts of the world. It is generally recognized as one of the most efficient ways to increase effectiveness, reduce operating costs and preserve energy. Growth has been rapid over the past few decades, and the provision of adequate radiofrequency spectrum to meet the increasing demand for licences is of primary concern to national radio regulatory administrations both in the long and short term. For many years, however, the continued demand for channels has not been matched by a corresponding increase in the amount of spectrum available. The allocation of a frequency band near 900 MHz for cellular mobile radio and the release of part of Band III (174–225 MHz) for trunked mobile radio systems in the UK has helped to ease the short-term problem, but there is evidence that even these bands will soon become congested, particularly in the major conurbations, and further action will be necessary.

The history of mobile radio is a story of continuously increasing usage, particularly so since the 1950s when, following the invention of the transistor, it became possible to design and manufacture reasonably light, compact equipment with a much improved reliability. The first use of land mobile radio as we now know it was in 1921 by the Detroit police department for police car dispatch purposes, with the New York police department following suit in 1932. These early systems operated at around 2 MHz, but as the technology improved and the demand increased during the next decade, higher frequencies came into use. In 1933 the Federal Communications Commission (FCC) in the USA authorized the use of four channels in the 30–40 MHz band on an experimental basis, and in 1945, immediately after World War II, during which radio communication technology advanced rapidly, experimental work started at Bell Telephone Laboratories with the object of improving mobile radio services and investigating the use of higher frequencies. In 1964 a new 150-MHz radiotelephone system was made available and this was followed in 1969 by a similar service at 450 MHz.

Rapid advances have also been made in other countries. In Japan extensive use is made of mobile radio for business and commercial purposes. Frequency allocations exist for communication between mobiles and their base, from railway trains to the normal subscriber telephone network, and for radio paging. The total number of mobile transceivers in use exceeds 5 million and new services are being introduced. In Scandanavia, where a fully automatic public mobile telephone system has existed for many years, the demands for the service and the number of calls made have far exceeded expectations. When the Nordic Mobile Telephone system was introduced in 1981 it was estimated that there would be about 50 000 subscribers in Sweden by 1991. That figure was reached in 1984, and the estimate for 1988 is 120 000 subscribers. This story of sustained and substantial growth is repeated in every country where systems have been introduced.

The extent of interest in mobile communications in the UK is apparent from the fact that from small beginnings in 1947 its use expanded rapidly so that 30 years later, in 1977, there were over 200 000 licensed mobile transceivers in the private mobile radio service. Mobile radio usage continues to expand at a rate of about 10% per annum and it is forecast that by the end of the century the number of licensed transceivers may well be approaching 2 million.

1.2 Mobile radio system fundamentals

At first sight, mobile radio systems appear to employ a very wide range of techniques in equipment design, operating principles and system configuration. However, this apparent dissimilarity results largely from relatively minor variations in system implementation which arise from local differences in government regulations relating to land mobile radio equipment, from differences in the operating environment, population density and terrain, and from the differing demands on systems made by a wide variety of users. Beneath these variations lie fundamentals which are common to the vast majority of all operating systems, and this introductory chapter is concerned with describing the types of service available and discussing some of the primary parameters such as choice of operating frequency and type of modulation. In later chapters we return to these and other topics and give them more detailed consideration.

1.2.1 The operating frequency

It is instructive to examine the requirements which an operational mobile radio system must fulfil, so that the constraints which influence subsequent design philosophy can be clarified.

The majority of mobile radio systems exists in and around centres of population, as is to be expected in view of the uses to which systems are put.

Emergency services such as police, fire and ambulance all use mobile radio, as do taxi operators, those who provide maintenance functions of various kinds, public authorities with responsibilities for power distribution, water and associated services, public transport and haulage organizations, and members of the public. Two features of these operations are apparent immediately: firstly, a large number of users must be able to operate at the same time without interfering with each other, and secondly, the distances over which the system must provide communication need not in general be great, since much of the system's operation will be either within the confines of a conurbation or along well-defined major vehicle trunk routes or railway lines. Additionally, it is evident that if mobiles are to communicate freely with their base or each other throughout a given area (which may or may not be the total service area of the system), all the transmitters involved must be able to provide an adequate signal strength over the entire area concerned. Operating frequencies must be chosen in a region of the radio spectrum where it is possible to use efficient, compact antennas that radiate uniformly in the horizontal plane. Since the signal is rapidly attenuated with increasing distance from the transmitter, this also implies that the frequencies chosen must be such that substantial power can be easily and efficiently generated.

The part of the electromagnetic spectrum which includes radio waves extends from 30 kHz (wavelength 10 km) to 300 GHz (0.1 cm). Electromagnetic energy in the form of radio waves propagates outwards from a transmitting antenna, and there are three main ways in which these waves travel, much depending on the frequency of transmission.

The ground wave is that portion of the radiation directly affected by the surface of the Earth. It has two components, an Earth-guided surface wave and a space wave. The surface wave travels in contact with the Earth's surface and is the type of wave that provides reception over large distances in the medium-wave broadcast bands during the daytime. The attenuation of this type of wave is rather high, so the intensity dies off rapidly with distance from the transmitter. The attenuation also increases rapidly with frequency, with the result that the surface wave can be neglected as a means of reliable communication at VHF and higher frequencies.

When the transmitting and receiving antennas are within line of sight of each other, the received signal strength is the combination of a direct and a ground-reflected ray, forming what is commonly known as a space wave. This space wave is a negligible factor in communication at low frequencies, but is the dominating factor in ground-wave communication at VHF and UHF.

The tropospheric wave is that portion of the radiation kept close to the Earth's surface as a result of refraction (bending) in the lower atmosphere. This tropospheric refraction can sometimes make communication possible over distances far greater than can be covered with the ordinary space wave. The amount of bending increases with frequency, so tropospheric communication improves as the frequency is raised. It is relatively inconsequential below 30 MHz.

The ionospheric or sky wave arises from radio waves which leave the antenna at angles somewhat above the horizontal. Under certain conditions, these waves can be sufficiently refracted by the ionospheric layers high in the Earth's atmosphere, to reach the ground again at distances ranging from zero to about 4000 km from the transmitter. By successive reflections at the earth's surface and in the upper atmosphere, communications can be established over distances of thousands of miles, and on occasions radio signals can travel completely around the Earth. Propagation by the regular layers of the ionosphere takes place mainly at frequencies below 30 MHz.

For two-way mobile radio, it is seldom that the vehicle or portable antenna has a direct line-of-sight path to the base station. Radio waves will penetrate into buildings to a limited extent and, because of diffraction, appear to bend slightly over minor undulations or folds in the ground. Fortunately, due to multiple scattering and reflection, the waves also propagate into built-up areas, and although the signal strength is substantially reduced by all these various effects, sensitive receivers are able to detect the signals even in heavily built-up areas and within buildings.

However, for point-to-point links and particularly for broadband links carrying many channels, a true line-of-sight path between antennas is generally required. Tall masts or towers carry the antennas well clear of local obstructions. The role of point-to-point links in mobile radio systems will be discussed later, in section 1.4.1. They are usually operated at high UHF, and at these frequencies it becomes possible to design directional antennas which can concentrate the radiofrequency energy in the wanted direction.

Two-way radio has chiefly developed around VHF, and latterly UHF, for very good reasons. In a city police force, for example, the central police station receives reports from incidents in the city area—often by emergency telephone calls. The police are obliged to investigate the incident, so the control-room operator puts out a call to a police officer believed to be in the area. He may be on foot with a personal radio, or in a car fitted with mobile radio. In either case, on receipt of the call, he acknowledges it, investigates the incident, and reports back by radio on the situation he has found. Because the police use frequencies in the VHF and UHF bands that are specifically allocated for their use only, they do not interfere with other users. However, all police officers on duty who carry a receiver tuned to appropriate frequency will hear the calls as they are broadcast.

If the message had been radiated on HF, then it is possible that the signals could have been detected at distances of several hundred kilometres, which of course is unnecessary and not the intention. The VHF and UHF bands are therefore used for mobile radio because of their relatively short-range propagation characteristics. The same frequency can be used in another city a few tens of kilometres away, if geographical conditions permit, without one user interfering with another.

In each country, use of the radiofrequency spectrum is controlled by an authority, frequently the PTT (in the UK it is the Department of Trade and Industry, and in the USA it is the Federal Communications Commission) whose job is to divide the usable radiofrequency spectrum into individual channels, and to license users for the use of a particular channel. Because of the short-range propagation characteristics of VHF and UHF, it is possible to allocate the same channel to different users in areas separated by distances of 50–100 km with a substantial degree of confidence that, except under anomalous propagation conditions, they will not interfere with each other.

1.2.2 Modulation

Having decided on an operational frequency it remains to determine the modulation method that will be used to convey the information over the radio network. Returning to the basic constraint of the need to operate a number of channels simultaneously, the simplest technique is to divide the available spectrum into a number of channels, each wide enough to carry the information but as narrow as possible consistent with that need. At the present time in the UK, channels in the VHF band are 12.5 kHz wide, whilst at UHF the spacing is 25 kHz.

For conventional mobile radio systems principally intended to carry speech, the choice lies between amplitude modulation (AM) and angle modulation. The latter can take two forms: either the instantaneous frequency or instantaneous phase of the carrier can be made to vary in sympathy with the modulating signal giving rise to frequency or phase modulation (FM or PM). These systems are discussed in Chapter 4.

1.3 A simple mobile radio system

We are now in a position to visualize the configuration of a basic system which would comprise a number of transceivers (transmitter/receivers), either hand-held or vehicle-mounted, communicating with, and to some extent controlled by, a base station. Such systems are often termed 'dispatch systems' and are used extensively by police forces, taxi companies and service organizations who operate fleets of vehicles. One further parameter remains to be determined, and that is the mode in which the system will operate. This amounts to a choice between operating the base–mobile and mobile–base paths, at the same frequency (but sequentially), termed 'single-frequency simplex'; at two frequencies sequentially, termed 'two-frequency simplex' (sometimes 'semi-duplex'); or at two frequencies simultaneously, known as 'duplex' operation.

Simplex implies the use of a press-to-talk switch, since stations cannot transmit and receive simultaneously. If all the stations use the same frequency for transmit and receive, then the system is known as single-frequency simplex. It has the advantage that mobile stations can speak directly to each other, provided that they are within radio range. Because vehicle antennas are only a few metres above ground, vehicle-to-vehicle working is very variable depending on the terrain between the cars, and can vary from a few hundred metres to several kilometres. Although it might seem a good thing for vehicles to have the facility for direct communication, it can be a serious disadvantage because of the risk of interference to other users of the same channel. A further disadvantage is that two vehicles can engage the whole system and make it difficult for the base station to exercise control. For two-frequency operation, the fixed station transmits on one frequency and the mobile stations on a second frequency, generally at least 5 MHz removed from the first. The mobile stations cannot hear each other. This system is normal practice in the UK on private mobile radio systems.

'Semi-duplex' is a somewhat ambiguous term sometimes used for two-frequency simplex. Another interpretation implies the use of a duplex base station with mobiles operating two-frequency simplex.

'Duplex' implies the use of a two-frequency system where the transmitter and receiver are both operated simultaneously (as in a telephone or cellular radio system). Frequency separation is essential between the 'outward' and 'return' channels to avoid the transmitter breaking in or desensitizing the receiver. Other precautions are also needed to prevent breakthrough, such as the use of two antennas, or alternatively one antenna with a duplexer (a special arrangement of RF filters protecting the receiver from the transmit frequency). Duplex is an essential feature of cellular radio systems, and in the UK 900-MHz system, known as TACS (Total Access Communication System), the transmit and receive frequencies are separated by 45 MHz. Otherwise, duplex mobiles are not common, but are sometimes required as a part of a system involving communication through a telephone network or exchange. The mode of operation is illustrated in Figure 1.1.

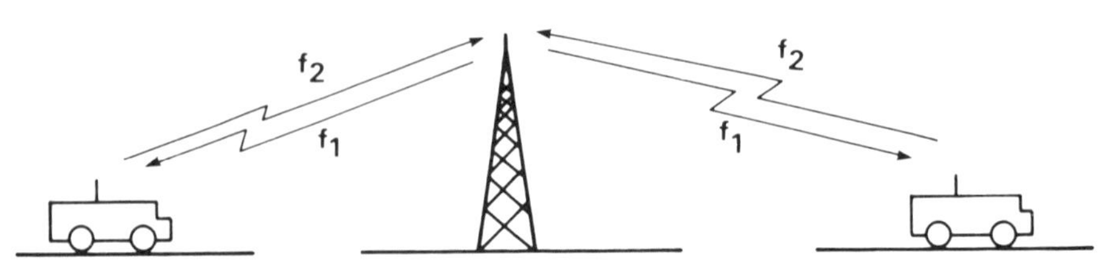

Figure 1.1 Two-frequency simplex, with f_1 used for the 'outward' path and f_2 for the 'return' path. For duplex operation both frequencies are used simultaneously.

1.4 Practical communication systems

The components of a simple mobile-radio system are now defined although, as we have seen, a number of alternative combinations of facilities exist, such as AM or FM, simplex (single- or two-frequency) or duplex operation. However, a number of additional requirements exist; these must be identified and implemented before really useful and flexible systems can be assembled.

The first and most essential requirement is for a means of operating the base station from a point remote from the site where it is installed. The need for this facility can be readily appreciated when considering the propagation characteristics of VHF transmissions. In some mobile radio systems, it is desirable to achieve maximum range and to minimize variations in received signal strength over the operating area of the mobiles, and in such cases it is essential that the base station transmitter and receiver should be located at a point which has maximum elevation relative to the surrounding terrain. Almost invariably the control centre of the mobile radio system is an office which is unfavourably situated in relation to the required coverage area, and some form of link equipment is therefore required.

1.4.1 Links

There are two ways in which a link can be established between the system control point and the remote site housing the transmitter and receiver, either by land lines or by a radio link (or a combination of both). In localities where a comprehensively developed line telephone system exists, it has been common in the past to employ the former alternative, since a very high degree of reliability and circuit continuity is available. In remote rural areas where few lines are available, or because there are local problems with difficult terrain, there may be no alternative to radio links. From the technical point of view, there are some problems in interfacing radio systems with line communication facilities, since it is inevitable that stringent regulations must be observed and specifications met by any equipment connected to the public-switched telephone network. However, these requirements can usually be met with equipment which is less expensive than the alternative radio link equipment.

As far as radio links are concerned, it is immediately apparent that the requirements imposed on fixed-link equipment differ markedly from those relating to base and mobile equipment. In particular, the link path remains fixed at all times, and since it is usually possible to establish a line-of-sight path, the more demanding propagation characteristics of UHF and microwave frequencies can be accommodated. Further, omnidirectional coverage is not required, so that highly directional antennas can be utilized to concentrate the transmitted energy along the required path, resulting both in high gain and the avoidance of interference with other users. Figure 1.2 shows a basic system which may be regarded as a starting point adequate in itself for local operations and capable of much further extension.

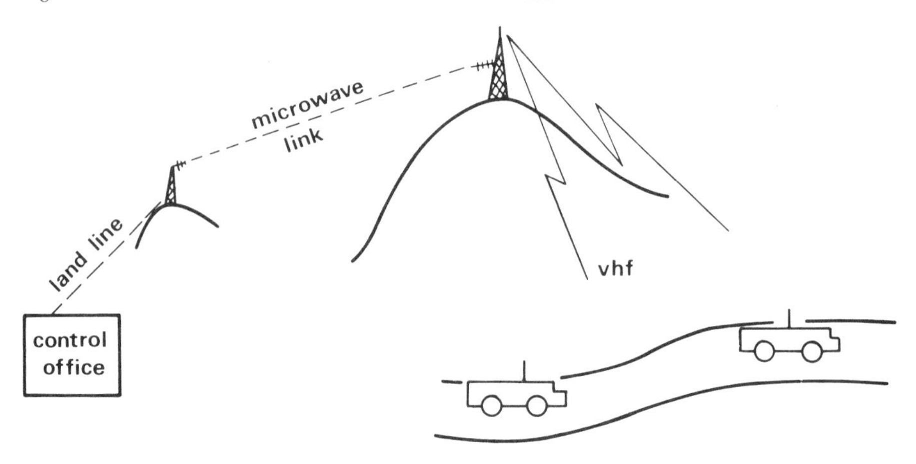

Figure 1.2 Simple link system in which a VHF base station on a high site is operated from a control office some distance away using a land line and radio link.

1.4.2 Repeaters

We have just discussed a situation in which a radio link is used to replace a telephone line, and it is equally to be expected that other features of the line communication network will have their radio counterparts. The repeater (amplifier) is common in line communication systems, but the constraints imposed by the radio system differ from those in line communications, principally because it is very difficult to obtain the high isolation between the receive and re-transmit antennas that is vital to avoid oscillation in 'on-frequency' repeaters, i.e. those using only one frequency. Problems can be minimized if the receive and re-transmit frequencies are different, but great care must be exercised in applying 'on-frequency' repeater techniques to extend range.

1.4.3 Talk-through

Arguments for and against two-frequency simplex operation have already been advanced, but a disadvantage that remains is that, whereas mobiles can communicate readily with base, direct contact between mobiles is impossible. Clearly an operator at a base station can relay messages from one mobile to another, but it is much more desirable that this should be achieved automatically. The requirement therefore is that when one mobile wishes to communicate with another, the message received at base must be simultaneously re-transmitted for the other vehicle to receive. In other words, the output of the base-station receiver must be applied to the input of the base transmitter, which must be energized while the incoming message is being received. Such operation is referred to as 'talk-through' and is a feature of many practical systems.

There are advantages and disadvantages to the use of talk-through, one obvious advantage being that mentioned above. A second is that, in the usual situation of line or radio link operation of a remote base, the system can be made to automatically adopt the talk-through mode of operation in the event of link failure, so that a measure of useful operation can continue. Without talk-through the system would be totally inoperative under link-failure conditions.

The disadvantages of the system are associated with the problem of controlling the talk-through function at the remote site, since excessive use of the base facility in this mode clearly interferes with the normal function of the system in conveying information from base to mobiles. Suitable strategies must therefore be adopted to permit adequate control of the system and avoid unwanted triggering of the talk-through condition at base.

1.4.4 Communication with selected mobiles

Thus far we have considered the simple case of a base station communicating with many mobiles on a single channel, i.e. using a single frequency for all transmissions in the case of single-frequency simplex, or in the case of two-frequency simplex or duplex, one frequency for base transmitter and mobile receivers and a second for base receiver and mobile transmitters. This implies that whenever the base-station operator has a message for one mobile, all other mobiles in the system also receive the same message, although they do not need to respond. It is clearly desirable in many cases to confine the message transmission to the single mobile or subset of mobiles to which the information is directed, without calling other vehicles in the system. One solution would be to allocate different channels to different groups of mobiles operating from the same base, but this is far from ideal both on the grounds of equipment cost, and more particularly because of the resulting inefficient use of the already crowded radio spectrum. Moreover, if it were possible to call one group of mobiles without interfering with a second group on the same channel, then more than one user could share the same frequency or pair of frequencies in the same area. Such operations can, in fact, be achieved, and the techniques employed are generally referred to as 'selective calling' or 'selective signalling'. They utilize a common feature of the mobile radio receiver usually referred to as the 'mute' or 'squelch' circuit.

1.4.4.1 The receiver mute facility. In normal system operation, neither base nor mobiles have any foreknowledge of the arrival of a message, so that receivers (base and mobile) must be in continuous operation. Further, since signal strength varies greatly over the normal operating area, as we will see in Chapter 2, the receiver must have high gain capability, with comprehensive automatic gain control (AGC) circuits, so that the operator does not need to continually adjust the audio output level control. It can be readily appreciated, therefore, that in the absence of transmissions from base, when the mobile receiver has no input other than noise, the AGC circuits increase the gain to maximum and a high level of audio noise is produced, particularly in the case of FM receivers. This noise quickly becomes offensive and a circuit is therefore introduced which detects the presence of a carrier from the base transmitter and only allows the receiver to produce an audio output when a transmission is being received. This 'mute' circuit silences the receiver until a message is received and the level of signal required to 'open the mute' can be adjusted to avoid unwanted noise outputs from local interference.

Here, therefore, is a convenient means of achieving selective signalling since if it can be arranged that the mute opens only on the required 'selected' receiver, other receivers on the same channel produce no output.

1.4.5 Continuous tone controlled signalling system (CTCSS)

This system is conceptually the simplest method of obtaining selection among mobiles on the same channel. The transmitters (base and mobile) transmit carriers which are continuously modulated by a tone at a frequency too low to produce an audible response from the receiver, but which can be recognized by the appropriate receiver as corresponding to a wanted transmission. Consider two bases A and B, A having CTCSS tone 77 Hz, B having tone 123 Hz, but sharing a common carrier frequency. When A transmits, all the mobiles of both the A and B systems receive the signal, but only the mobiles of A have their mutes opened, the mobiles of B having detected a 77 Hz modulation rather than 123 Hz and remaining therefore in the muted condition. The tone frequencies available for assignment lie in the range 67–250 Hz.

Clearly, if base station B attempted to transmit at the same time as A, confusion would result, so that operators must first check the status of their allocated channel before attempting a transmission.

1.4.6 Selective calling

In common usage, the term 'selective calling' has been associated with a more elaborate system than that described above, but one which has much greater capability to identify a particular mobile or group of mobiles in the presence of others on the same channel. In essence, the basis of selective calling is that of coding the carrier in such a way that the receiver can recognize transmissions intended for it and reject others. It is a vital and essential facility in cellular radio systems where the receiver that has been called must be selectively and uniquely addressed. The CTCSS method is restricted in the number of separate codes available by the requirement for continuous transmission—there can only be as many identifiable codes as there are allocated tone modulation frequencies. The alternative is to begin each transmission by modulating the carrier with coded information identifying the required receiver, which must then open its mute and remain in the active condition during the subsequent message transmission. In this way a great many separate codes can be constructed.

At present three separate systems of code formation are in use in private mobile radio. These are known by initial letters ZVEI, CCIR and NATEL, but all have a great deal in common in that they consist of a sequence of five tones each transmitted for a short interval (10–50 milliseconds), the correct sequence of tones identifying the required receiver. Taking the three coding systems together, a total of 11 tones is available for each of the five 'time slots' which form the code. Since accurate timing is not available, no tone can be repeated in two consecutive time slots, so that a total of $11 \times 10 \times 10 \times 10 \times 10 = 110\,000$ separate codes (or addresses) is possible.

Inevitably, a complicated circuit is required to decode the received information, to form the code initially and to maintain the receiver in the correct condition during message transmission, but integrated circuits which perform the necessary functions accurately and reliably are widely available. Once such a circuit has been developed, many additional features are readily incorporated, and two such features are particularly attractive.

First, it is often desirable to communicate with several mobiles simultaneously (group calling). This is accomplished by adding an additional tone to the eleven required for the address codes, the occurrence of the 'group tone' in various positions in the last three digits of the address code enabling group calling to be accomplished over 10, 100 or 1000 mobile units.

An example of group calling might occur as follows:

Input Code	Activates receivers coded	Groups of
25784	25784 only	1
2578G	25780–25789	10
257GR	25700–25799	100
25GRG	25000–25999	1000

G is the group tone, R is the 'repeat tone' used because of the restriction mentioned previously that the same tone cannot be repeated in consecutive code digits.

Second, it is often convenient to have confirmation from the mobile that a message has been received and this is easily achieved by making the mobile transmitter respond to the received code corresponding to its own address by transmitting its own code in reply. This process is frequently referred to as 'vehicle identification'.

1.5 Paging

We now have at our disposal all the necessary components for assembly of a range of systems of varying degrees of complexity, but one additional special case of selective calling techniques deserves separate consideration.

A paging system is a limited form of mobile radio in that two-way communication is not normally available. The objective of a paging system is to notify personnel working within range of the base station that they are required to report by telephone to a known fixed location, often their office, to receive further instructions. The requirement therefore is for a selective calling system which will initiate a response in an individual receiver carried by the user. Usually paging systems are required to have only limited range, 2–5 km, so that UHF transmission is appropriate and the response is indicated by an audio tone or a vibration. There are several levels of complexity in paging systems.

1.5.1 Simple paging systems

Figure 1.3 shows a simple paging system. The operator at the base station transmits the code of the person required by a simple automatic process, his receiver responds if within range, and contact is established.

1.5.2 Voice paging system

In the system shown in Figure 1.4, the simple system is expanded to permit a voice or alphanumeric message to be transmitted from base to the person required, but there is no facility for response. In this way, simple instructions can be conveyed without the necessity to establish telephone contact, such contact being needed only if the person paged is unable to carry out the instructions passed.

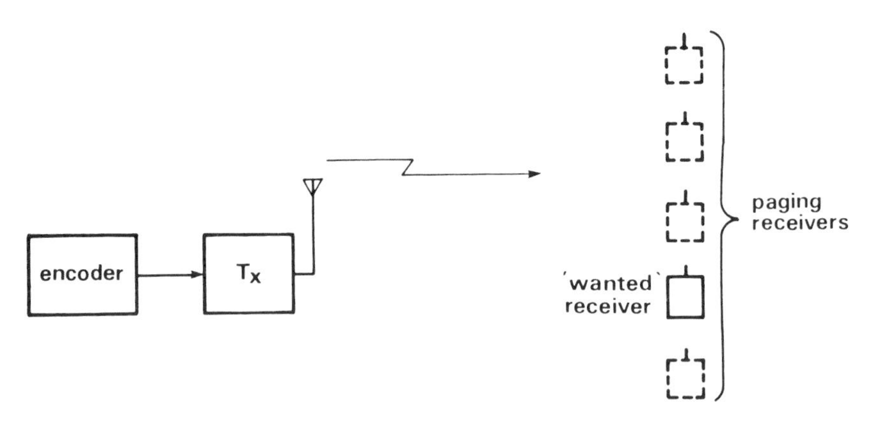

Figure 1.3 Simple paging system.

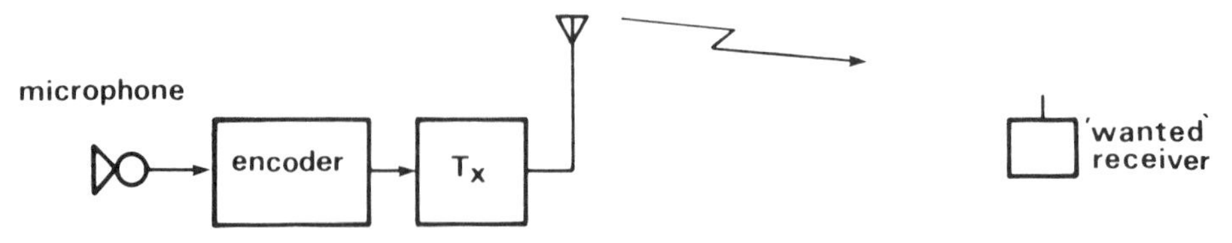

Figure 1.4 Expanded paging system in which the 'wanted' receiver can accept a voice message (or some alphanumeric characters).

1.5.3 Overlay paging system

In Figure 1.5, a paging facility is superimposed on a conventional two-way mobile radio system to extend the range of access to personnel who are frequently out of range of a local paging transmitter, working instead from a vehicle, and engaged typically on a service and maintenance operation. This is described as overlay operation.

1.5.4 Wide-area paging

Local-area paging grew as an extension of the need for companies occupying large sites to contact personnel who were not at their desks or normal places of work. However, the type of facilities offered by such systems were soon in demand by organizations operating over much larger areas, and paging was seen as having very great advantages in connection with calling firemen or doctors in the case of an emergency. Initially, the requirement of the system was for very sensitive receivers to improve the range whilst maintaining the low-cost transmission networks, and wide-area paging was soon perceived by manufacturers as an extension of the mobile radio business.

The concept of a paging service operational over areas much larger than those relevant to a private or professional organization was first introduced in North America in 1963 and subsequently in other countries, notably in Europe and Japan. Because of the unconnected ways in which these systems were designed and developed, various incompatible signalling systems were developed [1]. However, during the 1970s the British Post Office (now British Telecom) took the lead in proposing and eventually implementing a paging

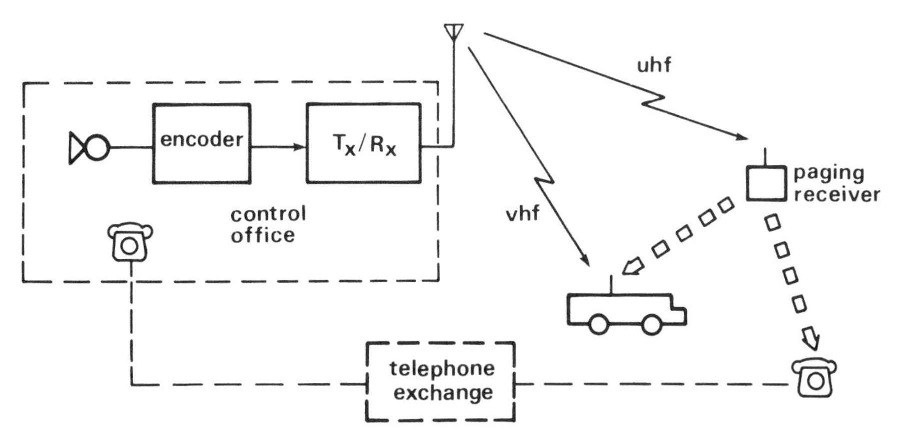

Figure 1.5 Overlay paging system used to contact drivers who are away from their vehicles. Contact can be made either via the vehicular VHF radio system or by means of a convenient telephone.

code and signalling format that would meet performance requirements on a national basis and could be internationally accepted. This was initially known by the initials POCSAG from the name of the group (Post Office Code Standardisation Advisory Group) which was formed to establish the concept. The POGSAG code [2] was initially developed at Philips Research Laboratories, and has now gained international recognition by the CCIR. It has the capacity to cope with eight million pagers, provides the facility for alphanumeric messages, and can operate in 12.5 or 25 kHz channels. Up to the present time the data rate has been fixed at 512 bits/second, but because of the increased paging traffic in the UK, especially resulting from the growing use of message pagers and the restriction of available spectrum, BT have led the introduction of enhanced POCSAG in which the bit rate will be increased to 1200 bits/second with a corresponding increase in capacity. Paging systems are extensively described in reference 3.

1.6 Portables

The next obvious stage in the development of the paging type of system is to include in the equipment carried by people a transmitter of adequate power output to permit direct voice response from the person paged back to base. Such a complete transceiver is termed a 'portable' to distinguish it from the 'mobile' equipment which is permanently attached to a vehicle. Many compromises must be accepted in the design and operation of portables, but ranges of similar order to that achievable with paging systems can be obtained from equipment having output transmitter powers of around 500 mW. This can be generated readily without the need for excessively large and heavy batteries in the portables.

1.7 Dialling systems

A large portable system employing a comprehensive selective calling strategy approaches in many respects the address-accessing capacity of a conventional line telephone system, in that the operator at base effectively 'dials the number' of the portable and then holds a conversation with the person required. It is therefore a small step conceptually to systems in which these facilities are extended to include a local PAX (Private Automatic Exchange). In practice, the interfacing of radio and local line systems is readily achieved, since the only additional function required is to convert the received coded address information from the radio system into dialling impulses for the PAX and, conversely, to convert normal telephone extension signalling into selective calling address codes for transmission to portables and mobiles. Systems having this facility are readily available in the UK; local variations between alternatives operating in other countries are of relatively minor significance.

1.8 Radiophone services

Once a system capable of accessing a local line network has been developed, it is a small step in principle to extend the operation to the national telephone communications systems. In practice, however, such an extension is a major undertaking for reasons which are historical as much as technical. The development of line and mobile radio systems have been totally separate and unrelated for so long that completely different codes of practice, engineering philosophies and standards, terminology and administrative procedures have become established in the two fields, and the development of totally automatic interfacing facilities between radio equipment and the telephone network has been a slow process.

The obvious and simplest way to circumvent the automatic interface problem was to provide telephone operators at each base station to make connection with the line system manually. Each user had a multichannel mobile in his vehicle and contacted the operator at the base station in the vicinity on any free channel. The operator then dialled the required number and made the connection with the mobile. This system was clearly inefficient in the utilization of radio channels, since an extended 'air-time' was required to establish the call before conversation could commence. Nevertheless, the system proved to be popular with the public, despite the high cost of the equipment hire and the high cost of operator-initiated calls. In the UK, a service of this kind was provided for several years by British Telecom, and was replaced subsequently by a fully automatic dial service. Nowadays cellular radio systems have been introduced; these provide a much increased capacity and are compatible with hand-portable equipment.

1.9 Channel sharing

Traditional private mobile radio services have developed on the basis of an exclusively allocated radio channel for each user, although for some years the radio regulatory authorities have not guaranteed the majority of users this exclusive facility. As far as the user is concerned, exclusive use of a channel gives a large measure of privacy, since there are no other users on the same channel, and with two-frequency simplex operation a single radio receiver can only be tuned to one half of the conversation. On the other regulatory point of view since, with certain exceptions, the channels are actually in use only for a very small percentage of the time. In the search for more efficient ways of using the available spectrum it is therefore vital to find methods by which channels can either be shared, or can be allocated in a more satisfactory manner.

We have already discussed, in section 1.4, CTCSS and selective calling which allow some channel sharing while retaining a measure of privacy. However both these methods have some disadvantages since with CTCSS for example, base station operators have to check the status of the channel to ascertain whether it is already in use. A much more advanced technique, which has been made possible by modern technology, is known as dynamic channel allocation. Because of its similarity with standard practice in telephone systems it is also commonly known as trunking. Dynamic channel allocation (trunking) allows a much improved utilization of channels—it is an essential feature of cellular schemes which will be described later and the trunked radio systems to be used in the UK in Band III [4].

1.9.1 Trunking

The basis of trunking is that if a set of channels is available for use in a certain geographical area, then each user is allocated exclusive use of one of those channels on demand, but only when he needs to make a call. When the call is completed the channel is returned to the pool so that it can be allocated to another user. This means that each mobile has to be able to tune to all the available channels, but modern frequency synthesized oscillator circuits allow this to be done at reasonably low cost. In addition, mobiles have to be able to signal to the central control that they wish to initiate a call so that a channel can be allocated. In large systems such as cellular radio, one or more channels within the available set is designated as a control channel. Quiescent mobiles automatically tune to this channel and use it for communication with the control so that they can initiate instructions or respond to them. However, in small systems this would be uneconomical, and communication with the central base is usually carried out on one of the channels that is not currently in use for voice transmission.

The reason why a dedicated signalling channel becomes desirable in larger systems is because of the importance of signalling time. The problem is that the time which elapses between channels becoming free and their being allocated again must be short compared with the rate at which channels become available. Otherwise a queue of unused but available channels builds up merely because the signalling is not fast enough. For example, in a ten-channel system with an allocation time of 2 seconds, almost half of the system capacity is lost in this manner. The time taken to assign a channel includes not only the time for the exchange of messages but also, in a small system where the mobile has to scan all the channels, the time taken for the mobiles to locate the signalling channel. This factor assumes more importance as the number of channels is increased, and eventually it becomes more efficient to fix the signalling channel on a permanent basis. The point at which the loss of capacity due to signalling efficiency is greater than the loss due to underutilization of the signalling channel, is about six channels.

A further problem affecting signalling systems is that of overcoming interference between two or more mobiles which try to signal at the same time. There are two obvious solutions. The first is a 'polling' system in which each mobile is assigned a particular time slot in which to signal if it wishes to send a message. This slot may be derived by counting from some initial timing signal, or each user may be invited to transmit in turn by the base station. The disadvantage of polling is that the signalling load depends on the number of users who have to be polled rather than on the traffic offered. This is clearly inefficient when there are a large number of users, each offering traffic only occasionally, although it may be effective when the number of users is small. It is also difficult to increase the number of users on the system after the initial design has been completed. A second solution is the contention or 'aloha' type of system. In this type of system, every message contains error check bits allowing the base station to determine whether the message was corrupted by a clash with a simultaneous transmission from another source. If the signal is received successfully an acknowledgement is sent; if it is not, the mobiles concerned repeat their messages but with a random delay between the first and second attempts. The process continues until the message get through. There are many different kinds of aloha systems [5], with efficiencies ranging from 18% for simple systems to over 80% for the more complicated schemes.

A detailed treatment of trunking is well beyond the scope of this book. However, any consideration involves two quantities, load and grade of service. Load is a basic concept in traffic theory and is the product of call rate and call duration time (holding time). It is a dimensionless quantity and is measured in Erlangs. One Erlang is the traffic intensity in a channel that is continuously occupied, so a channel occupied for 50% of the time carries a load of 0.5 Erlang, etc. Grade of service is a term used to quantify the extent to which congestion occurs in any system and is typically expressed as the probability of finding congestion or the probability of exceeding a certain delay before a connection can be established. In practical systems, the values quoted for load and grade of service relate to the 'busy hour', i.e the continuous one-hour period in the day during which the highest average traffic density is experienced.

Load and grade of service are related through various formulae developed by Erlang [6], who laid the foundations for modern telephone traffic theory. The Erlang C formula gives the probability of a call experiencing a delay as

$$p_{\mathrm{d}} = \frac{A^{C}}{A^{C} + C!\left(1 - \dfrac{A}{C}\right)\sum\limits_{k=0}^{C-1} A^{k}/k!} \tag{1.1}$$

where A is the total traffic load in Erlangs and C is the number of trunked channels.

The probability that delayed calls will have to wait more than a specified time t is

$$P(\mathrm{W} > t) = \exp\left[-\frac{(C-A)t}{H}\right] \tag{1.2}$$

where H is the mean call duration time (holding time) in the busy hour. Thus the probability that any call will be delayed for a time greater than t can be obtained from eqns (1.1) and (1.2) as

$$P(W > t) = p_{\mathrm{d}} \exp\left[-\frac{(C - A)t}{H} \right] \tag{1.3}$$

Derivation of the Erlang C formula is based on a number of assumptions:

(i) The number of users is infinite
(ii) Intervals between calls are random
(iii) Call duration times (holding times) are random
(iv) Call set-up times are negligible
(v) Delayed calls are placed in a queue and serviced on a 'first-come-first-served' basis.

A practical radio system will not fulfil all these criteria, but nevertheless, the formula has been shown to provide a good estimate of the global channel requirements in a system, and eqn (1.3) can be taken as an expression for the grade of service available.

As an example, Figure 1.6 shows the percentage probability that a call will experience a delay greater than 20 seconds as a function of the load per channel, $a(= A/C)$, if the average holding time per call is also 20 seconds [4]. The benefits of trunking are immediately obvious, since Figure 1.6 shows that for any probability of delay the traffic load per channel, a, increases, as the number of trunked channels increases. The increase continues as the number of trunked channels becomes greater, but the rate of increase becomes progressively smaller. Systems with more than about 20 channels in a group exhibit very little extra benefit over 20-channel groups. The information in Figure 1.6 can be used to estimate the mean waiting time and the total number of mobiles that the system can accommodate. For the latter calculation it is necessary to make an assumption about the number of calls each mobile makes during the busy hour. If one call is assumed, then a single-channel (untrunked) system offering a 30% grade of service can accommodate 90 mobiles with a mean waiting time of 20 seconds, whereas a 20-channel trunked system with the same grade of service can accommodate 3430 mobiles with a mean waiting time of 16.4 seconds. Although the example shows that the number of mobiles that can be accommodated far exceeds the pro-rata increase in the number of channels, it does assume that all the channels are used solely for message purposes. In practice, an allowance is necessary for call set-up and the mechanism of channel allocation which will reduce some of the benefits of trunking.

Finally, it is worth pointing out the non-linear nature of Figure 1.6 and the fact that the 'knee' becomes sharper as the number of channels is increased. This affects the relationship between the number of mobiles per channel and the grade of service. Calculation shows that, for a single-channel system, a

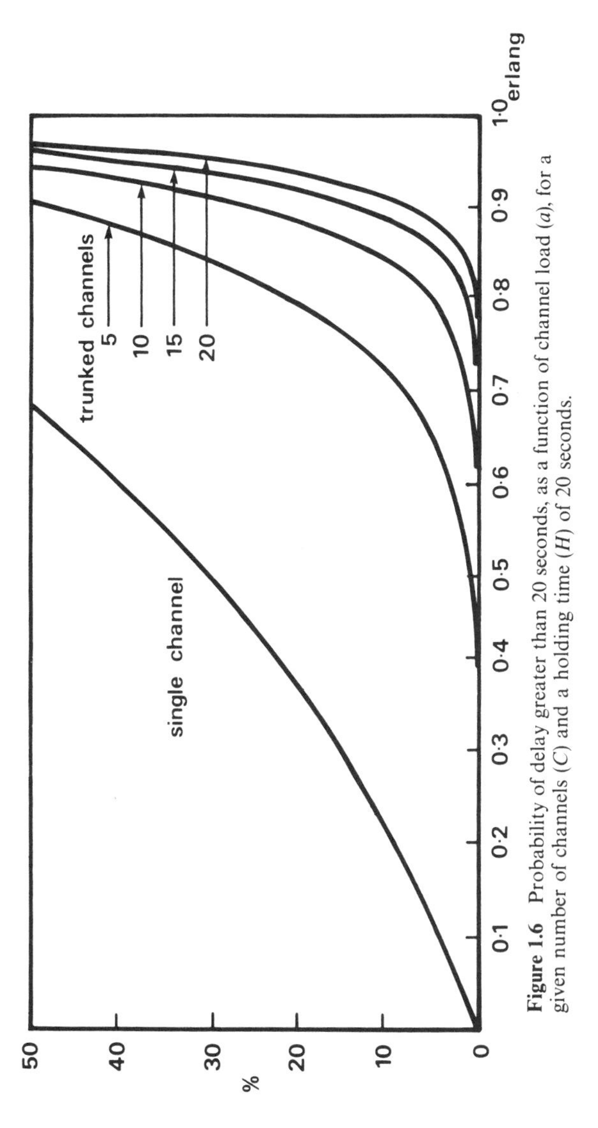

Figure 1.6 Probability of delay greater than 20 seconds, as a function of channel load (*a*), for a given number of channels (*C*) and a holding time (*H*) of 20 seconds.

reduction in the number of mobiles from 90 to 22 improves the grade of service from 30% to 5%, i.e. a single channel can accommodate four times as many mobiles at a 30% grade of service as at a 5% grade of service. On the other hand, a 20-channel system which provides a 5% grade of service to 3185 mobiles can tolerate an increase to only 3430 (about 8% more) before the grade of service is reduced to 30%.

The basic configuration of a trunking system is shown in Figure 1.7. The central base station *B* is able to transmit and receive on any channel available in the system. It receives requests for service from users in vehicles, who gain access by radio and from users at fixed locations *F* who may gain access by radio or land line. If only radio is required the system serves as a multichannel repeater; if line connections are also made to the fixed stations, a switching matrix *M* will be needed to interface between the land lines and the radio channels. The central processor *C* has three principal functions: to receive and

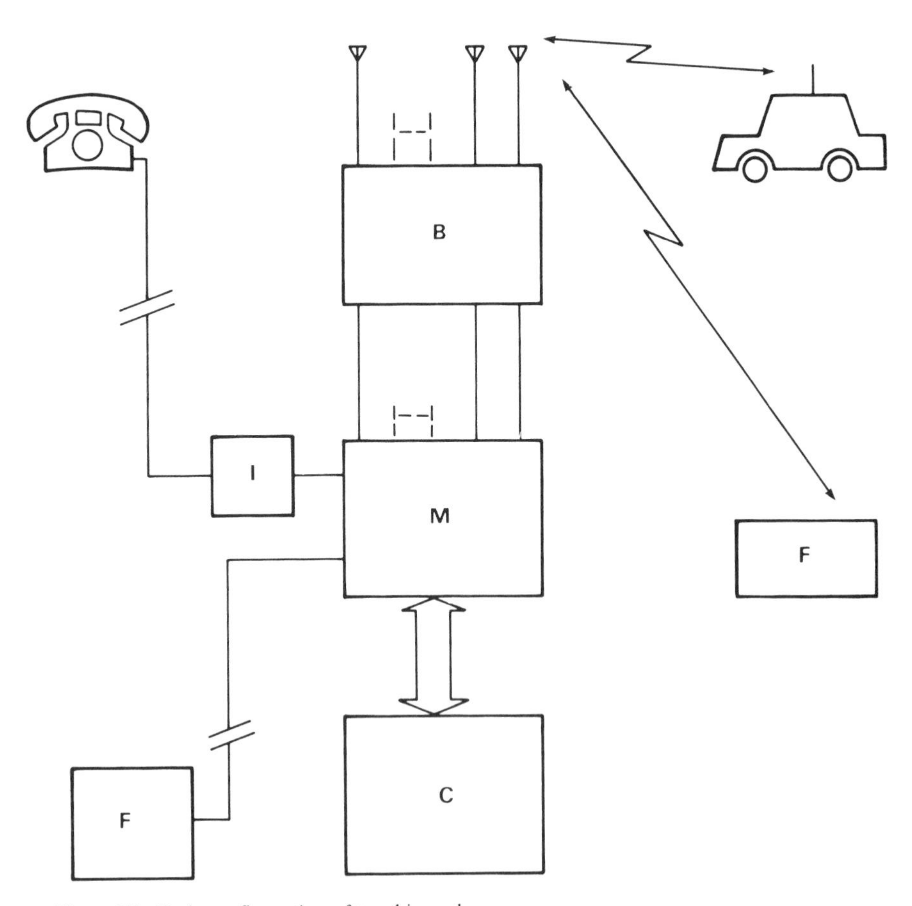

Figure 1.7 Basic configuration of trunking scheme.

process requests for service, to form an orderly queue (if queuing is allowed in the system) and to allocate free channels to the mobiles concerned, making the necessary connections through the switching matrix. Various protocols have been established for the necessary signalling [5.7].

1.10 Area coverage techniques

Many mobile radio telephone and paging systems are required to operate over areas too large to be economically covered by a single base station transmitter, or there may be topographical features which restrict the coverage from a given base station. The requirement therefore exists for some kind of 'area coverage technique' using a number of transmitters. It is desirable that if more than one transmitter is used to cover a given area, the different transmitters should all operate within the same radiofrequency channel, partly for reasons of spectrum utilization, but also to ensure that any receiver within the overall coverage area can be paged without the necessity for channel switching.

There are a number of ways in which area coverage can be achieved using multiple transmitters, and various advantages are claimed for the different methods. However, it is generally accepted that simultaneous operation of all transmitters is better than sequential transmitter operation, provided certain conditions are met, and a technique known as 'quasi-synchronous operation' has been evolved as a solution to this particular problem. Quasi-synchronous operation is possible using either AM or FM transmitters.

The first co-channel AM area coverage scheme was proposed in the late 1940s. In those days channel separations were wide (100 kHz), and transmitters were operated with carrier frequency offsets of several kHz so that the heterodyne beats between the various transmitters fell at frequencies higher than the maximum frequency of the recovered audio signal. In order that a multistation scheme can operate entirely within the confines of a modern 12.5 kHz channel, the transmitter frequency spacing has had to be reduced so that the heterodynes between carriers occur at a frequency lower than the minimum frequency of the received audio signal, and co-channel area coverage schemes with this sub-audio carrier spacing are generally termed quasi-synchronous.

There are some significant differences between the techniques used in AM and FM receivers, which lead to substantial differences in performance under quasi-synchronous conditions. For example, a FM receiver will normally produce a high level of audio noise at its output when the input signal level is low, whereas under the same conditions an AM receiver will produce a very low level of audio noise. In addition, FM receivers normally rely on hard limiting to cope with variations of the input signal level, and a limiter is an essential part of such a receiver. An AM receiver is normally equipped with an automatic gain control circuit. It is apparent that in FM systems the modulation parameter (phase or frequency) and the signal level (amplitude) are quite different. However, in an AM system, changes in either the modulation or in the signal level affect the same parameter, amplitude. Therefore, under conditions when the mobile receiver is subjected to signal components from more than one transmitter, modulation will be additive in the AM case whereas the situation with FM receivers is much more complex.

Quasi-synchronous AM schemes have been found to operate most successfully when the frequency offsets between the transmitters are a few Hz. The AGC system in an AM receiver is designed to compensate only for variations in the mean signal level and in practice, as will be seen later, these occur relatively slowly. It is considered undesirable to increase the speed of response of the AGC system, as this would have a detrimental effect on any sub-audio signalling systems such as those used for selective calling. Careful matching of the transmitter modulation indices is employed in quasi-synchronous schemes, but the accuracy of matching is not highly critical. Deterioration of performance occurs gradually with mismatch in AM schemes and this significantly eases the task of installation. In addition, with AM schemes, it is relatively simple to employ 'fill-in' base stations to improve coverage in areas where reception would otherwise be very poor. In an area where the receiver is subject to weak signals, the effect of quasi-synchronous operation is to cause a periodic increase in the background noise level at a rate equal to the offset frequency. This has only a minor detrimental effect on speech transmissions, but is potentially more serious when high-speed data is being transmitted over the link.

Quasi-synchronous FM (also known as Simulcast) schemes were initially set up with carrier frequency offsets of a few tens of Hz, which in the early days represented the performance limitation of quartz crystal oscillators. It was assumed that, due to the capture effect inherent in FM receivers, the mobile would respond only to the strongest signal present and that periods of apparently equal signal strength, when capture would not occur, would be infrequent and would produce few problems. In practice, it was difficult to obtain good performance unless the modulation matching was very carefully undertaken with respect to both amplitude and phase. In addition, steps needed to be taken to avoid even approximately equal strength signals in areas where intelligible communication was required. The difficulties in establishing a successful quasi-synchronous FM scheme were considerable. It was marginally easier at UHF than at VHF, and the overall quality of reception was improved significantly if the frequency offsets were reduced from the few tens of Hz of the early schemes down to a few Hz as in the schemes most recently installed. The careful and accurate modulation matching that was necessary in FM schemes was much more easily achieved when the various transmitters were controlled by radio links rather than by land lines. Generally speaking, dedicated lines are not available, and it is relatively straightforward to match the modulation delay and modulation index over a radio link with adjustment needed only on rare occasions.

1.10.1 Alternative techniques

The difficulties of setting up a quasi-synchronous scheme can be avoided if all base station transmitters use the same channel but are operated sequentially rather than simultaneously. However, this apparently simple technique has several disadvantages, since the system operator will not be aware of the location of the mobiles, and a simple vehicle location scheme must therefore be incorporated into the system. This requires selective calling facilities, which have been described earlier, and a return path receiver 'voting system', to be incorporated. When the operator in a system of this type wishes to initiate a call, each base station in turn transmits the appropriate selective call sequence for the required mobile. On receipt of its call sequence, the mobile automatically re-transmits its own sequence (identity) and alerts the driver to the fact that a call is imminent. Each base station receiver that is within range of that particular mobile will then inform the central control point of the system of the strength or quality of the signal that has been received from the mobile so that the optimum base station site can be selected for that particular mobile, i.e. the receivers 'vote' as to which is the best base station site. For calls initiated by the mobile, the driver will normally transmit his selective call sign before initiating a message, and the system will select the optimum base station site in the same manner as before. This technique, which is in common use by message handling organizations who need a selective calling system anyway, provides satisfactory performance in almost all circumstances. In general terms, complexity is its major disadvantage, and its success relies very largely on the skill of the fixed-station operators.

Finally, in UHF schemes it is possible to extend the range of coverage, or to fill in areas within the overall coverage area where reception is poor, by the use of 'on-frequency' repeaters. These are essentially high-gain amplifiers which receive the signal on one antenna and then re-transmit it in a different direction on another antenna. Clearly there are problems, since any feedback around the system caused by pick-up between one antenna and the other are likely to cause oscillation, and an isolation of about 90 dB is commonly required. This can be achieved at UHF, but is not possible at lower frequencies. Generally speaking, on-frequency repeaters are not favoured where other techniques can be used.

1.10.2 Cellular schemes

The most sophisticated technique in current use for area coverage is a cellular scheme [8]. The cellular technique will be described in detail in a later chapter; at this point we will merely identify the principle of frequency re-use by which a large area is covered.

If a fixed number of radio channels are available for use in a given radio telephone system, they can be divided into a number of sets, each set being allocated for use in a given small area (a cell) served by a single base station. It is apparent that the greater the number of channels available in any cell, the more simultaneous telephone calls that can be handled, but the smaller the total number of cells in a cluster that uses all the channels. For example, if the total number of channels is 56, then these can be split into four groups of 14 or seven groups of 8, after which the frequencies have to be re-used. Some of the ways in which a fixed number of channels can be grouped to cover a given area are illustrated in Figure 1.8. Frequency re-use is a fundamental concept in cellular radio systems, but such systems need careful planning to avoid degradation by co-channel interference, i.e. interference with calls in one cell caused by a transmitter in another cell that uses the same set of frequencies.

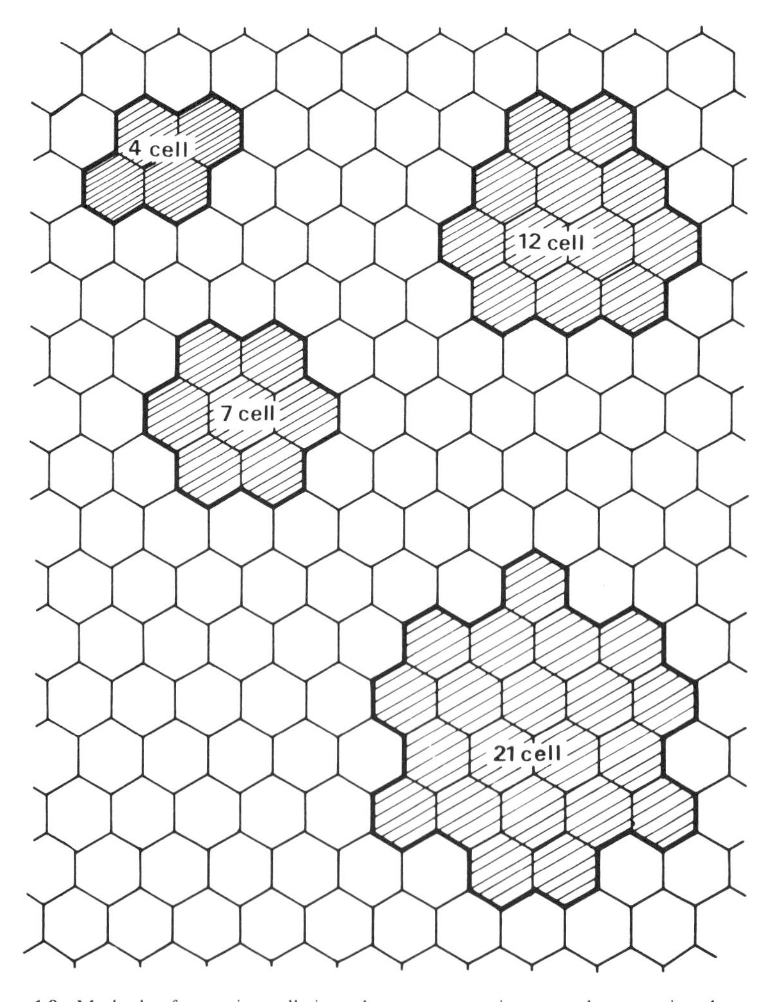

Figure 1.8 Methods of grouping cells in order to cover a given area by repeating the pattern.

1.11 Present and future use of mobile radio

As we have seen earlier in this chapter, land mobile radio services were on a very small scale until the end of World War II. However, the invention of the transistor, together with other important technological advances, provided an enormous boost, and from a situation where in 1946 there were less than 1000 civilian mobile units in the UK, we have advanced to the point where no policeman, fireman, taxi-driver or security guard would be able to fulfil his designated role efficiently without suitable radio equipment. Growth has been rapid and continuous, and projections for the future show a consistent upward trend. For many years mobile radio was dominated by dispatch services, but the use of radio paging and radio telephony have continued to increase.

Research into mobile radio, as distinct from the development of mobile radio systems to meet specific operational and economic requirements, was very small (with one or two notable exceptions) until the 1960s. Until that time there had been no pressing need for such research. The demand for mobile radio had increased, but advances in technology were able to meet the requirement for narrower channel allocations. Although not good by line communication standards, the quality of service was at least adequate. However, there was no sign of a decrease in the demand for licences; channel separations had reached the limit that the technology could support with existing modulation systems, and the amount of spectrum available for mobile services was not being increased to support the continuing demand. The situation was the same in other countries, particularly in the USA, where the problem was exacerbated in the big cities by much more widespread use of mobile radio telephones in addition to dispatch service operations. However research in the USA, and in the laboratories of several major European companies and universities, has led to significant additional knowledge in this field over the past twenty years or so.

The provision of adequate radiofrequency spectrum to meet the increasing demand for mobile radio licences has been a problem to national radio regulatory administrations for many years. Much of the research work related to mobile radio has been devoted to the search for spectrally efficient systems, and recent interest in digital modulation methods, which will be discussed later, is yet another step in this direction. A major step forward was the introduction of dynamic channel allocation (or trunking), which is an essential feature of cellular radio and has substantially increased spectrum utilization. Although the cellular concept which relies on frequency re-use is still in its infancy, it has the potential to provide a radio telephone service to individuals on the move, either through vehicular installations or by means of hand-portable equipment. The setting up of a UK national network (TACS) with access to the public switched telephone network and all its associated facilities has produced a demand which has exceeded all expectations.

It is the availability of low-cost microprocessors and other large-scale integrated circuits which has brought about the most recent advances in mobile radio. Selective addressing, signalling, frequency synthesis and other facilities can now be incorporated into transceivers at an economical cost. This technology is perhaps most obvious in cellular radio schemes, but is capable of being introduced into other new types of service to meet user demands. As an example, the new private advanced radio service (PARS) in the UK is intended to provide business users with a high-quality voice and data service over short ranges in the 900 MHz band. PARS is a two-way digital voice and data private mobile radio service which uses all-digital techniques, can accommodate hand-portable equipment, and uses advanced trunking methods. Cordless telephones, which have been common for some time, are increasing in popularity, and cordless telephone exchanges cannot be far behind. It seems that in the future we can look forward to increasing use of all types of services, a move towards digital modulation methods, and the introduction of other system techniques such as diversity reception on vehicles. This may become much more important when digital transmissions become the norm. There is no doubt that the frequency bands currently available for all types of service will soon become congested, and moves towards higher frequencies initially in the range 1-2.5 GHz, but perhaps in the long term up to 60 GHz, cannot be far away.

References

1. Sandvos, J.L. A comparison of binary paging codes. *IEEE. Conf. on Vehicular Technology*, May 1982.
2. British Telecom. A standard code for radio paging. Report of the Post Office Code Standardisation Advisory Group (POCSAG), June 1978 and Nov. 1980.
3. Sharpe, A.K. Paging systems. In *Land Mobile Radio Systems*, ed. R.J. Holbeche, Peter Peregrinus (1985).
4. Bands I and III. A consultative document. HMSO, London, Cmnd 9241, May 1984.
5. Kleinrock, L. *Queuing Systems*. Vol. 2, John Wiley, New York (1976).
6. Erlang, A.K. Solution of some problems in the theory of probabilities of significance in automatic telephone exchanges. *Post Office Elect. Engrs. J.* **X**, (Jan. 1918) 189–197.
7. Davis, C. and Mitchell, R. Traffic on a shared radio channel. *IERE Int. Conf. on Land Mobile Radio, 1979* [*IERE Conf. Proc.* **44**, 341–349].
8. Advanced Mobile Phone Service. Special issue, *Bell Syst. Tech. J.* **58**, January 1979.

2 Multipath characteristics in urban areas

2.1 Introduction

In all except very simple transmission conditions, radio communication links are subjected to conditions in which energy can travel from the transmitter to the receiver via more than one path. This 'multipath' situation arises in different ways depending upon the application. For example, in HF skywave transmissions, it arises because of reflections from two or more ionospheric layers, or from two-hop propagation. In mobile radio it arises because of reflection and scattering from buildings, trees and other obstacles along the path.

Radio waves therefore arrive at a mobile receiver from different directions with different time delays, and they combine vectorially at the receiver antenna to give a resultant signal which can be large or small depending upon whether the incoming waves combine constructively or destructively. A receiver at one location may experience a signal strength several tens of dB different from a similar receiver located only a short distance away. As a vehicle-borne receiver moves from one location to another, the phase relationship between the various incoming waves changes; hence there are substantial amplitude fluctuations and the signal is said to be subject to fading. It is worth noting that whenever relative motion exists there is a Doppler shift in the received signal, this being a manifestation in the frequency domain of the envelope fading in the time domain.

In the mobile radio case, the fading and Doppler shifts arise as a result of motion of the receiver through what is basically a spatially-varying field. However, whatever the mechanism, the effect of multipath propagation is to produce a received signal with an amplitude that changes quite substantially with location.

The service area of most mobile radio installations includes regions which can be generally classified as 'urban'. Here there are problems due to the fact that the mobile antenna is low (so there is no line-of-sight path to the base station), and is often located in close proximity to buildings. Propagation is

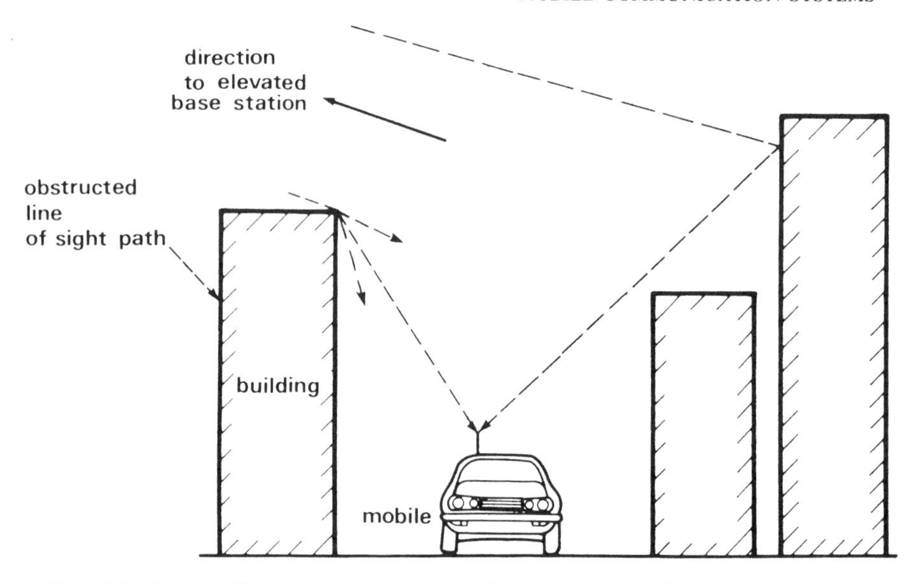

Figure 2.1 Diagram illustrating the mechanism of radio propagation in urban areas.

therefore mainly by means of scattering and multiple reflections from the surrounding obstacles, as shown in Figure 2.1. As we suggested earlier, the position of the antenna does not have to be changed very much to change the signal level by several tens of dB. Although the rapidly changing signal is basically a spatial phenomenon, a receiver mounted on a vehicle moving continuously through the field experiences a time-related variable signal which is further complicated by the Doppler shift mentioned above. The signal fluctuations are known as fading, and the rapid fluctuations caused by the local multipath are known as fast fading to distinguish them from the much longer-term variation in mean level which is known as slow fading. This latter effect is caused by movement over distances large enough to produce gross variations in the overall path between the base station and the mobile. The typical experimental record shown in Figure 2.2(*a*) illustrates these characteristics. Fades of a depth less than 20 dB are frequent, but deeper fades in excess of 30 dB, although less frequent, are not uncommon. Rapid fading is usually observed over distances of about half a wavelength, and at VHF and UHF, a vehicle moving at 50 km/h can pass through several fades in a second. If we consider the nature of the multipath and the shadowing effects, and their influence on the characteristics of radio wave propagation, it is apparent that it is pointless to pursue an exact or deterministic characterization and we must resort to the powerful tools of statistical communication theory.

Examination of Figure 2.2(*a*) shows that it is possible to draw a distinction between the short-term multipath effects and the longer-term variations of the local mean. Indeed, it is convenient to go further and suggest that the mobile

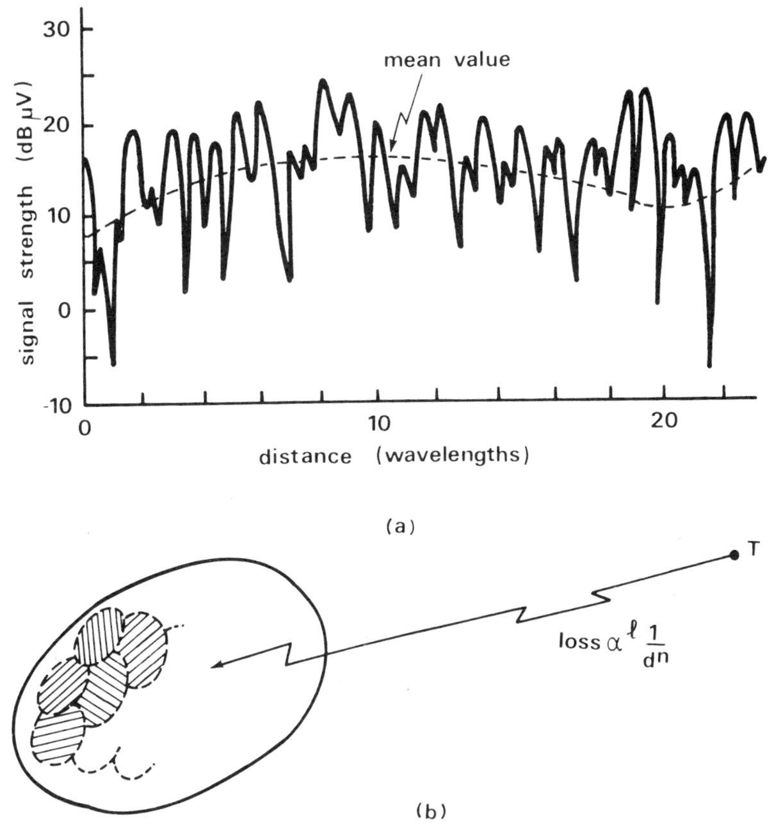

Figure 2.2 (*a*) Typical fading envelope in an urban area. (*b*) A model of mobile radio propagation showing small areas where the mean signal is constant within a larger area over which the mean value varies slowly as the receiver moves.

radio signal consists of a local mean value, which is sensibly constant over a small area, but varies slowly as the receiver moves, superimposed upon which is the short-term rapid fading as illustrated in Figure 2.2(*b*). In this chapter we consider only the short-term effects, both for narrowband and wideband channels; in other words, we consider the signal statistics within one of the small shaded areas in Figure 2.2(*b*), assuming the mean value to be constant. In Chapter 3 we consider methods of predicting the local mean value and statistically characterizing its variability.

2.2 The nature of multipath propagation

A multipath propagation medium contains several different paths by which energy travels from the transmitter to the receiver. If we consider firstly the case of a stationary receiver, we can imagine a 'static multipath' situation in which the different propagation paths are distinguishable from each other if the differences in their electrical path lengths are such that the various delayed versions of a signal radiated from the transmitter can be recognized sequentially at the receiver. This is illustrated in Figure 2.3, which shows the case of two resolvable paths where the differential time delay is greater than the reciprocal of the signal (pulse) bandwidth.

If we consider the transmission of a narrowband signal, for example an unmodulated carrier, then several versions still arrive sequentially at the receiver. However, the effect of the differential time delays will now be to introduce relative phase shifts between the component waves, and superposition of the different components that leads to either constructive or destructive addition (at one instant of time) depending upon the relative

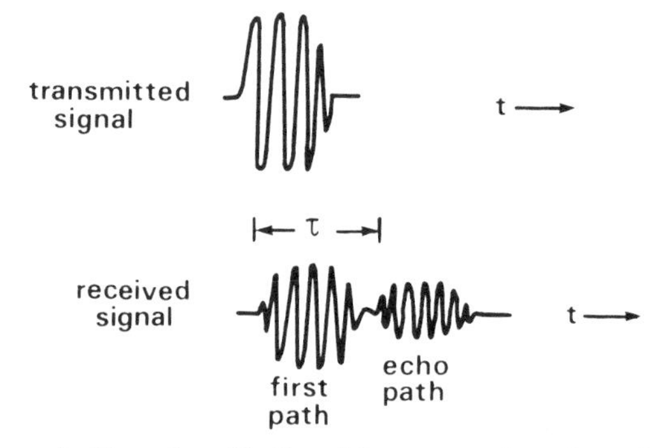

Figure 2.3 Two resolvable paths with time delay greater than the reciprocal of the signal bandwidth.

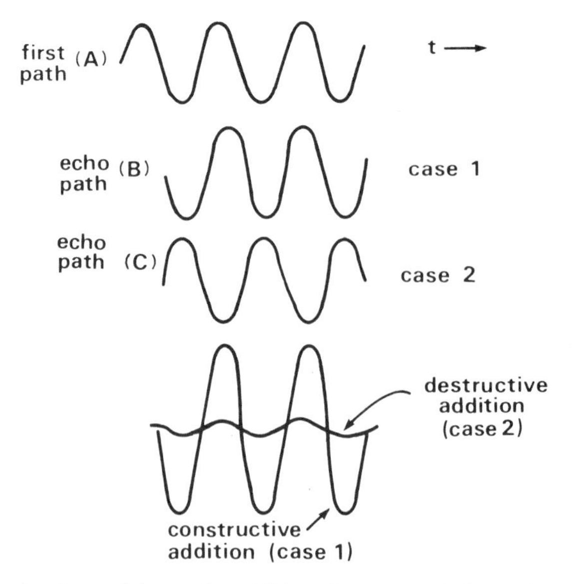

Figure 2.4 Constructive and destructive addition of two transmission paths.

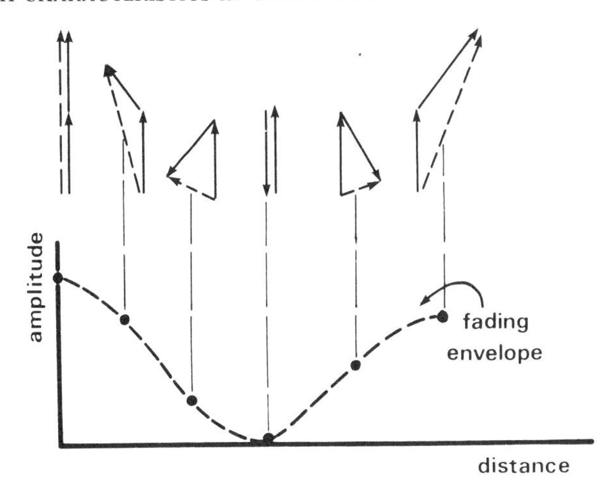

Figure 2.5 Illustrating how the envelope fades as two incoming signals combine with different phases.

phases. Figure 2.4 illustrates the two extreme possibilities. The resultant signal arising from two paths (*A* and *B*) will be large because of constructive addition, whereas that from paths *A* and *C* will be very small.

If we now turn to the case when either the transmitter or the receiver is in motion, we have a 'dynamic multipath' situation in which there is a continuous change in the electrical length of every propagation path, and thus the relative phase shifts between them change as a function of spatial location. Figure 2.5 shows how the received amplitude (envelope) of the signal varies in the simple case when there are two incoming paths with a relative phase that varies with location. At some positions there is constructive addition, whereas at others there is almost complete cancellation. In practice, of course, there are several different paths which combine in different ways depending on location, and this leads to the more complicated signal envelope function shown in Figure 2.2. The space-selective fading which exists as a result of multipath propagation is experienced as time-selective fading by a mobile receiver which travels through the field.

The time variations, or dynamic changes in the propagation path lengths, can be related directly to the motion of the receiver and indirectly to the Doppler effects that arise. The rate of change of phase, due to motion, is apparent as a Doppler frequency shift in each propagation path, and this arises because the changes of phase $\Delta\phi$ and the change in path length Δl are related by

$$\Delta\phi = \frac{2\pi}{\lambda}\Delta l \tag{2.1}$$

where λ is the carrier wavelength.

The change in path length will depend on the spatial angle between any component wave and the direction of motion, and it is apparent that waves arriving from directly ahead of, or directly behind, the vehicle are subjected to the maximum rate of change of phase.

In a practical case then, the several incoming paths will be such that their individual phases as experienced by a moving receiver will change continuously and randomly. The resultant signal envelope and RF phase therefore also will be random variables, and it remains to devise a mathematical model to describe the relevant statistics. Such a model must be mathematically tractable, and lead to results which are in accordance with the observed signal properties. For convenience, we will consider only the case of a moving receiver.

2.3 Short-term fading

Several multipath models have been suggested to explain the observed statistical characteristics of the electromagnetic fields and the associated signal envelope and phase. The earliest of these was due to Ossana [1], who attempted an explanation based on the interference of waves incident and reflected from the flat sides of randomly-located buildings. Although Ossana's model predicted power spectra that were in good agreement with measurements in suburban areas, it assumes the existence of a direct path between transmitter and receiver and is limited to a restricted range of reflection angles. It is therefore rather inflexible and inappropriate for urban areas where the direct path is almost always blocked by buildings or other obstacles.

A model based on scattering is more generally appropriate, and the most widely quoted and accepted is that of Clarke [2]. This assumes that the field incident on the mobile antenna is composed of a number of plane waves of random phase, these plane waves being vertically polarized with horizontal angles of arrival and phase angles which are random and statistically independent. Furthermore, the phase angles are assumed to have a uniform probability density function (PDF) in the interval $(0–2\pi)$. This is reasonable at VHF and above where the wavelength is short enough to ensure that small changes in path length result in significant changes in the RF phase. The PDF $p(\alpha)$ for the horizontal angle of the plane waves is not specified by Clarke, as it depends on the scattering characteristics of the environment and is likely to change over spatial distances greater than a few tens of wavelengths. A model such as this, based on scattered waves, allows the establishment of several important relationships describing the received signal, for example the first- and second-order statistics of the signal envelope and the nature of the frequency spectrum.

2.3.1 The scattering model

At every receiving point we assume the existence of N plane waves of equal amplitude. This is a realistic assumption in heavily built-up areas, since the

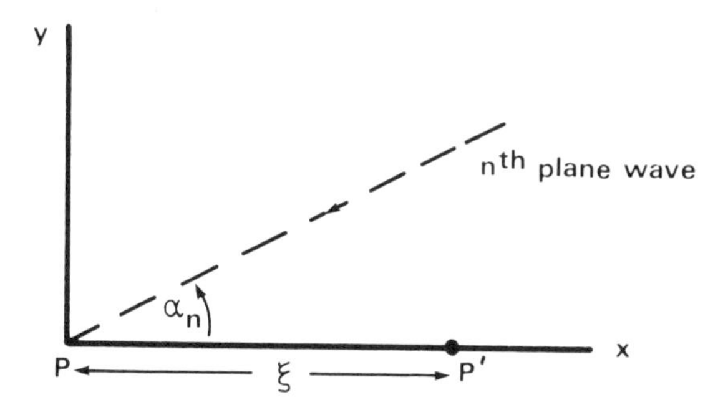

Figure 2.6 Path-angle geometry for nth plane wave.

scattered components are likely to experience similar attenuation and there will be no dominant component. In certain situations a direct line-of-sight path may contribute a steady non-random component. This will alter the nature of the fading envelope and its statistics, but in this discussion we restrict analysis to the case of N equal-amplitude waves. The path-angle geometry for the nth scattered plane wave is shown in Figure 2.6. If the transmitted signal is vertically polarized, i.e. the electric field vector is aligned along the z axis, the field components at the mobile are the electric field E_z, the magnetic field component H_x and the magnetic field component H_y. The expressions for the resultant fields at the receiving point P in Figure 2.6 are obtained in the complex equivalent baseband form [2] as

$$
\begin{aligned}
E_z &= E_0 \sum_{n=1}^{N} \exp(j\phi_n) \\
H_x &= -\frac{E_0}{\eta} \sum_{n=1}^{N} \sin\alpha_n \exp(j\phi_n) \\
H_y &= \frac{E_0}{\eta} \sum_{n=1}^{N} \cos\alpha_n \exp(j\phi_n)
\end{aligned}
\tag{2.2}
$$

In these equations ϕ_n is the phase relative to the carrier phase, E_0 is the real amplitude of the N plane waves, and η is the intrinsic wave impedance. The real and imaginary parts of each field component constitute the in-phase and quadrature components of the received RF fields when an unmodulated carrier has been transmitted. By applying the central limit theorem and observing that α_n and ϕ_n are independent, it follows [3] that E_z, H_x and H_y are complex Gaussian random variables for large N.

The assumption that the mobile received signal is of the scattered type, with each component wave being independent, randomly phased and having a random angle of arrival, leads to the conclusion that the probability density function of the envelope is

$$
p_r(r) = \frac{r}{b_0} \exp(-r^2/2b_0) \tag{2.3}
$$

which is a Rayleigh distribution. The corresponding cumulative distribution function is

$$\text{prob}[r < R] = P_r(R) = 1 - \exp(-R^2/2b_0) \tag{2.4}$$

where b_0 is the mean power $(= E_0^2/2)$.

2.3.2 Angle of arrival and signal spectra

If either the transmitter or receiver is in motion, the components of the received signal each experience a Doppler shift, the frequency shift being related to the spatial angle between the direction of arrival of that component and the direction of vehicle motion. For a vehicle moving at a constant speed v along the x axis in Figure 2.6, the Doppler shift f_n of the nth plane-wave component is

$$f_n = \frac{v}{\lambda} \cos \alpha_n \tag{2.5}$$

It can be seen that component waves arriving from ahead of the vehicle experience a positive Doppler shift (maximum value $f_m = v/\lambda$) whilst those arriving from behind the vehicle have a negative shift.

If N is large, the fraction of the incident power contained within a spatial angle between α and $\alpha + d\alpha$ for an omnidirectional antenna is $p(\alpha)d\alpha$; if the antenna has a gain $G(\alpha)$, the corresponding value is $G(\alpha)p(\alpha)\,d\alpha$. Equating this power to the incremental power determined through the relationship between Doppler shift and spatial angle, viz.

$$f(\alpha) = f_m \cos \alpha + f_c \tag{2.6}$$

we obtain the frequency spectrum $S(f)$ of the received signal as

$$S(f)|df| = \{G(\alpha)p(\alpha) + G(-\alpha)p(-\alpha)\}|d\alpha| \tag{2.7}$$

A practical case of interest is the omnidirectional* vertical monopole. We can obtain an expression for $S(f)$ by assuming $p(\alpha)$ to be uniformly distributed in the angular range $(-\pi, \pi)$. This need not actually be the case, as any other suitable function can be considered. In fact, experimental data confirms that such an assumption is simplistic, since the scattering model changes from one location to another, even within a fairly homogeneous locality. However, with the simplifying assumption, the RF Doppler spectrum becomes

$$S_{E_z}(f) = \frac{1.5}{\pi f_m}\left[1 - \left(\frac{f - f_c}{f_m}\right)^2\right]^{-1/2} \tag{2.8}$$

This, together with the corresponding baseband spectrum, is shown in Figure 2.7*a*.

* In common usage, the term omnidirectional is used to imply a uniform radiation pattern in the horizontal (azimuth) plane. It is not to be confused with 'isotropic' (all-directions).

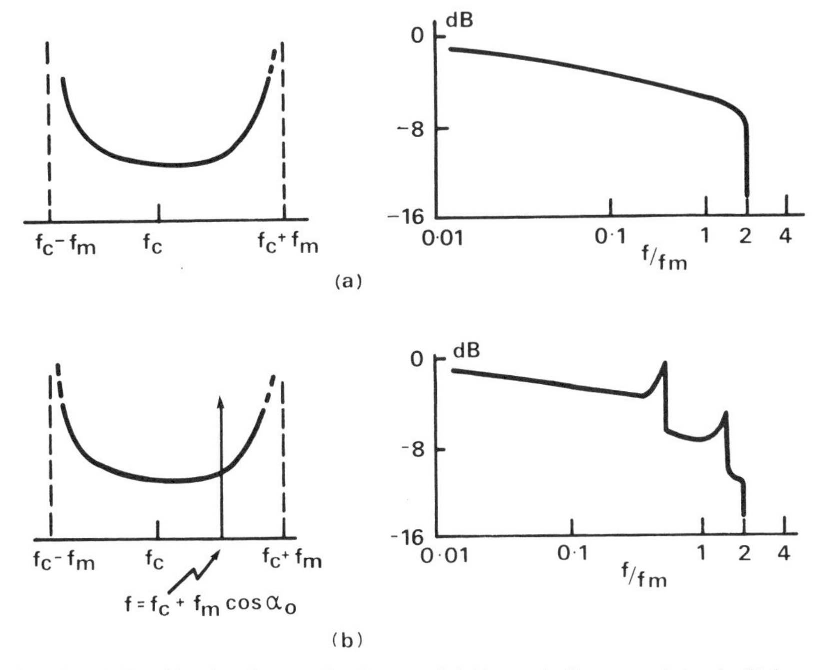

Figure 2.7 RF and baseband spectra for the case of: (*a*) isotropically scattered signals; (*b*) the case of a dominant component.

If there is a dominant component in the incoming signal, then this has a substantial influence on the signal spectrum. Such a component arriving at an angle α_0 gives rise to a spectral line at $f_c + f_m \cos \alpha_0$ in the RF spectrum, and two additional peaks at $f_m(1 \pm \cos \alpha_0)$ in the baseband spectrum, as shown in Figure 2.7*b*.

In the time domain, as we have already seen, the effects of the randomly-phased and Doppler-shifted multipath signals appear in the form of a fading envelope.

2.3.3 Fading envelope statistics

The fading envelope directly affects the performance of radio receivers. It is of interest to consider the average rate at which the envelope crosses a given level and how long it remains below that level. The Rayleigh fading envelope only occasionally experiences very deep fades; for example, 30 dB fades occur for only 0.1% of the time. To specify some of the constraints in a quantitative manner, we can derive some parameters of the fading envelope that have direct relevance to system performance. In particular, we are interested in how often the envelope crosses a specified signal level (the level-crossing rate), and the average duration of a fade below that specified level.

2.3.4 Level-crossing rate

The level-crossing rate at a specified signal level R is defined as the average number of times per second that the signal envelope crosses the level in a positive-going direction. This is illustrated in Figure 2.8. Mathematically this can be expressed [4] as

$$N(R) = \int_0^\infty \dot{r}p(R, \dot{r})\,\mathrm{d}\dot{r} \tag{2.9}$$

where $p(R, \dot{r})$ is the joint PDF of R and $\dot{r}$, and a dot indicates the time derivative. For a vertical monopole the level crossing rate becomes

$$N(R) = \sqrt{2\pi}\, f_m \rho \exp(-\rho^2) \tag{2.10}$$

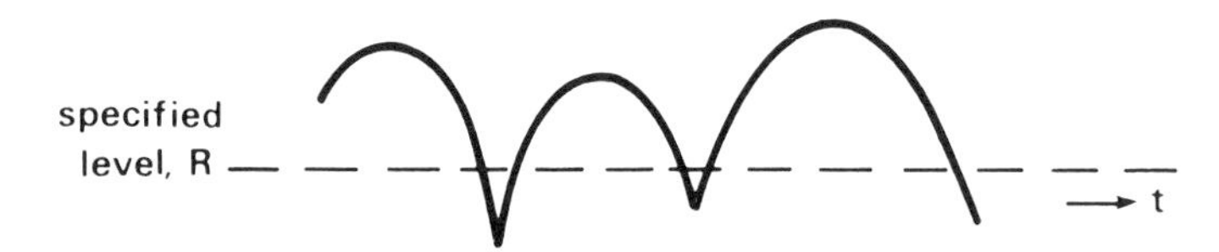

Figure 2.8 Level-crossing rate of Rayleigh-fading signal.

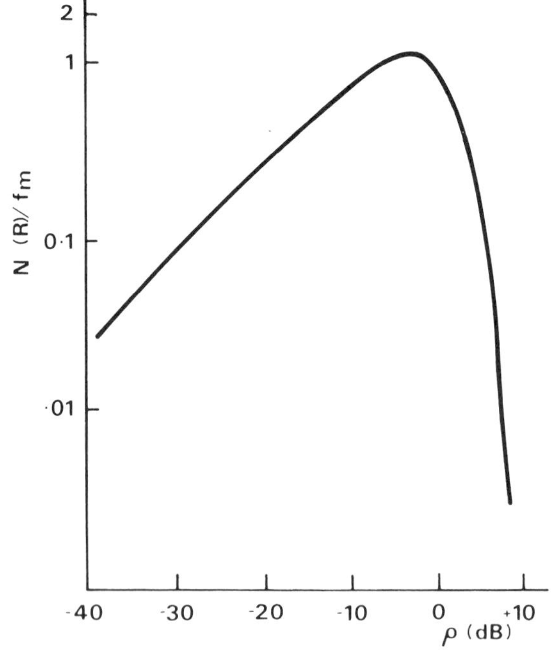

Figure 2.9 Normalized level-crossing rate for a vertical monopole under conditions of isotropic scattering.

where $\rho = R/R_{rms}$ is the ratio between the specified level and the rms amplitude of the fading envelope.

The normalized level-crossing rate for a vertical monopole is plotted in Figure 2.9. The maximum rate occurs at $\rho = -3$ dB, and decreases as the level is lowered or raised. We recall that, since the level-crossing rate is specified for a positive direction, then when a low level is set the signal remains above this level for most of the time; a similar argument applies when a level much greater than the rms value is selected. At $f_c = 900$ MHz and a vehicle speed of 48 kmh^{-1}, $f_m = 40$ Hz, thus the level crossing rate N_R is 39 per second at $\rho = -3$ dB.

2.3.5 Average fade duration

The average duration of fades below the specified level R can be found from the relationship

$$\text{average fade duration} = \frac{\text{prob}[r < R]}{N(R)} \qquad (2.11)$$

From eqn (2.4), prob$[r < R]$ can be evaluated as $1 - \exp(-\rho^2)$, so substituting for $N(R)$ from eqn (2.10) gives the average fade duration for a vertical monopole as

$$\tau(R) = \frac{\exp(\rho^2) - 1}{\rho f_m \sqrt{2\pi}} \qquad (2.12)$$

and this is plotted in Figure 2.10.

2.3.6 Spatial correlation of field components

In mobile radio systems, especially at VHF and above, the effects of fading can be combated by the use of space diversity techniques, at either the base station or the mobile, provided that antennas can be spatially separated far enough to ensure that the signal envelopes exhibit a low correlation. The principles of diversity, and diversity techniques will be discussed in Chapter 6. To obtain the spatial correlation function for the electric field, we consider the field incident at a point P' a distance ζ along the axis from the point P as shown in Figure 2.6. The phase of the nth component wave at P' will be $(\phi_n + k\xi \cos \alpha_n)$, where $k = 2\pi/\lambda$ is the phase shift factor in free space. The autocovariance function of the electric field is given in the form

$$R_{E_z}(\zeta) = \langle E_z^* E_z' \rangle = \left\langle E_0^2 \sum_{n=1}^{N} \exp(-j\phi_n) \sum_{n=1}^{N} \exp\{(\phi_n + k\zeta \cos \alpha_n)\} \right\rangle \qquad (2.13)$$

where E_z^* is the complex conjugate of the field component at P, and E_z' is the field component at P'. If we assume statistical independence of ϕ and α and also consider the case of isotropic scattering, i.e. assume that $p(\alpha) = 1/2\pi$, then

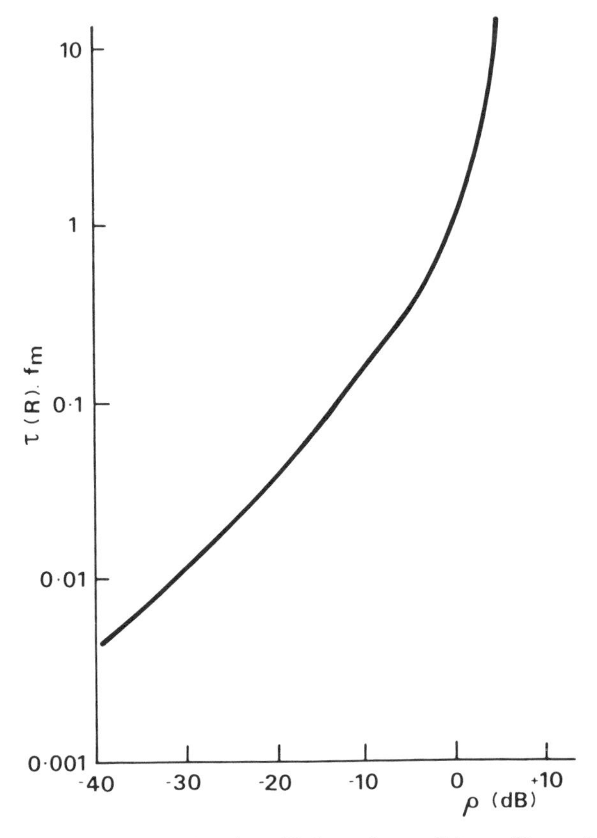

Figure 2.10 Normalized average duration of fades under conditions of isotropic scattering.

it follows that

$$R_{E_z} = NE_0 \int_{-\pi}^{+\pi} p(\alpha) \exp(jk\zeta \cos\alpha)\, d\alpha$$
$$= NE_0^2 J_0(k\zeta)$$

where $J_0(.)$ is the zero-order Bessel function of the first kind. It then follows, to a good approximation, that the normalized autocovariance function of the envelope is equal to the square of the normalized autocovariance function of the complex random field. Thus

$$\rho_{|E_z|} = J_0^2(k\zeta) \tag{2.14}$$

This function is plotted in Figure 2.11*a* for the case of an isotropically scattered field. There is rapid decorrelation, showing that space diversity can be implemented at the mobile end of the link, where the assumption of isotropic scattering is approximately true.

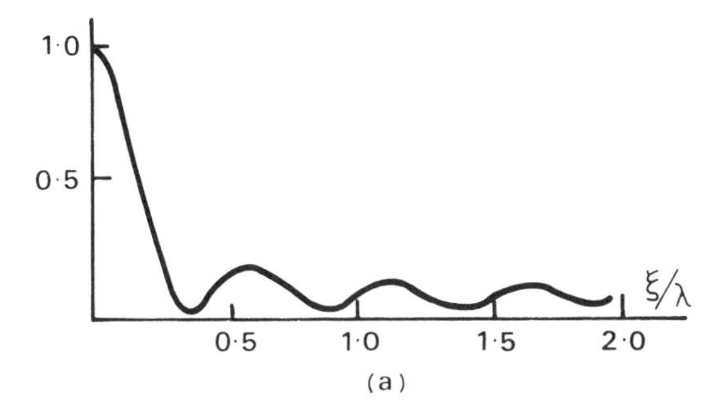

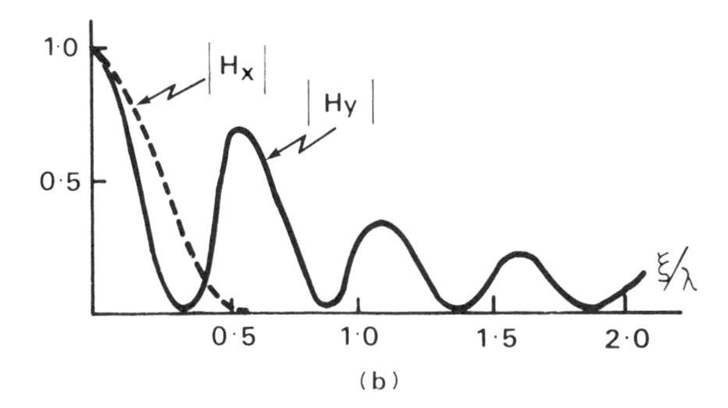

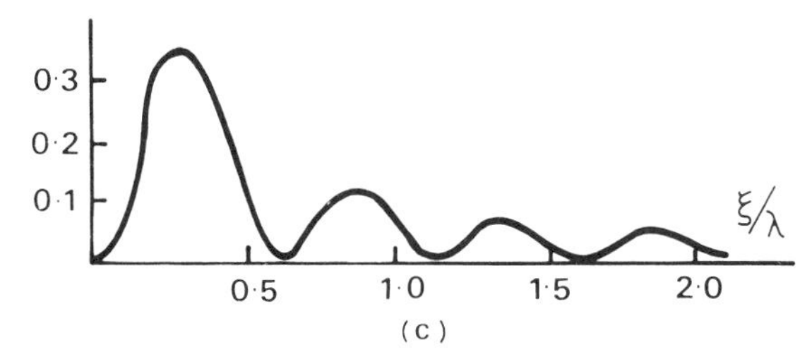

Figure 2.11 Covariance functions of the envelopes of the field components under conditions of isotropic scattering.

In a similar way, the parts of eqn (2.2) relating to the magnetic field components H_x and H_y can be used to show that the autocovariance functions of the envelopes of these components are

$$\begin{aligned} &[J_0(k\zeta) + J_2(k\zeta)]^2 \text{ for } H_x \\ &[J_0(k\zeta) - J_2(k\zeta)]^2 \text{ for } H_y \end{aligned} \tag{2.15}$$

and these are shown in Figure 2.11*b*.

Also of interest are the cross-covariance functions between the envelopes of the three field components, since low values point the way towards other methods of implementing diversity systems. It can be shown that E_z and H_x are always uncorrelated provided $p(\alpha)$ is an even function; thus, their envelopes will also be uncorrelated, and a similar argument applies to H_x and H_y. It can be shown that, to a high degree of accuracy, the cross-covariance between the envelopes of E_z and H_y is given by $J_1^2(k\zeta)$, and this is shown in Figure 2.11c. It is interesting to note from this diagram that, in a scattered field of the type described, the electric and magnetic field envelopes are uncorrelated at the same point in space, which is perhaps a rather surprising result.

On the other hand, it is clear that at base station sites the assumption of isotropic scattering is unlikely to hold. Base station sites are deliberately chosen to be well above local obstructions to give the best coverage of the intended service area, and the scattering objects which produce the multipath effects are located principally in a small area surrounding the mobile. The reciprocity theorem applies, of course, in any linear medium, but this should not be taken to imply that the spatial correlation distance at one end of the radio path is the same as it is at the other.

Figure 2.12 shows a number of scattering objects which surround the mobile, and a base station some distance away. If we assume that the distance d from the base station to the mobile is much greater than the distance a from the mobile to the significant scatterers, then it is possible to calculate the spatial correlation distance at the base station [4]. It seems intuitively obvious that antennas have to be further apart at base station sites than at mobiles to obtain the same correlation, and indeed this is the case. It is also apparent that, whereas at mobiles the assumption of isotropic scattering (from scatterers

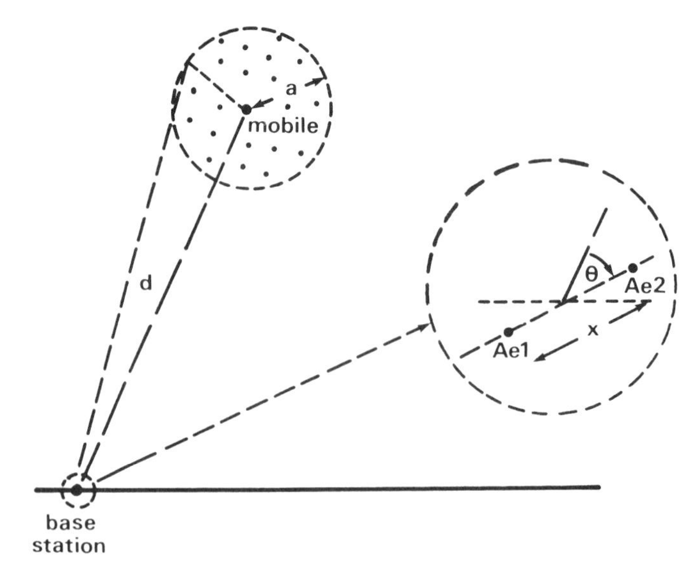

Figure 2.12 Path geometry relevant to base-station diversity showing the relationship between separation x and orientation θ.

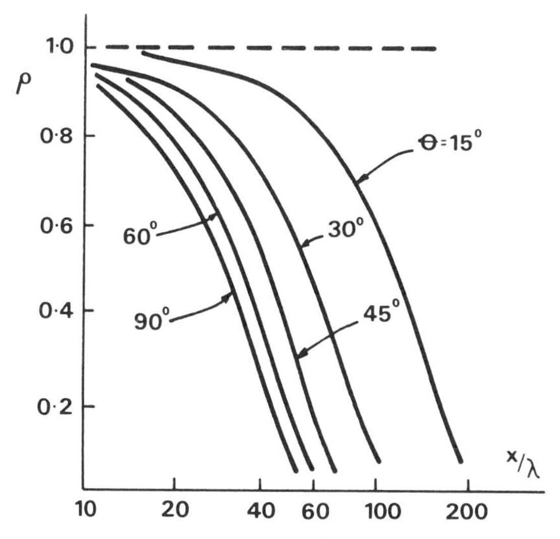

Figure 2.13 Base station correlation as a function of separation for various values of θ.

which surround the mobile uniformly) leads directly to the conclusion that the correlation between the electric field at two receiving points is a function of their separation only, this is not the case at base station sites. Here, scattering is not isotropic and the correlation between the electric field at two receiving points is a function of both their separation and the angle between the line joining them and the direction to the mobile. Figure 2.13 shows theoretical curves for correlation at base station sites and a comparison between this and Figure 2.11 shows how much more rapidly the signals at the mobile become decorrelated with antenna separation. It also shows that, at base stations, the antenna separation required along the line-of-sight direction to the mobile is greater than that along the perpendicular direction.

A further possibility for obtaining decorrelated signals is to separate the receiving points vertically rather than horizontally. Experimental investigations of this possibility [5, 6] have demonstrated its feasibility at base station sites. The angle subtended at the mobile by two antennas depends not only on their separation but also on the distance to the mobile, and the separation required to produce a certain correlation therefore depends on the distance from mobile to base. In practice, at 900 MHz the required separation is about 12λ when the mobile-to-base distance is about 1.5 km [6].

2.3.7 Random FM

The discussion so far has concentrated on the properties of the signal envelope, but it was mentioned in section 2.1 that the field components E_z, H_x and H_y are complex, Gaussian variables. The complex nature gives rise to time-varying in-phase and quadrature components which, in turn, means that the apparent frequency of the signal varies with time. The in-phase and quadrature components vary in a random manner and the signal therefore exhibits random frequency variations known as random frequency modulation. The characteristics of this random FM can be described in terms of its probability density function and power spectrum, and appropriate expressions have been obtained in reference 4.

For the E_z component, the probability density function of the random FM can be written as

$$p(\dot{\theta}) = \frac{1}{\omega_m\sqrt{2}}\left[1 + 2\left(\frac{\dot{\theta}}{\omega_m}\right)^2\right]^{3/2} \tag{2.16}$$

and similar expressions can be obtained for the other two field components. We note that, in contrast to the well-defined power spectrum of the signal (the Doppler spectrum, Figure 2.7) there is a finite (non-zero) probability of finding the frequency of the random FM at any value, although the larger-frequency excursions are caused by the deep fades and therefore occur only rarely. For a 20 dB fade, the significant frequency deviations occupy a bandwidth of about $10 f_m$, and this decreases to f_m when the signal level is at its rms value.

The power spectrum of the random FM may be derived as the Fourier transform of the autocorrelation function of $\dot{\theta}$. The mathematics is rather complex but leads to the one-sided power spectrum shown in Figure 2.14.

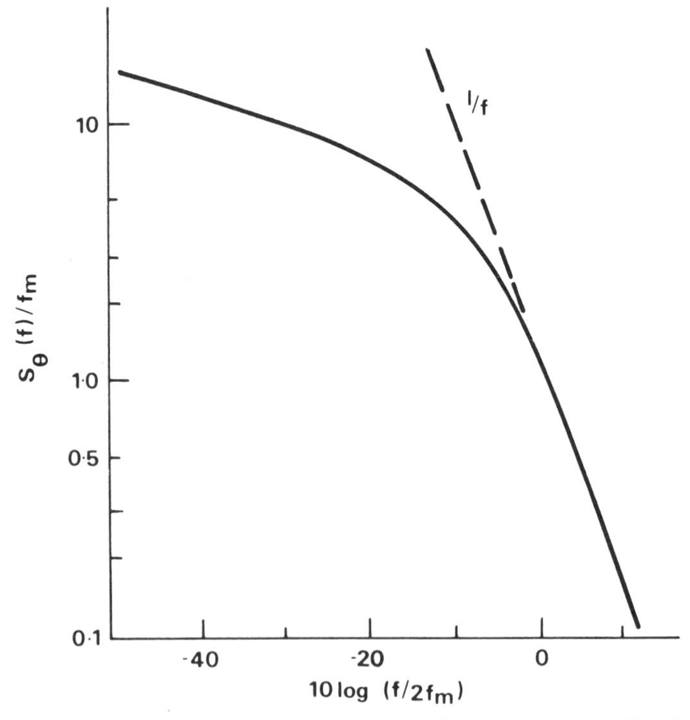

Figure 2.14 Power spectrum of random FM plotted as relative power on a normalized frequency scale.

Above an angular frequency of $2\omega_m$ the spectrum falls off as $1/f$. Thus, $2\omega_m$ can be regarded as a 'cut-off' frequency, both for the random FM and for the signal spectrum in Figure 2.7.

2.4 Frequency-selective fading

The discussion in the earlier sections has concentrated in general on describing the envelope and phase variations of the signal received at a moving vehicle when an unmodulated carrier is radiated by the base station transmitter. The question now arises as to the adequacy of this description of the channel when real signals, which occupy a finite bandwidth, are radiated. It is clear that in practice we need to consider the effects of multipath propagation on these signals, and we consider firstly the case of two frequency components within the message bandwidth. If these frequencies are close together, then the different propagation paths within the multipath medium have approximately the same electrical length for both components, and their amplitude and phase variations will be very similar. However, as the frequency separation increases, the behaviour at one frequency tends to become uncorrelated with that at the other frequency, because the differential phase shifts along the various paths are quite different at the two frequencies. The extent of the decorrelation depends on the spread of time delays, since the phase shifts arise from the excess path lengths. For large delay spreads, the phases of the incoming components can vary over several radians even if the frequency separation is quite small. Signals which occupy a bandwidth greater than that over which spectral components are affected in a similar way will become distorted, since the amplitudes and phases of the various spectral components in the received version of the signal are not the same as they were in the transmitted version. The phenomenon is known as frequency-selective fading, and the bandwidth over which the spectral components are affected in a similar way is known as the coherence, or correlation bandwidth.

The fact that the lengths of the individual propagation paths vary with time due to motion of the vehicle gives us a method of gaining further insight into the propagation mechanism, since the changing time of arrival suggests the possibility of associating each delayed version of the transmitted signal with a physical propagation path. Indeed, if a number of distinct physical scatterers are involved, it may be possible to associate each scatterer with an individual propagation path. However, it is not possible to distinguish between different paths merely by considering the difference between the times of arrival; the angular direction of arrival has also to be taken into account. If we consider only single-scattered paths, then all scatterers with the same path length can be located on an ellipse with the transmitter and receiver at its foci. Each time delay between transmitter and receiver defines a confocal ellipse, as shown in Figure 2.15. If we consider scatterers located at A, B and C, then we can distinguish between paths TAR and TBR (which have the same angle of

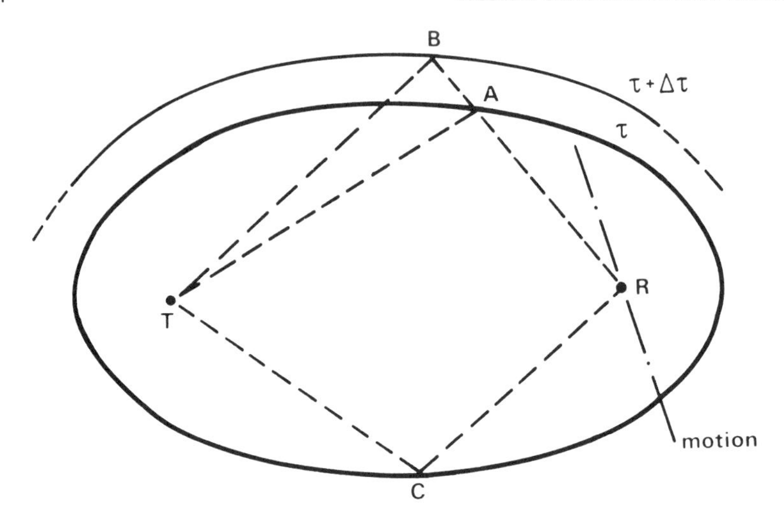

Figure 2.15 Path geometry for single scattering.

arrival) by their different time delays, and between TAR and TCR (which have the same time delay) by their different angles of arrival.

The angles of arrival can be determined by means of Doppler shift. As we have already seen, whenever the receiver or transmitter is in motion the received RF signal experiences a Doppler shift, the frequency shift being related to the cosine of the spatial angle between the direction of arrival of the wave and the direction of motion of the vehicle. If, therefore, we transmitted a short RF pulse and measured both its time of arrival and Doppler shift at the receiver, we could identify the length of the propagation path and the angle of arrival. Of course, there is a left/right ambiguity inherent in the Doppler shift measurement but this could be resolved, if necessary, by the use of directional antennas.

An important and instructive feature of Figure 2.15 is that, for a particular receiver location, a suitably scaled diagram with several confocal ellipses can be produced in the form of a map overlay. Co-ordinated use of this overlay together with experimental results for the location in question, allows the identification of significant single scatterers or scattering areas, and gives an indication of the extent of the contribution from multiple scattering.

2.5 Channel characterization

Signals that have suffered multipath propagation constitute a set of randomly attenuated, delayed, and phased replicas of the transmitted RF signal. The resultant sensed at the receiver is therefore the superposition of contributions from the various individual paths, and on this basis it is reasonable to describe the radio channel in terms of a two-port filter with randomly time-varying transmission characteristics. Since the signals at the input and output of the equivalent filter can be represented in either the frequency or time domains, there are four possible input–output functions which can be used to describe the behaviour of the radio channel [7].

Consider the time-domain description of a deterministically time-variant channel, expressed in terms of the impulse response of the equivalent filters. For a time-varying channel, the impulse response is also time-variant. In the notation of complex envelopes of real bandpass waveforms, the output $z(t)$ is related to the input $y(t)$ and the impulse response $h(t, \tau)$ by the expression

$$z(t) = \int_{-\infty}^{\infty} y(t-\tau)h(t,\tau)\,d\tau \tag{2.17}$$

A circuit model of the complex convolution represented by eqn 2.17 is shown in Figure 2.16, in the form of a tapped delay-line transversal filter. From a physical viewpoint, the multipath is seen to arise from a continuum of stationary elemental scatterers distinguished in path lengths by their elemental delays $d\tau$. Each scatterer produces gain fluctuations $h(t, \tau)\,d\tau$ which are time-dependent. Modelling the time-variant impulse response of the channel in the form of a densely-tapped delay line provides added insight into the physical mechanisms causing multipath, since it explicitly shows that the received signal comprises delayed, attenuated replicas of the input signal, and it demonstrates the nature of multipath arising from scatterers with different path lengths.

Although the impulse response explicitly shows the contributions from scatterers with different path lengths, the Doppler shift phenomenon is embedded in the complex envelope of the time-varying waveform. Direct illustration of the Doppler shift phenomenon is provided by characterization in terms of $H(f, \nu)$, a dual function of $h(t, \tau)$ in the frequency/frequency-shift domain. This is defined in terms of the complex spectra at input and output by

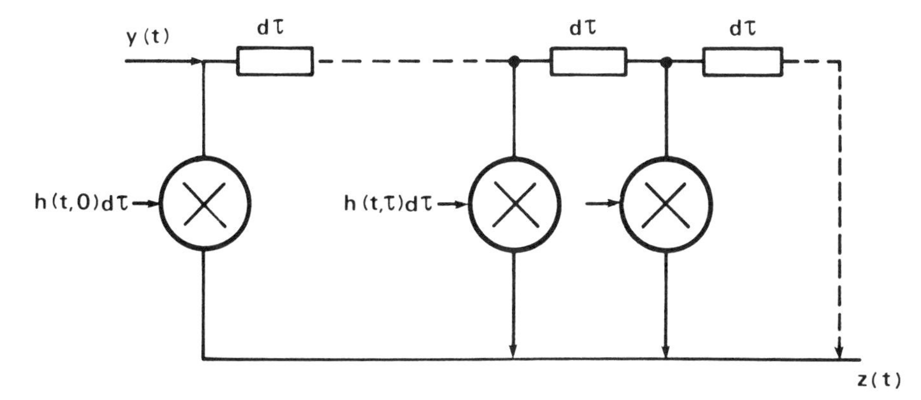

Figure 2.16 Tapped delay-line model of the channel.

the superposition in Doppler shifts ν

$$Z(f) = \int_{-\infty}^{\infty} Y(f-\nu)H(f-\nu, \nu)\,d\nu \tag{2.18}$$

Characterizing a channel using $H(f, \nu)$ necessitates measurement over a suitably wide frequency range in order to study the frequency-selective effects caused by paths of varying lengths.

Another characterizing function $T(f, t)$ describes the response of the channel to cissoidal excitations in the frequency band of interest. In effect, this is the time-variant transfer function of the channel, defined by

$$z(t) = \int_{-\infty}^{\infty} Y(f)\,T(f, t)\exp(j2\pi f t)\,df \tag{2.19}$$

The effects of multipath propagation are now observed as time-selective fading of $T(f_1, t)$ for transmission of a single frequency f_1. $T(f, t)$ is the function we used earlier to characterize narrowband radio channels, the underlying assumption being that the fading character at a single frequency is representative of the actual signals received. However, $T(f, t)$ does not give direct physical insight into the multipath phenomenon, since it only displays the manifestation of multipath as a fading envelope of the received signal. It can therefore be seen that while $h(t, \tau)$ directly distinguishes paths of different lengths, $H(f, \nu)$ can directly identify different Doppler shifts, each shift being associated with the angle of arrival of a scattered component path.

A system function which can simultaneously provide an explicit description of the multipath in both the time-delay and Doppler-shift domains can be introduced on the following basis. It has been observed above that the Doppler spectra are embedded in the time-varying complex envelope of $h(t, \tau)$. Therefore, a system function $S(\tau, \nu)$ can be envisaged such that, while it retains explicitly the time delays of $h(t, \tau)$, the Doppler spectra emerge more directly from the complex envelope by defining $S(\tau, \nu)$ as the spectrum of $h(t, \tau)$:

$$h(t, \tau) = \int_{-\infty}^{\infty} S(\tau, \nu)\exp(j2\pi\nu t)\,d\nu \tag{2.20}$$

The input–output relation then becomes

$$z(t) = \int_{-\infty}^{\infty}\int_{-\infty}^{\infty} y(t-\tau)S(\tau, \nu)\exp(j2\pi\nu t)\,d\nu\,d\tau \tag{2.21}$$

and represents the sum of delayed, weighted and Doppler-shifted signals. Thus, signals arriving during the infinitesimal delay range τ to $\tau + d\tau$ and having Doppler shifts in the range ν to $\nu + d\nu$ have a differential gain of $S(\tau, \nu)\,d\tau\,d\nu$.

Significant symmetric relationships emerge on inspection of the four system functions [8] and are shown in Figure 2.17. The Fourier transform and inverse Fourier transform relationships between these functions are indicated by

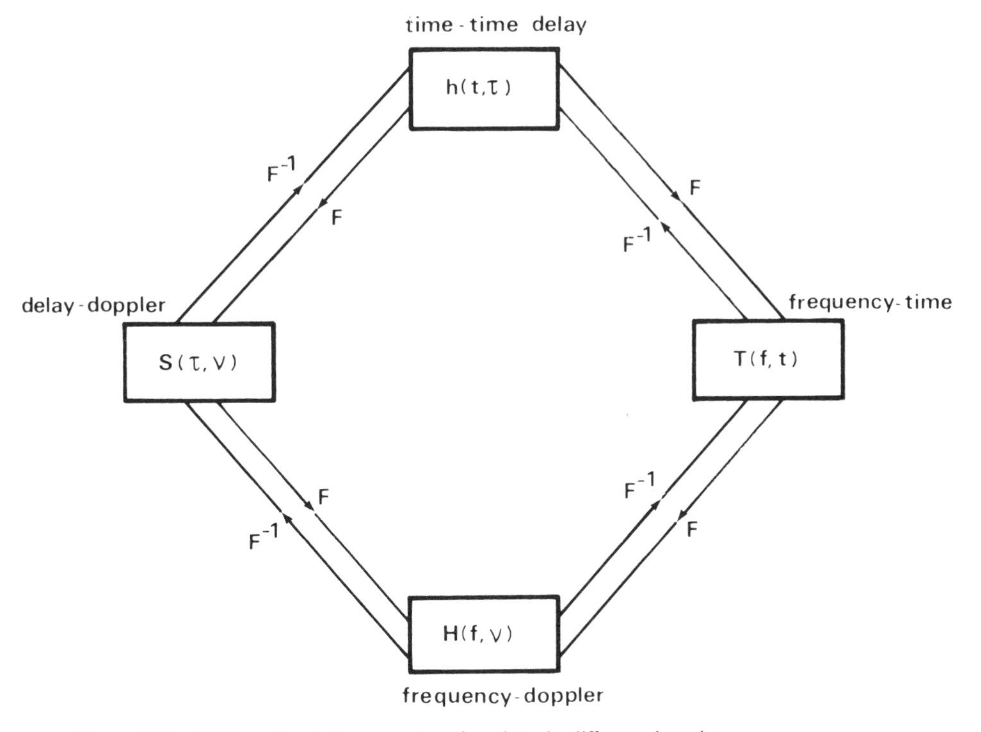

Figure 2.17 Relationships between system functions in different domains.

arrowheads with F and F^{-1} respectively. Thus measurements in one domain can be related to parameters derived from a system function in another domain.

Scattering of radio waves in built-up areas depends on the topography and terrain, and the radio channel between a base station and mobile therefore has randomly time-varying characteristics. Thus, an exact description using system functions is not possible unless the joint multidimensional probability density functions can be specified for all the random variables. In mobile radio communications, the statistics of the received signal exhibit temporal stationarity over spatial distances of only a few tens of wavelengths. The grossly non-stationary character of multipath propagation in changing environments does not easily allow a global characterization. However, experimental observations [9, 10] have shown that it is reasonable to devise a two-stage model in which certain characterizing functions are regarded as statistically stationary over a specific locality of the order of a few tens of wavelengths, and variations over a larger area are studied through the behaviour of moments of these statistical channel descriptors. The study can therefore be conveniently divided into small-area and large-area characterizations.

In practice, of course, multipath radio channels have random rather than deterministic characteristics, and it is then necessary to consider the system functions as stochastic processes which can be specified by their statistical moments. In all practical cases, this is confined to a study of the mean and correlation functions of the two-dimensional system functions; for channels with Gaussian statistics the mean and correlation functions provide an exact description. As extension of the system functions to include the random behaviour has a twofold advantage. Firstly, we retain the description in terms of a two-port filter, thereby allowing the autocorrelation function of the channel output to be obtained from a knowledge of the autocorrelation of the input–output system function. Secondly, the behaviour of the correlation functions aids understanding of the physical scattering of radio waves. Statistical classification of the channel on the basis of the time and frequency statistics, i.e., correlation functions, provides vital clues about the nature of the scattering sources. This aspect of channel characterization is covered in references 7 and 11.

2.6 Channel sounding techniques

The symmetrical relationships between the various input–output channel functions via the Fourier and inverse-Fourier transforms imply that a characterization in either the time or frequency domain is adequate. For narrowband channels, sounding by means of an unmodulated CW carrier leads to characterization via the $T(f,t)$ function. However, for wideband (frequency-selective) channels other methods have to be used. One possibility is to simultaneously (or sequentially) transmit several spaced carriers and to receive each one via a narrowband receiver at the mobile. This leads directly and conveniently to a measurement of the coherence bandwidth via the fading characteristics of the different frequencies, but measurements of time delay and delay spread are less easily extracted from the results. This technique has been used [12], and although there are some advantages, in general it is not as flexible as the alternative technique in the frequency/time domain of using a swept CW carrier, i.e. a 'chirp' signal.

Channel sounding techniques in the time/time-delay domain using a wideband RF pulse, or its equivalent, amount to measuring the channel impulse response, and explicitly show the time-delay multipath characteristics. Coherent detection of the in-phase and quadrature components of the RF pulse also allow the Doppler shifts associated with each time delay to be extracted after suitable signal processing, and hence the angles of arrival can be determined. Measurements in the frequency–time domain using a swept CW carrier provide essentially the same information, but not explicitly in terms of the time-delay or Doppler-shift multipath characteristics.

In the implementation of a wideband RF pulse-sounding system, pulse compression is the key feature. Inherent advantages then include reduced interference to users sharing the same RF band, better immunity from certain types of interference, and enhanced detectability within the peak power limitations of the transmitter. In channel sounders of the pulse-compression type, pseudorandom binary sequences are usually used to angle-modulate an RF carrier. Such maximal length sequences have excellent correlation properties and, as a result, their use has become widespread. With this form of pulse compression, the overall sequence length determines the system sensitivity, and the clock rate determines the limit of resolution, as far as time delay is concerned. At the receiver, pulse compression can be obtained either by convolution using a matched filter, or by processing the signal using correlation. Evidently, there is a subtle difference between the two types of signal processing. The output of the matched filter is a series of snapshots, in real time, of the channel response. The time delays therefore appear in time, and this amounts to a 1:1 mapping of time delays in the time domain. To obtain equivalent correlation processing, a bank of correlators with infinitesimally different time-delay lags would be needed, and in this case the time delays would be related to the correlation lags. Practical correlation processing is usually achieved with a single correlator, using a swept time-delay correlation technique in which the incoming signal is correlated with a pseudorandom sequence identical to that transmitted, but clocked at a slightly slower rate. Inherent in this method is a time-scaling operation, the extent of the scaling being determined by the difference in the clock rates used at the transmitter and receiver. Alternatively, the sequence can be run at the same speed, but reset digitally at appropriate times to achieve the same effect. The time delays in the channel response are therefore observed sequentially in time, in order to reconstruct the snapshot available directly from the matched filter. The snapshots from the matched filter have the same bandwidth as the transmitted signal, and so in practice, although they are observable in real time, in the absence of expensive wideband recording equipment they cannot be directly recorded for subsequent signal processing, and some form of bandwidth compression has to be used. Bandwidth compression is, of course, identical to time scaling. Instantaneous real-time snapshots are therefore available from matched-filter processing, but bandwidth compression normally has to be used before recording, whereas, with swept time-delay correlation, instantaneous snapshots are not available, and the bandwidth compression is an integral part of the signal processing.

2.7 Practical channel sounders

A simplified block diagram of a wideband channel sounder is shown in Figure 2.18. All oscillator frequencies are derived from a highly stable frequency source, which is often a rubidium standard. Frequency synthesizers are used to provide the clock for the pseudorandom sequence generator and the input to the mixer. Typically, a time delay resolution capability of 0.1 μs is available from a 10 MHz clock, and this allows propagation paths with a

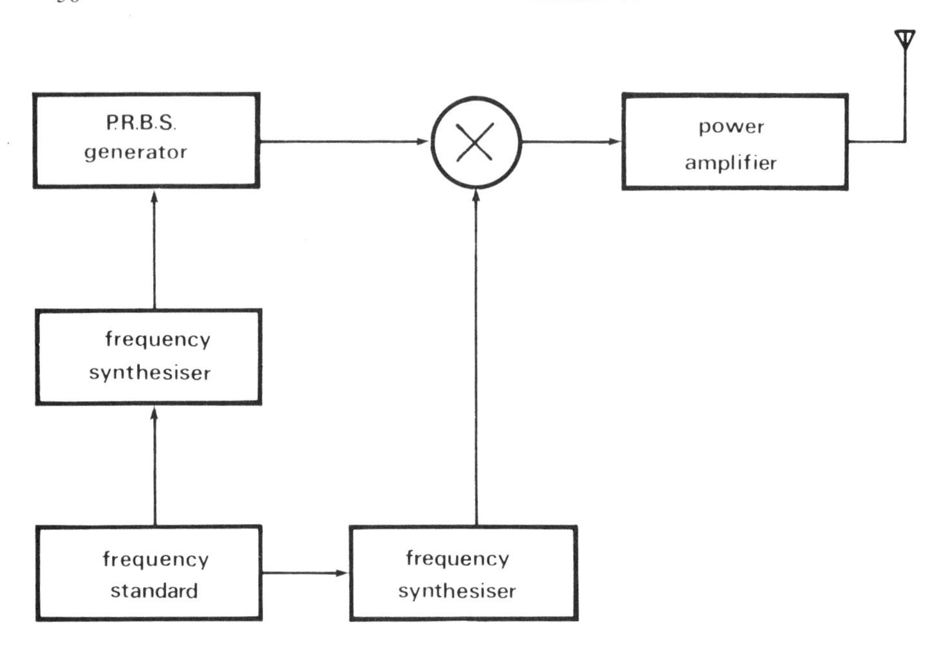

Figure 2.18 Simplified diagram of wideband transmitter.

differential length of 30 m to be distinguished. The pseudorandom sequence is commonly of length 127, 511 or 1023 bits, giving a maximum unambiguous time delay resolution capability (with a 10 MHz clock) of between 12.7 and 102.3 μs. If the longer time delays are of no interest, then the sequence can be reset at any convenient time. The inherent dynamic range of the system is limited by the residual value of the autocorrelation function of the pseudorandom sequence, so that longer sequences give a higher dynamic range. For a 127-bit sequence the range is $1/127 = 42$ dB. Finally, the pseudorandom sequence is used to phase-reversal modulate a carrier so that the transmitted signal is of constant amplitude with the pseudorandom code embedded in the 180° phase transitions.

As we have already seen, there are two basic forms of receiver in which pulse compression is obtained, using either convolution or correlation processing [9, 10]. Figure 2.19 shows a system using a swept frequency correlator in which the multiplication part of the correlation process is achieved at an intermediate frequency by superimposing the pseudorandom sequence on to the local oscillator. If the difference in the clock rates is 5 kHz, then this is the bandwidth of the resulting signal, and the circuits which follow the mixer can be relatively narrowband, improving the noise performance of the receiver. After demodulation to baseband in two quadrature channels, the signal is subjected to further low-pass filtering, and the in-phase (I) and quadrature (Q) outputs are recorded on an instrumentation tape recorder. Other recorder

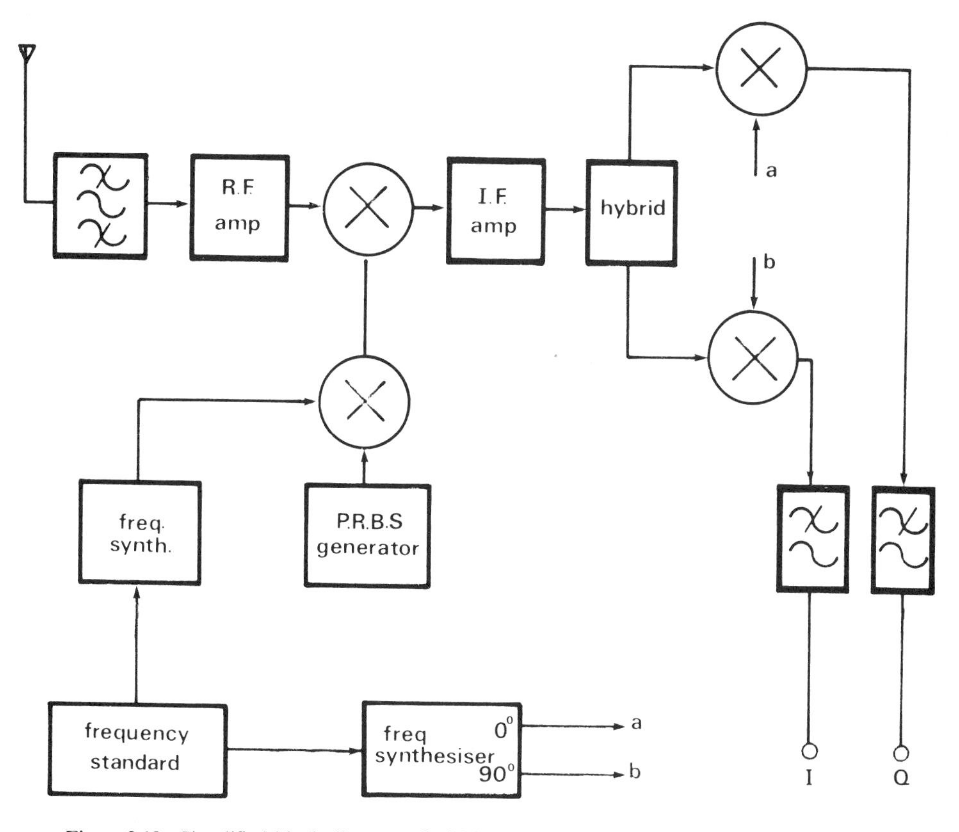

Figure 2.19 Simplified block diagram of wideband receiver.

channels are often used for other relevant information, such as time reference pulses which mark a precisely-defined point in the pseudorandom sequence.

2.7.1 Data processing

Data processing is based on the symmetrical relationships between the statistical characterizing functions discussed in section 2.5. These functions include the time-delay/Doppler scattering function, the average power-delay profile and the frequency correlation function. Data available in the form of the complex impulse response $h(t, \tau)$ is processed to obtain the characterizing functions. To explain the signal processing, the in-phase and quadrature components of $h(t, \tau_k)$ are visualized as time-varying signal phasors for each time-delay bin τ_k, as shown in Figure 2.20*a*.

As the amplitudes of $I(t, \tau_k)$ and $Q(t, \tau_k)$ change, the resultant phasor varies in magnitude and phase relative to the carrier phase angle. The phasors for successive time-frames are shown in Figure 2.20*b*. As time t progresses, the elapsed phase advances or retards. The illustration shows the phase retarding

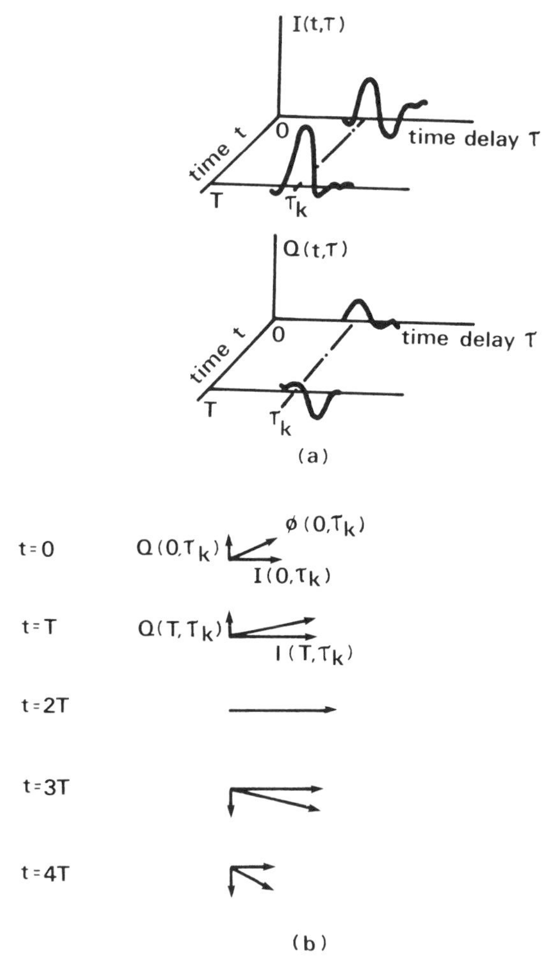

Figure 2.20 Illustrating the data processing. (*a*) In-phase *I* and quadrature *Q* demodulated outputs in successive time frames. (*b*) Phasor representation of resultant demodulated output in one time-delay cell.

at a constant rate, this being caused by a negative Doppler shift which appears as an offset in the carrier frequency. A number of Doppler-shifted components, each with different amplitude, are present in the experimental results, so it is essential that a Fourier analysis is invoked, because the time variations of the phasors represent waveforms that can be described only by the orthogonal Fourier basis-set. By applying the Discrete Fourier Transform method, Doppler spectra can be obtained for each time-delay to effectively assemble the time-delay/Doppler scattering function $S(\tau, \nu)$. The scattering function is a powerful aid in the interpretation of the data from a viewpoint of the scattering mechanics, and in most instances the interest focuses on multipath as a source of distortion in radio systems. If the time-delay dispersion is insignificant compared with, say, the bit-rate in data communication, then the Doppler-shifted components will appear as fading amplitudes of the resultant phasors. When time-delay dispersion is appreciable, the phasors corresponding to the different delayed echoes combine in a destructive or constructive manner as the carrier frequency is varied, the resultant phasor becoming frequency-selective in amplitude. Time-delay dispersion is studied directly from observations of the echo envelope $|h(t, \tau)|$, i.e. the magnitude of the phasor for each delay bin. Averaging a sufficient number of echo profiles, by summing the envelope power for the same delay bin in consecutive frames of the impulse response, produces the statistical distribution $P(\tau_k)$; $k = 0, 1, \ldots (\max)$. Parameters such as the average delay d and delay spread s are the first and second central moments respectively of the average power–delay profile $P(\tau)$. For wideband communication systems, the frequency selectivity relates to the frequency coherence of the spectral content of the signal. This is specified by the frequency correlation function $R_T(\Omega)$ obtained by using the Wiener Khinchin relationship, and Fourier transformation of $P(\tau)$ yields $R_T(\Omega)$. We have already stated that decorrelation of signal components in the frequency domain is quantitatively expressed in terms of the coherence bandwidth, which is commonly defined as the frequency separation which corresponds to 0.5 correlation, i.e. $|R_T(\Omega)| = 0.5$, although other values such as 0.9 have been used.

2.8 Small-area characterization

The characteristics of radio propagation in urban areas depend largely on factors such as the nature of the terrain, the height and density of buildings and the street orientation, and these characteristics can be studied by observing the features of measured scattering functions. For example, Figure 2.21 shows a scattering function measured in a suburban area of a city, the particular road in this case being almost radial with respect to the transmitter. Two-storey detached houses are evenly spaced along this street on one side, while the opposite side has fewer but larger dwellings. Echoes in the scattering function are identified below by the coordinates (x, y), where x is the excess time delay in μs and y is the Doppler shift in Hz. Thus the large-amplitude echo at (0.1, -3.8) represents the shortest path, in the absence of a line-of-sight (LOS) path, in this mid-block section of the street, and this scattered component is expected in view of the near-radial street orientation with respect to the transmitter. The presence of strong echoes with near-maximum Doppler shifts leads to the conclusion that the majority of the scattered waves arrive from ahead and behind the vehicle rather than from the 90° angular directions. Scattering from the flat sides of houses in the immediate vicinity of the mobile is indicated by the presence of echoes within the rectangle defined by the co-ordinates (0.1, 0) to (0.3, 2.5). The elevated houses lying ahead contribute

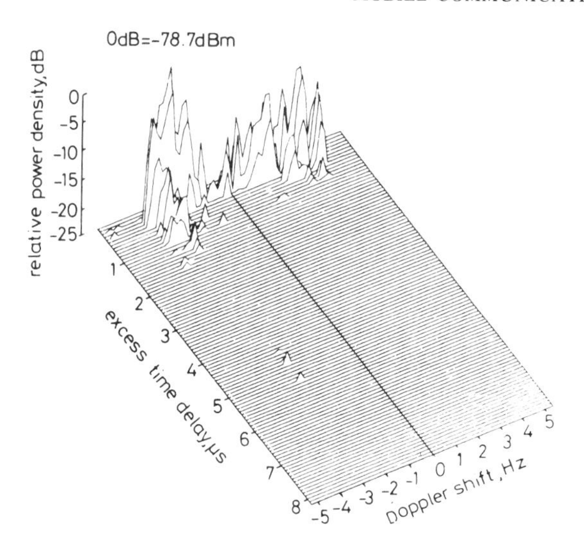

Figure 2.21 Measured scattering function in a radial suburban street.

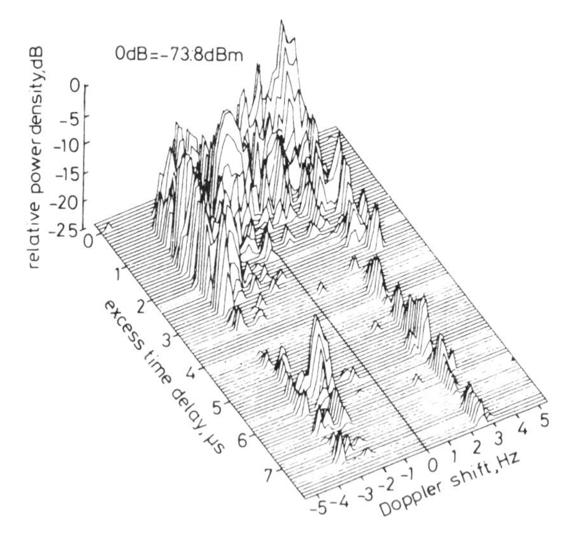

Figure 2.22 Measured scattering function in an area having extreme multipath conditions.

scattered components with excess delays up to 0.5 μs. Weaker echoes with longer delays and negative Doppler shifts are due to non-local reflections from large buildings some distance away. Clearly, the scattering in this residential area is localized to the immediate environment and depends on the street orientation and sloping terrain.

In certain environments, isolated large buildings can play a significant role in determining the multipath characteristics. Figure 2.22 shows a scattering function for another road and this represents a more extreme case of suburban multipath. The direct, LOS, path is blocked by a large building, and other non-local features that influence propagation in this area include some high-rise residential blocks. A significant feature at this location is the wide spread of time delays due to contributions from the large non-local scatterers. Due to the complex topography, it is not possible to explain all echoes on a basis of single scattering. In general, the local scattering includes all angles of arrival for the radiowaves, due to the circumferential orientation of the street. Furthermore, the scattering is uncorrelated for different time delays, this being evident from the dissimilar Doppler spectra at different delays. The presence of large scatterers in suburban areas gives multipath radio propagation an inhomogeneous character, and large reflected components, when shadowed due to motion of the vehicle relative to the local topography, can drastically alter the resultant scattering.

In densely built-up city areas, propagation is mainly by means of scattering. The environment is more uniform than in suburban areas, although echoes from large buildings can sometimes be identified. Nevertheless, significant differences can be observed between data collected at street intersections and those collected in mid-block areas. Street orientation influences the scattering pattern in mid-block areas, but at intersections, radio waves scattered by buildings along both streets arrive unobstructed at the mobile receiver, and street orientation with respect to the transmitter does not influence the angular distribution of scattered paths. In certain situations, the non-local scattering can also change.

2.8.1 Statistical characterization

The discussion in the previous section has illustrated the role of the scattering function in the study of multipath radio propagation. Assessment of communication systems can be aided by viewing this, and other system functions, as statistical input–output descriptors of the two-port channel model. Because the power-delay profiles represent a statistical function, it is possible to extract parameters that indicate the severity of the multipath. These parameters, the first and second central moments of the average power delay profile, have been discussed in section 2.7.1, and the manner in which they can be determined from the experimental results has been described.

In suburban areas, multipath is predominantly due to local scattering from residential houses, and the time–delay dispersion is usually less than 1 μs. The power–delay profile and frequency correlation function for a typical suburban street are shown in Figure 2.23, the average delay being 0.64 μs and the delay spread 0.81 μs. There is a low degree of delay dispersion, and this leads to a smoothly varying frequency correlation function and a reasonably large coherence bandwidth of 2 MHz. To obtain some indication of the likely overall performance of communication systems in suburban areas, it is necessary to consider a more extreme multipath situation, such as that

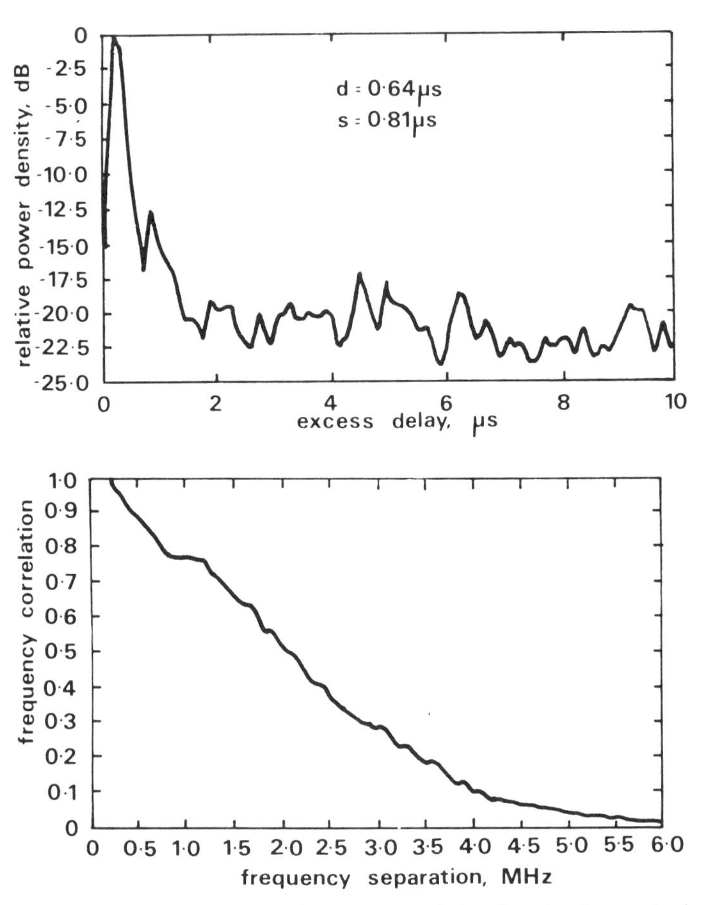

Figure 2.23 Power delay profile and frequency correlation function in a suburban area. (*a*) Typical average envelope power against time delay for a suburban area. (*b*) Typical frequency correlation function for a suburban area.

represented by the power delay profile and frequency correlation function in Figure 2.24, which correspond to the scattering function of Figure 2.22. The average delay and the delay spread are more than double the 'typical' values, and the frequency correlation function, although it decreases rapidly at first, contains a number of oscillatory variations. The periodicity in the fine oscillatory structure in the correlation function is determined by the echo at an excess delay of 5.5 μs in the power delay profile, and a coherence bandwidth of 200 kHz indicates that the radio channel is highly frequency-selective. The difference between typical and extreme multipath conditions highlights the variability of propagation conditions in suburban areas.

The multipath characteristics in the dense high-rise localities are representative of typical urban areas. In these localities, the scattering is generally stable due to the homogeneous environment, and consequently the multipath

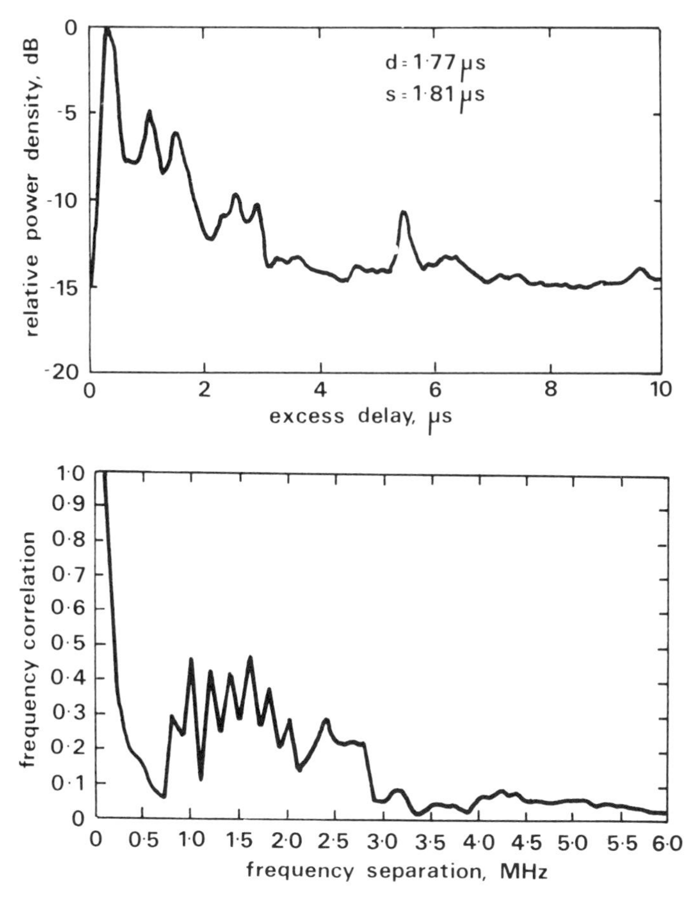

Figure 2.24 Power delay profile and frequency correlation function in a suburban area under extreme multipath conditions.

parameters are less variable. A power-delay profile taken close to an intersection is shown in Figure 2.25*a*. The echoes due to local scattering decrease in strength, except for the large echo at an excess delay of 2 μs. The presence of such an echo appears in the form of oscillations in the frequency correlation function of Figure 2.25*b*. Using a definition of coherence bandwidth in which the frequency separation for 0.5 correlation is taken, it is clear that Figure 2.25*b* indicates three different values. Obviously the smallest frequency separation indicates the actual coherence bandwidth, but nevertheless it demonstrates that measurement of coherence bandwidth using spaced tones is not a satisfactory method unless measurements are carefully repeated for different frequency separations. In the literature [4, 13], a smoothly decreasing echo power distribution is often assumed, implying a smooth decorrelation due to the inherent relationship between these functions [8].

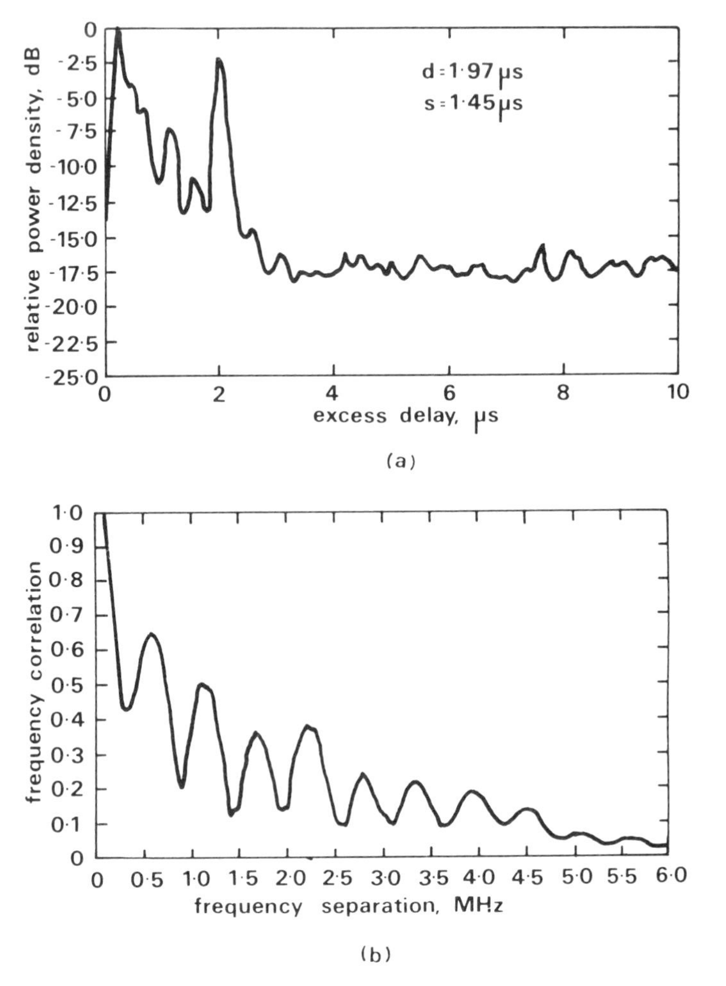

Figure 2.25 Power delay profile and frequency correlation function in an urban area. (*a*) Typical average envelope power against time delay for an urban area. (*b*) Typical frequency correlation function for an urban area.

Clearly these assumptions are unrealistic for multipath propagation in certain suburban, and most urban, areas. More extreme multipath characteristics in urban areas can be found [10], particularly in industrial areas. Echoes with large excess delays are often a salient feature of the power delay profile, these echoes being due to non-local scattering from large buildings some distance away. It is interesting to note the resemblance of these multipath characteristics to those in suburban locations influenced by remotely sited high-rise buildings, because in both areas this environment represents an extreme case. The first minimum in the frequency correlation function often indicates significant decorrelation, although not enough to reach the value of 0.5.

However, it could be argued that this is indicative of the actual coherence bandwidth and that the value obtained by considering decorrelation to 0.5 is deceptively optimistic. The very substantial influence of strong delayed components is obvious in these cases, since components at a certain excess delay produce corresponding oscillatory variations in the frequency correlation function.

2.9 Large-area characterization

In the previous sections we have discussed parameters suitable for channel characterization and have described experimental techniques and data processing methods which yield these parameters. It is evident that the multipath characteristics in urban areas are grossly non-stationary in nature, so that experimental data collected at a given location provide parameters that, strictly, describe propagation only at that location, that is, provide a small-area characterization. Extension from local to global characterization is therefore not a trivial matter. Nevertheless, it is very important, because designers of radio communication systems intended to operate in the presence of multipath wish to know, given a certain performance criterion, at what percentage of locations within the service area the equipment will be satisfactory and what can be done to improve matters.

We therefore turn briefly to the question of large-area characterization and attempt to provide a bridge between the local and the global situation. It has been observed that many radio channels have fading characteristics that are stationary over short periods of time (or over short distances of travel). These channels, although not stationary in the strict sense, can be regarded as stationary in the wide sense (weakly stationary). Wide-sense stationary (WSS) channels have the property that the channel correlation functions are invariant under a translation in time.

Over spatial distances of the order of a few tens of wavelengths the descriptive statistics of the mobile radio channel are wide-sense stationary. As the mobile receiver moves from one locality to another, changes in the dominant features of the environment are manifested in the form of non-stationarity in the statistical characteristics of the multipath. In general, mobile radio channels can be described as wide-sense stationary uncorrelated scattering (WSSUS) and, since certain parameters adequately describe the dispersiveness of such channels, it is reasonable to study the non-stationary behaviour by observing the variability of these parameters over a large, but characteristically similar, area. As far as an assessment of communication systems is concerned, this two-stage model of the multipath represents a practical and adequate approach to the large-scale characterization of mobile radio channels [10, 14]. Variability in the homogeneity of the environment influences the dispersiveness of the channel, and changes in the channel parameters reflect these characteristics.

2.9.1 Distribution of the channel parameters

The approach therefore is to obtain WSSUS channel descriptors in the form of average power time-delay profiles for a large number of relatively small areas and to observe the variation in the parameters of these descriptors. In general, the average echo power versus time-delay represents a statistical distribution of echo power over the small area. Therefore, the average excess delay and delay spread are relevant statistical parameters of the WSSUS channel. These parameters are insensitive to the shape of the power-delay profile for homogeneous environments due to the predominantly scattered mode of radio propagation. However, this is not the case in the presence of strong reflected components that have large excess delays, this situation being evidenced in practice by spikes in the power-delay profile at the longer values of time delay.

2.9.1.1 Suburban areas. Histograms of the average delay and delay spread in suburban areas are shown in Figure 2.26. For a significant percentage of locations, the values of these two parameters greatly exceed the values typical of the areas under consideration. For example, at 22% of locations the average delay is more than $\mu_d + \sigma_d$ (1.4 μs). The extreme values, which correspond to the 'tails' of the statistical distribution of the parameters, occur predominantly in areas with high-rise buildings. These 'tails' exhibit the same behaviour for the two parameters, since they consistently constitute more than 20% of the locations. The bimodal nature of the histograms, emphasizing low and high values of delay and delay spread, reflects the grossly inhomogeneous composition of the suburban environment. System designs based on typical or mean values of the parameters are therefore unlikely to provide the desired overall performance due to the large variability of the channel parameters over the service area. In general, the more typical residential areas display less time-delay dispersion than suburban areas influenced by non-local high-rise buildings.

The scatter plot in Figure 2.27 shows the correlation of the average excess delay and the delay spread. It is clear that the two parameters tend to vary in a related manner. The concentration of the points about the regression relationship $d = 0.143 + 0.88s$ is indicative of the high correlation. Thus, low values of d are usually associated with low values of s, and vice versa. Furthermore, the low values of s and d are usually associated with a multipath situation in which the scattered power decreases smoothly, whereas high values are found in locations where strong specular components exist with large excess delays.

A histogram of the coherence bandwidth of the channel (as defined by a 0.5 correlation coefficient) is shown in Figure 2.28. The lower extreme signifies the worst multipath dispersion which, of course, manifests itself in the form of frequency selectivity. For multipath resulting mainly from scattering, the coherence bandwidth B_c is inversely proportional to the delay spread s. The

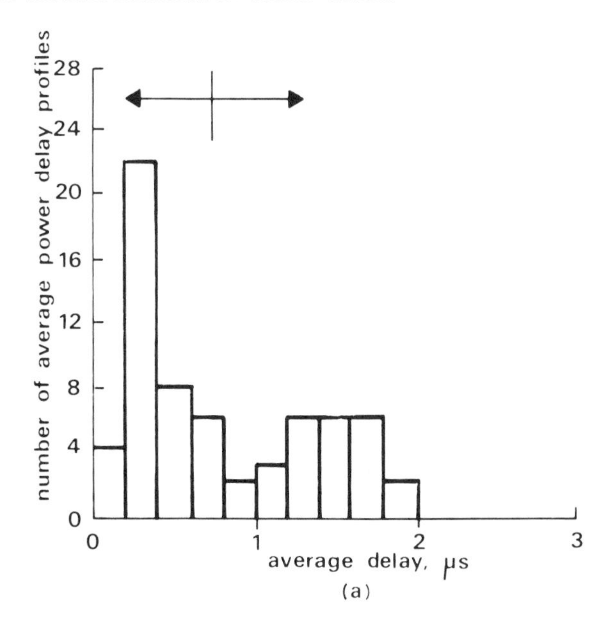

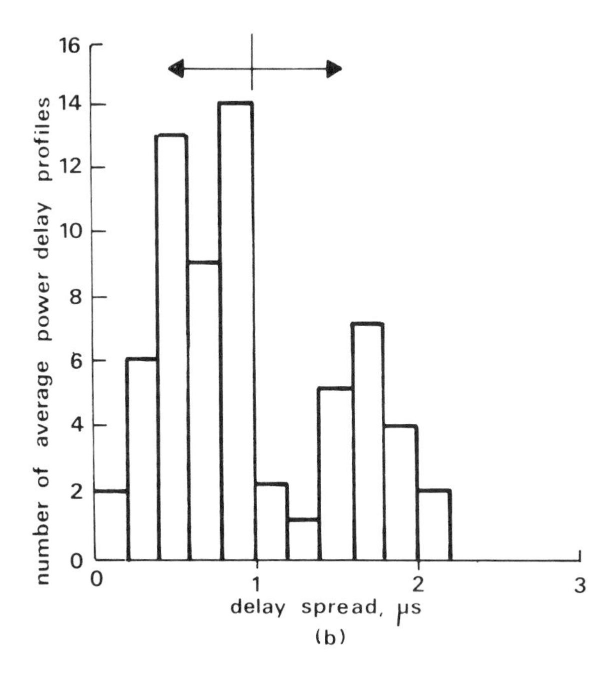

Figure 2.26 Histograms typical of suburban areas. (*a*) Average delay: mean $\mu_d = 0.78\,\mu s$, standard deviation $\sigma_d = 0.56\,\mu s$; (*b*) delay spread: mean $\mu_s = 1.04\,\mu s$, standard deviation $\sigma_s = 0.57\,\mu s$.

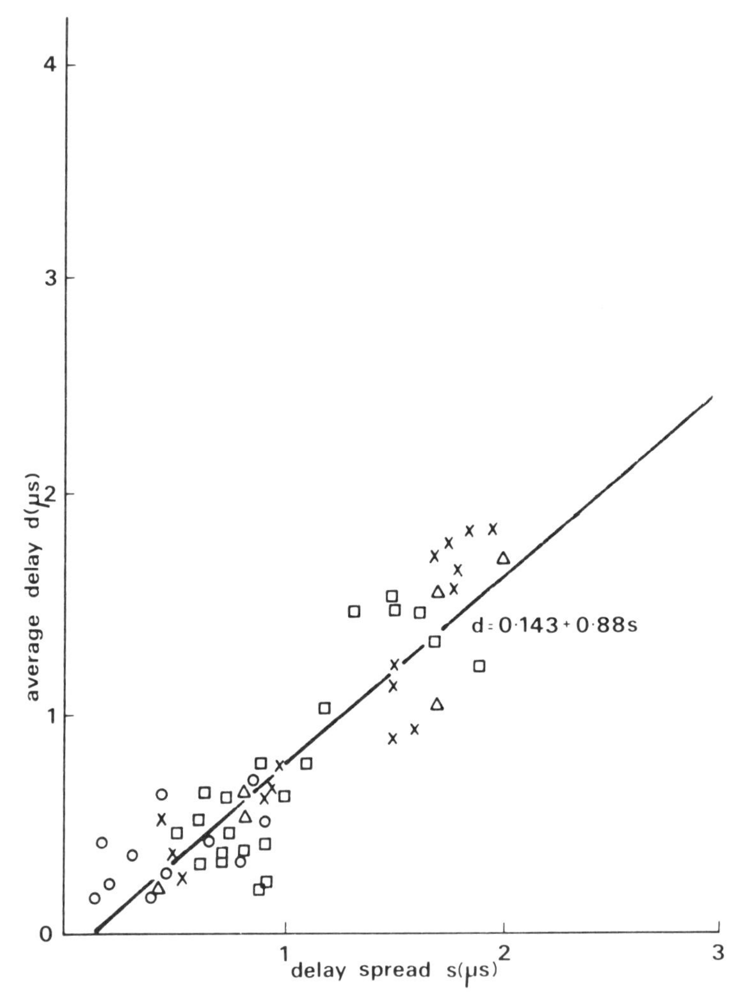

Figure 2.27 Scatter plot of average delay against delay spread in suburban areas. ○ Residential (typical); □ residential (sparse); × residential (high-rise); △ residential (sparse/high-rise).

scatter plot of the coherence bandwidth in Figure 2.29 shows the departures of experimental data from this relationship, caused by the differences in modes of propagation at different locations. From a linear regression of the logarithms, the empirical relationship is determined as $B_c = 1420s^{-1.57}$, where s is expressed in μs and B_c in kHz. It is clear that the coherence bandwidth decreases more sharply as a function of delay spread than would be the case in a purely scattered mode. The presence of large specular components determines the empirical relationship, and these specular components also induce an oscillatory variation in the frequency correlation function, the first minimum being set by the inverse of the excess time-delay of this component in

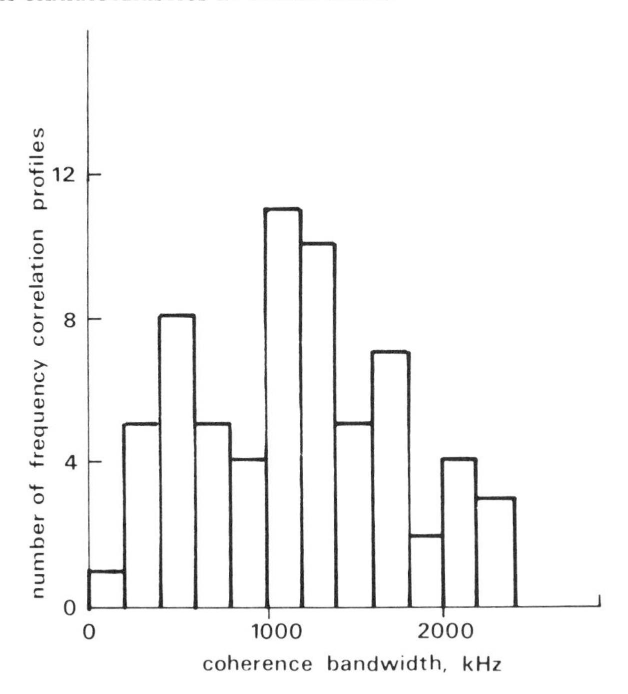

Figure 2.28 Histogram of typical coherence bandwidth in suburban areas.

the echo profile. Moderate to low values of B_c are usually associated with locations which exhibit such correlation functions, and these locations are often found in the high-rise residential areas.

2.9.1.2 Urban areas. Histograms of average excess delay and delay spread for urban areas are shown in Figure 2.30. As far as average delay is concerned the very low values in Figure 2.30*a* are found to a greater extent in dense high-rise areas. Moderate values for the excess delay have a tendency to occur in all types of urban environment.

The low to moderate values of delay spread in Figure 2.30*b* are mainly derived from observations in dense high-rise surroundings, while the much higher values occur only in commercial and industrial urban areas. The variance of the delay spread is smaller than in suburban areas, due to the greater homogeneity of the urban environment, and the extreme value, $\mu_s + \sigma_s = 2\,\mu s$, is exceeded in only 7% of the locations. This is comparable to the fluctuations induced by the statistical sample size. However, the average excess delays exceed $\mu_d + \sigma_d = 2\mu s$ in 20% of the locations and will therefore require consideration in the assessment of automatic vehicle location systems.

In urban areas, the average excess delay and the delay spread show a low degree of correlation, and, in contrast to suburban areas, the trend in the dependence of d and s is different for low and high values of s. In urban areas,

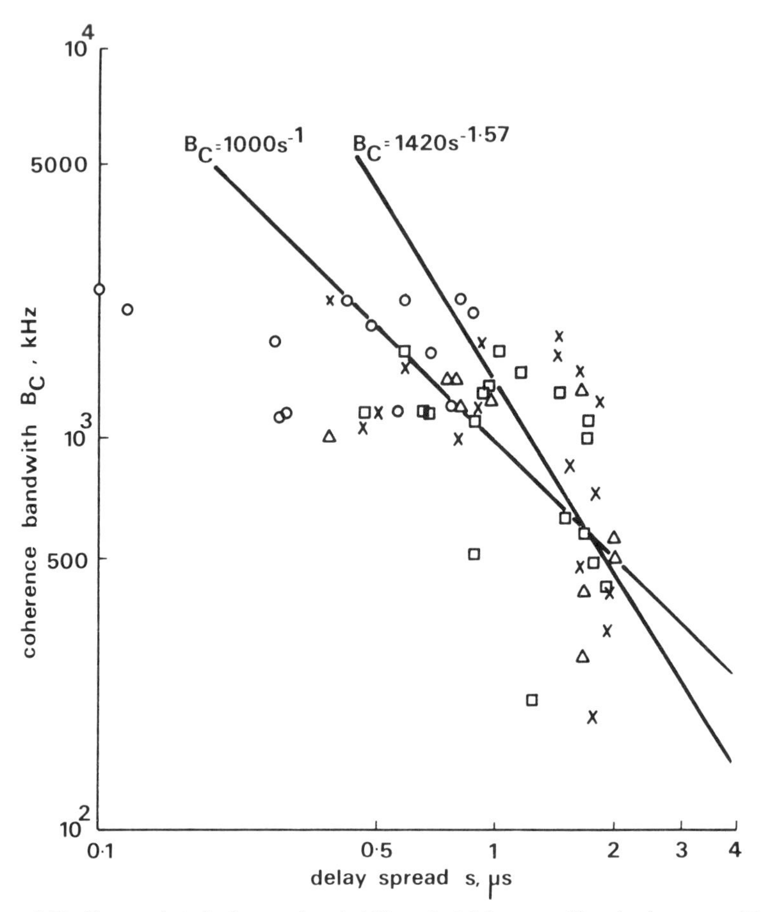

Figure 2.29 Scatter plot of coherence bandwidth against delay spread in suburban areas. (Key as in Figure 2.27).

average delays are smaller than the delay spreads for low to moderate values, whereas the reverse is the case for high values. Power delay profiles for dense high-rise areas depict smoothly scattered multipath components and consequently moderate values.

A histogram for the coherence bandwidths in an urban environment is shown in Figure 2.31. Apart from the 300 to 400 kHz quantile which dominates, all the other quantiles below 1 MHz exhibit a similar probability of occurrence. Scatter plots drawn on the same basis as that in Figure 2.27 show a smaller spread in urban areas, and a typical relationship between B_c and s is $B_c = 730\ s^{-1.42}$. By comparison with suburban data, it appears that the coherence bandwidth is less dependent on s because of the smaller index in the inverse relationship. However, the decrease is still more rapid than in a simple

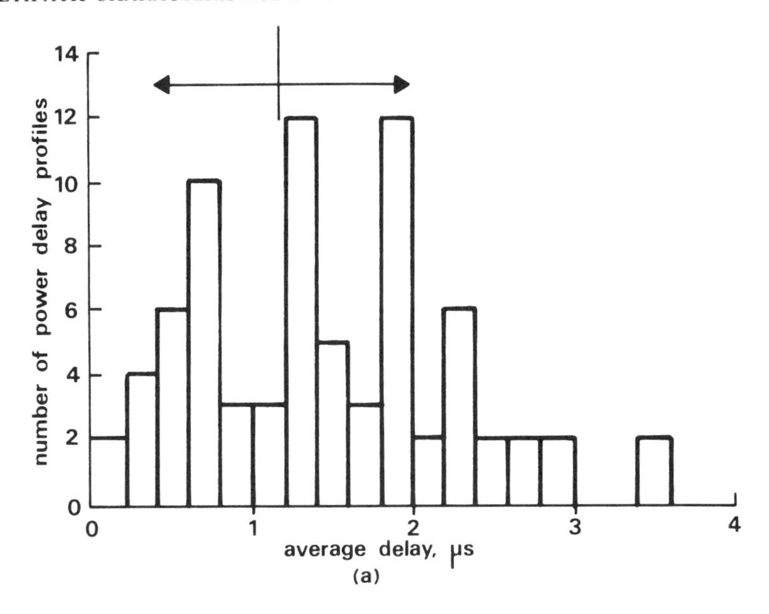

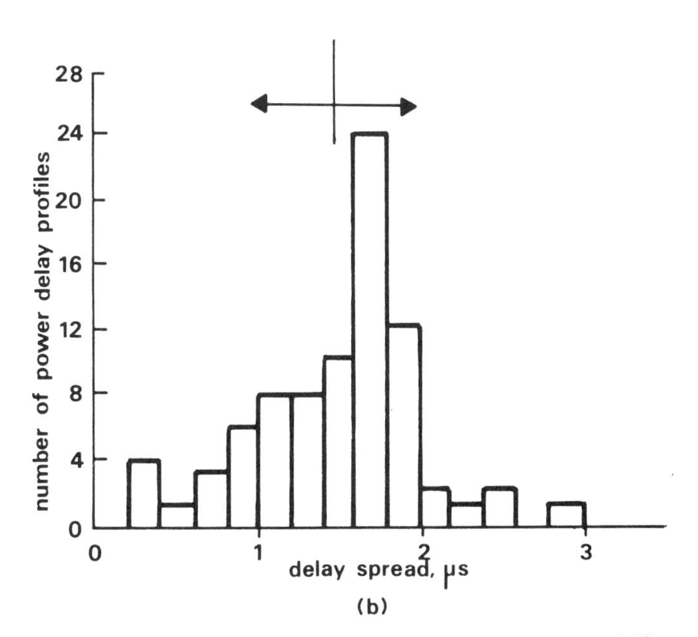

Figure 2.30 Histograms typical of urban areas. (*a*) Average delay: mean $\mu_d = 1.18\,\mu s$, standard deviation $\sigma_d = 0.79\,\mu s$; (*b*) delay spread: mean $\mu_s = 1.44\,\mu s$, standard deviation $\sigma_s = 0.5\,\mu s$.

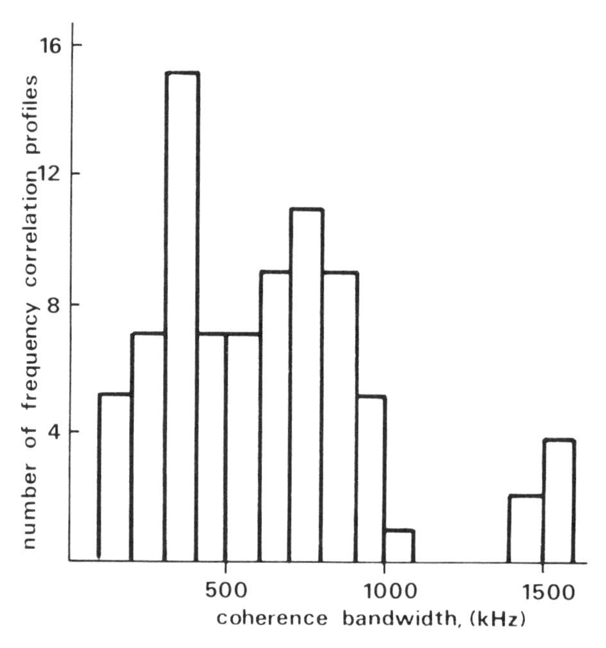

Figure 2.31 Histogram of typical coherence bandwidth in urban areas.

inverse law, due to the presence of large-amplitude echoes at longer excess delays.

2.10 Conclusions

Small-scale channel descriptors in the form of the average power-delay profile provide useful information about mobile radio channels, and the variability of the statistical parameters, each extracted from a WSSUS channel representation, is an appropriate large-scale descriptor in a two-stage characterization of the transmission medium. Extreme values of average delay and delay spread occur quite frequently in suburban areas; they are most common in inhomogeneous environments with high-rise residential buildings, irrespective of street orientation. Urban areas show comparable extreme values, though less frequently in the case of the delay spread. The outer fringe of the urban environment often represents the worst multipath, being influenced to a greater extent by remote but tall buildings some distance away. The effects of inhomogeneous scattering are also evident in the statistical distributions for the coherence bandwidths. Regression analysis shows the assumed inverse dependence of B_c on the delay spread to be optimistic, the measured inverse law with s having exponents greater than unity in both urban and suburban areas. Variability of the parameters is greater in suburban areas, although the extreme values are comparable. This large-scale characterization can be applied to an evaluation of radio systems [15] and the performance bounds then emerge in terms of the percentage of locations, where the given performance measure is not exceeded.

References

1. Ossana, J.F. Jr. A model for mobile radio fading due to building reflections: theoretical and experimental fading waveform power spectra. *Bell Syst. Tech. J.* **43** (Nov. 1964) 2935–2971.
2. Clarke, R.H. A statistical theory of mobile radio reception. *Bell Syst. Tech. J.* **47** (July–Aug. 1968) 957–1000.
3. Rice, S.O. Statistical properties of a sine wave plus random noise. *Bell Syst. Tech. J.* **27** (1948) 109–157.
4. Jakes, W.C. Jr. (ed.) *Microwave Mobile Communications.* Wiley–Interscience, New York (1974).
5. Rhee, S.B. and Zysman, G.I. Results of suburban base-station spatial diversity measurements in the UHF band. *IEEE Trans.* **COM-22** (Oct. 1974) 1630–1634.
6. Adachi, F., Feeney, M.T., Williamson, A.G. and Parsons, J.D. Correlation between the envelopes of 900 MHz signals received at a mobile radio base-station site. *Proc. IRE, Part F* **133** (1986) 506–512.
7. Bello, P.A. Characterisation of randomly time-variant linear channels. *IEEE Trans.* **CS-11** (Dec. 1963) 360–393.
8. Parsons, J.D. and Bajwa, A.S. Wideband characterisation of fading mobile radio channels. *Proc. IEE, Part F* **129** (April 1982) 95–101.
9. Cox, D.C. 910 MHz urban mobile radio propagation: multipath characteristics in New York City. *IEEE Trans.* **COM-21** (Nov. 1973) 1188–1194.
10. Bajwa, A.S. Wideband characterisation of UHF mobile radio propagation in urban and suburban areas. Ph.D. thesis, Department of Electronic and Electrical Engineering, University of Birmingham (1979).
11. Kennedy, R.S. *Fading Dispersive Communication Channels.* Wiley–Interscience, New York (1969).
12. Matthews, P.A., Molkdar, D. and Rashidzadeh, B. Measurement, description and modelling of the UHF terrestrial mobile radio channel. *Proc. 3rd Int. Conf. on Land Mobile Radio*, Cambridge, Dec. 1985 (*IERE Conf. Publ.* **65**, 119–125).
13. Bello, P.A. and Nelin, B.D. The effect of frequency fading on the binary error probabilities of incoherent and differentiability coherent matched filter receivers. *IEEE Trans.* **CS-11** (June 1963) 170–186.
14. Cox, D.C. and Leck, R.P. Distribution of multipath delay spread and average excess delay for 910 MHz urban mobile radio paths. *IEEE Trans.* **AP-23** (March 1975) 206–213.
15. Bajwa, A.S. and Parsons, J.D. Large area characterisation of urban UHF multipath propagation and its relevance to the performance bounds of mobile radio systems. *IEE Proc., Part F* **132** (April 1985) 99–106.

3 Propagation and signal strength prediction

3.1 Introduction

We saw at the beginning of Chapter 2 that in general the mobile radio signal is extremely variable. The mean signal strength will be constant only over relatively small areas, and will vary slowly as the receiver is moved. Superimposed on this slowly-varying mean is the fast fading which is caused by multipath propagation in the immediate vicinity of the receiver, and this was also discussed in Chapter 2.

In this chapter we describe methods by which the mean signal strength in any small area can be predicted, and introduce statistical techniques for expressing the variability of the mean over larger areas. This is arguably the most important task in system planning, since the most essential requirement in any radio communication system is that there should be an adequate signal strength in areas where receivers are intended to operate. Indeed, it is crucial for efficient use of the allocated frequency band and for the success of cellular systems, which rely heavily on frequency re-use, to be able to predict the minimum power of transmission required from a given base station at a given frequency to provide an acceptable quality of coverage over a predetermined service area, and to estimate the likely effect of such transmissions on existing adjacent services. In this context, an understanding of the influence of various urban and terrain factors on the mobile radio signal and its variability is essential. Therefore, before describing methods that are useful for predicting propagation over irregular terrain and in built-up areas, we briefly review some of the fundamentals which are the foundation on which the prediction techniques are based.

Traditional mobile radio services have been based on the concept of an elevated base station located on a good site, communicating with a number of mobiles in the surrounding area. However, the expanded services which are now being implemented use hand-portable equipment and cordless telephones, so in considering radio propagation it is necessary not only to take into account the effects of irregular terrain and built-up areas, but also propagation into buildings and in the areas immediately surrounding buildings. For cordless telephones, of course, there is also a need to study propagation totally within buildings. The frequencies in current use lie in the range 70–900 MHz, but the increasing congestion within these bands makes it almost certain that in the future even higher frequencies will be used.

Propagation studies are required not only in connection with narrowband transmissions of the traditional type, but also in connection with wideband systems which may be used in the future. The way in which local multipath propagation affects various systems will differ, as illustrated in Chapter 2, but no system can operate without an adequate signal. Indeed, it is worth pointing out that adequate reception conditions require not only an acceptable signal, but also an acceptable carrier-to-interference ratio and carrier-to-noise ratio. In considering the suitability of a particular band of frequencies for mobile radio use it is therefore important to take into account not only the signal characteristics, but also the magnitude and nature of the ambient noise. This latter aspect will be considered in Chapter 5.

3.2 Fundamentals of VHF and UHF propagation

Electromagnetic waves are launched into space by energizing a suitable antenna. At frequencies below 1 GHz the antenna normally consists of a wire or wires of a suitable length coupled to the transmitter via a transmission line. At these frequencies it is relatively easy to design an assembly of wire radiators which form an array, in order to beam the radiation in a particular direction. For distances large in comparison with the wavelength and the dimensions of the array, the field strength in free space decreases with an increase in distance, and a plot of the field strength as a function of angle is known as the radiation pattern of the antenna.

We can design an antenna to have a radiation pattern which is not omnidirectional, and it is convenient to have a figure of merit to quantify the ability of the antenna to concentrate the radiated energy in a particular direction.

The directivity D of an antenna is defined as

$$D = \frac{\text{power density at a distance } d \text{ in the direction of maximum radiation}}{\text{mean power density at distance } d}$$

This is a measure of the extent to which the power density in the direction of maximum radiation exceeds the average power density at the same distance.

The directivity involves knowing the power actually transmitted by the antenna; this differs from the power supplied at the terminals by the losses in the antenna itself. From the system designer's point of view, it is more convenient to work in terms of this quantity, and a power gain G can be defined as

$$G = \frac{\text{power density at a distance } d \text{ in the direction of maximum radiation}}{P_T/4\pi d^2}$$

where P_T = power supplied to the antenna.

Note that G/D is the efficiency of the antenna.

So, given P_T and G, it is possible to calculate the power density at any point in the far field that lies in the direction of maximum radiation. A knowledge of the radiation pattern is necessary to determine fields at other points.

The power gain is unity for an isotropic antenna, i.e. one which radiates uniformly in all directions, and an alternative definition of power gain is therefore the ratio of power density in the direction of maximum radiation to that of an isotropic antenna radiating the same power. We can calculate the power gain of an antenna or an array by integrating the outward power flow and relating it to the mean power flux. For example, the power gain of a $\lambda/2$ dipole is 1.64 (2.15 dB) in a direction normal to the dipole, and is the same whether the antenna is used for transmission or reception.

The concept of effective area is useful in considering receiving antennas. If an antenna is placed in the field of an electromagnetic wave, the received power available at its terminals is the effective area × power per unit area carried by the wave, i.e. $P = AW$.

It can be shown [1, Chapter 11] that the effective area of an antenna and its power gain are related by

$$A = \frac{\lambda^2 G}{4\pi} \tag{3.1}$$

3.2.1 Propagation in free space

Radio propagation is a subject that lends itself to deterministic analysis only in a few, rather simple cases. These cases represent idealized situations, but can be useful in giving insight into the basic propagation mechanisms.

If we consider two antennas separated by a distance d in free space, i.e. remote from any obstruction, then, if the transmitting antenna has a gain G_T, the power density at a distance d is

$$\frac{P_T G_T}{4\pi d^2} \tag{3.2}$$

The power available at the receiving antenna, which has an effective area A, is therefore

$$P_R = \frac{P_T G_T}{4\pi d^2} A$$

$$= \frac{P_T G_T}{4\pi d^2} \frac{\lambda^2 G_R}{4\pi}$$

So that

$$\frac{P_R}{P_T} = G_T G_R \left(\frac{\lambda}{4\pi d}\right)^2 \tag{3.3}$$

This is known as the free-space or Friis equation.

The propagation loss is usually expressed in dB, and from this equation we can write

$$\begin{aligned} L_F &= 10\log_{10} P_R/P_T \\ &= 10\log G_T + 10\log G_R - 20\log f - 20\log d + K. \end{aligned} \tag{3.4}$$

where

$$K = 20\log\frac{3\times10^8}{4\pi} = 147.56.$$

For unity-gain (isotropic) antennas we can define a 'basic transmission loss' L_B as

$$L_B(\text{dB}) = -32.44 - 20\log f_{\text{MHz}} - 20\log d_{\text{km}} \tag{3.5}$$

This is an inverse square law with distance, so the received power decreases by 6 dB for every doubling of distance. The equation also shows that the transmission loss increases with the square of the frequency, and so high-gain antennas are necessary to make up for this loss. The question to be answered is to what extent this formula is modified by atmospheric effects, the presence of the Earth and the effects of trees, buildings and hills which exist in, or close to, the transmission path.

3.2.2 Propagation over a reflecting surface

The situation of two mutually-visible antennas above a smooth reflecting surface is shown in Figure 3.1. It is convenient to draw ray paths as straight lines, and if the surface is that of the Earth (assumed spherical), then the radius r_e is often taken to be 4/3 times the actual Earth radius in order to account for refraction in the lower atmosphere. It can be seen that the total electric field at the receiving antenna is made up of two components: a direct wave and another wave that has been reflected from the surface. The attenuation along these individual paths is given by eqn (3.3), and the resultant received field will therefore depend upon the difference in length of the two paths and the reflection coefficient of the surface.

A situation of interest in mobile radio is when the antenna heights are small compared with their separation, but nevertheless the separation is only a few tens of kilometres, so that the reflecting surface (the Earth) can be considered to be flat (or plane). In these circumstances, it is usual to assume that the difference in path length is negligible, as far as attenuation is concerned. However, the difference in path length, as it affects the phase of the two received waves, cannot be neglected. Furthermore, although the reflection coefficient of the Earth depends upon the ground constants, such as

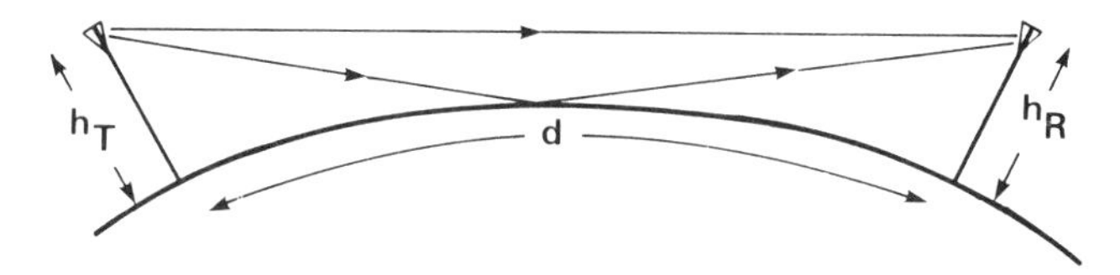

Figure 3.1 Two antennas above a smooth reflecting surface.

permittivity and conductivity, for h_T, $h_R \ll d$ i.e. grazing incidence, the Earth appears to be perfectly reflecting, so that the reflection coefficient $\rho = -1$, independent of the polarization. In these circumstances it can be shown [2, Chapter 4] that the received and transmitted powers are related by the expression

$$\frac{P_R}{P_T} = G_R G_T \left(\frac{h_T h_R}{d^2}\right)^2 \tag{3.6}$$

This is often termed the plane-Earth propagation equation, and can be written as

$$L_p(\text{dB}) = 10 \log G_T + 10 \log G_R + 20 \log h_T + 20 \log h_R - 40 \log d \tag{3.7}$$

Again, for the case of isotropic antennas this becomes the basis loss, which is

$$= 20 \log h_T + 20 \log h_R - 40 \log d \tag{3.8}$$

Here, the units of h_T, h_R and d must be the same.

We note that in this case there is an inverse fourth-power law with distance, so the received power decreases by 12 dB when the distance is doubled. However, surprisingly perhaps, the frequency does not appear in eqn (3.6); this is a consequence of the assumptions that h_T and h_R are very much smaller than d, and that the surface is perfectly reflecting.

3.2.3 The effect of surface roughness

The surface of the Earth is not really smooth, and some criterion is required to judge whether, in any given situation, the surface will give an appreciable specular reflection. Generally, if the undulations, as viewed from the angle of incidence, are small compared with the wavelength of the radio waves, the surface appears smooth; if not, it appears rough. It is common to define a factor C, known as the Rayleigh criterion, and given by

$$C = \frac{4\pi\sigma}{\lambda} \sin\theta \tag{3.9}$$

where σ is the standard deviation of the terrain irregularities relative to the mean height of the surface as shown in Figure 3.2. For grazing incidence, this can be simplified to

$$C = \frac{4\pi\sigma\theta}{\lambda} \tag{3.10}$$

Figure 3.2 Illustrating the Rayleigh criterion of surface roughness.

Experimental evidence suggests that, if $C < 0.1$, the surface can be considered smooth, so eqn (3.6) is valid. For values of $C > 10$, the reflected wave is almost insignificant.

3.3 Propagation over terrain obstacles

In the previous paragraphs we have assumed the Earth to be a smooth reflecting surface, but in practice of course this is far from the true situation. Obstacles such as trees, hills and buildings often exist in the radiowave path and these cause shadows within which there are large reductions of field strength, particularly at UHF. If the terminals of a path for which line-of-sight exists are low enough for the path to pass close to the surface of the Earth, then there may be a loss well in excess of the free-space loss, even if the path is not directly obstructed. In fact, a substantial clearance is required over terrain obstacles in order to obtain 'free-space' transmission.

A quantitative measure of the clearance required over terrain obstructions may be obtained in terms of Fresnel-zone ellipsoids drawn around the terminals, as shown in Figure 3.3.

3.3.1 Fresnel zones

Consider a transmitter T and a receiver R in free space, as in Figure 3.4. Imagine a plane perpendicular to TOR at some point along the transmission path. On this plane we sketch concentric circles which represent the loci of the origin of secondary waves travelling to R. The radii of the circles are such that the total path length from T to R via each circle is $n\lambda/2$ greater than that via TOR, where n is an integer. Thus, $TA_1 + A_1R$ is $\lambda/2$ longer than TOR, and $TA_2 + A_2R$ is λ longer than TOR, etc. The contribution of each successive zone to the field at R is thus alternately positive and negative. These zones are called Fresnel zones, and the radius of the nth Fresnel zone at a point defined by the geometry of Figure 3.4 is

$$R_n = \sqrt{\frac{n\lambda d_1 d_2}{d_1 + d_2}} \tag{3.11}$$

Since the contribution to the total field at R from successive Fresnel zones alternates, it is readily seen that if an obstructing screen were actually placed at the position shown, then as the radius of the central hole is increased, the field at R oscillates about the value that would exist in the absence of the screen (the

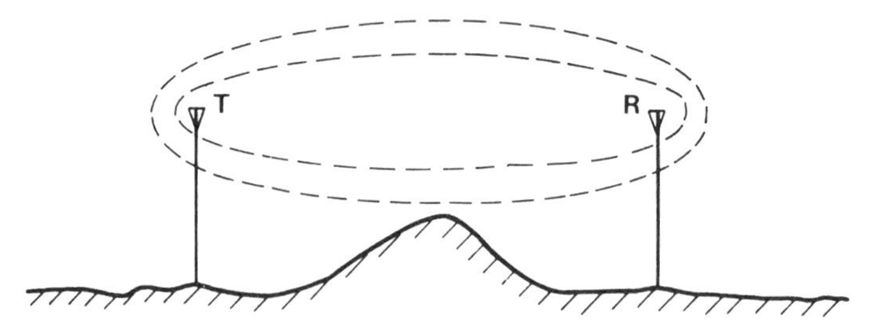

Figure 3.3 Estimating path clearance over obstacles, using Fresnel zones.

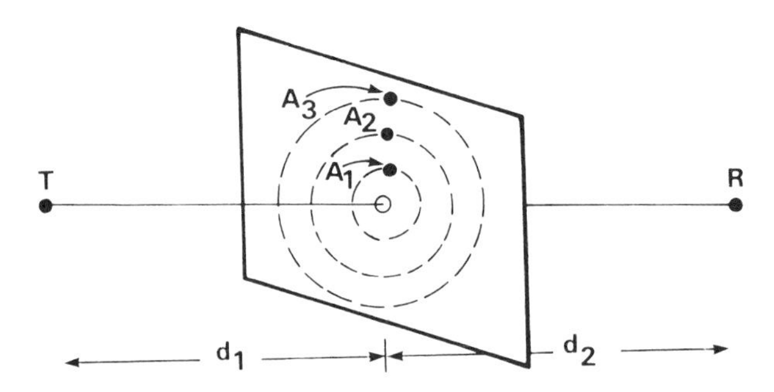

Figure 3.4 Illustrating Fresnel zones.

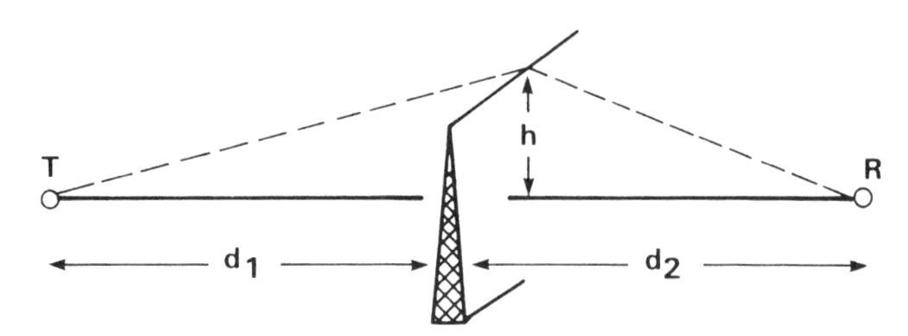

Figure 3.5 Knife-edge diffraction geometry.

free-space value). Only the first few zones are important, so the amplitude of the oscillation decreases rapidly as the radius is increased.

In radio propagation, the field at R is influenced not by a screen but by obstacles which lie in, or close to, the line-of-sight path as shown in Figure 3.5. If the straight-edge obstacle (considered as perfectly absorbing) interposed between T and R does not encroach into the first few Fresnel zones, then the field at R is unaffected. However, as the height is increased the field strength oscillates with increasing amplitude, in the manner described above. At the point where the obstructing edge is just in line with T and R, the field strength at R is 6 dB below the free-space value, and if the height of the obstruction is increased further, so that the line-of-sight path is actually blocked, the oscillation ceases and the field strength decreases steadily with height.

To express this in a quantitative way, we use classical diffraction theory (well documented in many books on optics) extended to radio propagation and we replace any obstruction along the path by an absorbing plane placed at the same position. The plane is normal to the direct path and extends to infinity in all directions except vertically, where it stops at the height of the original obstruction. All ground reflections are ignored.

The signal strength at a point R as shown in Figure 3.5 is determined as the sum of all contributions from the secondary Huygens sources in the plane above the edge. This yields an integral of the form

$$F(v) = \frac{(1+j)}{2} \int_v^\infty \exp\left(-j\frac{\pi}{2}t^2\right) \mathrm{d}t \tag{3.12}$$

which is known as the Fresnel integral. v is the Fresnel diffraction parameter given by

$$v = h\sqrt{\frac{2}{\lambda}\frac{(d_1+d_2)}{d_1 d_2}} \tag{3.13}$$

A graphical solution of this integral is shown in Figure 3.6 as a function of v.

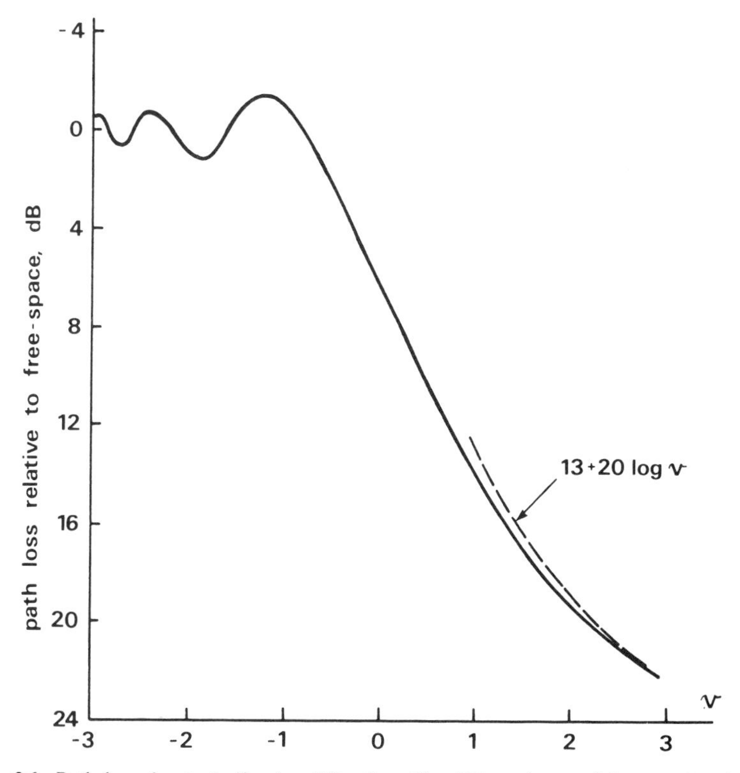

Figure 3.6 Path loss due to knife-edge diffraction. $13 + 20\log v$ is a useful approximation for $v > 2.4$.

It can be seen that if about 55% of the first zone around a radio path is kept free from obstructions, then we avoid the 6 dB loss associated with grazing incidence. In practice, designers of point-to-point links try to make the heights of their towers such that as much as possible of the first Fresnel zone is unobstructed, although there is not much to be gained from obtaining additional clearance once the 55% has been achieved.

For $v > 2.4$, the integral may be evaluated by series expansion of the integrand as

$$F(v) = \frac{2}{\sqrt{2\pi v}}$$

Approximations to the solution are available for $v < 2.4$ as:

$$-10\log_{10}|F(v)| = 6.02 + 9.11v - 1.27v^2 \quad \text{for } 0 \leqslant v \leqslant 2.4 \qquad (3.14)$$

$$-10\log_{10}|F(v)| = 6.02 + 9.0v + 1.65v^2 \quad \text{for } -0.8 \leqslant v \leqslant 0$$

3.3.2 Diffraction over rounded obstacles

Objects encountered in the physical world often have a shape far removed from that which can be approximated by a sharp knife-edge, and it is then necessary to consider modifications to the calculations made above [3]. The treatment normally given to the problem of a rounded obstacle is to replace it by a cylinder of radius equal to the crest, as illustrated in Figure 3.7. The solution is obtained in terms of a dimensionless parameter ρ, defined as

$$\rho = \left[\frac{\lambda}{\pi}\right]^{1/6} R^{1/3} \left[\frac{d}{d_1 d_2}\right]^{1/2} \qquad (3.15)$$

In terms of this parameter, the loss may be represented by a two-dimensional quantity $A(v, \rho)$, expressed in dB. This is related to the one-dimensional (ideal knife-edge) loss $A(v, 0)$ and a so-called curvature loss $A(0, \rho)$ by

$$A(v, \rho) = A(v, 0) + A(0, \rho) + U(v\rho)$$

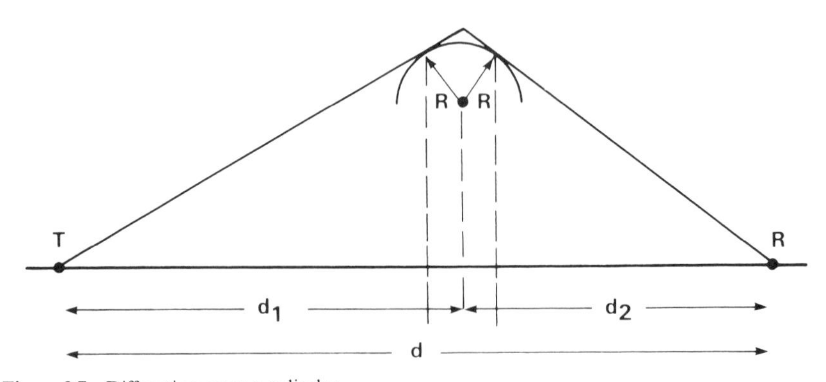

Figure 3.7 Diffraction over a cylinder.

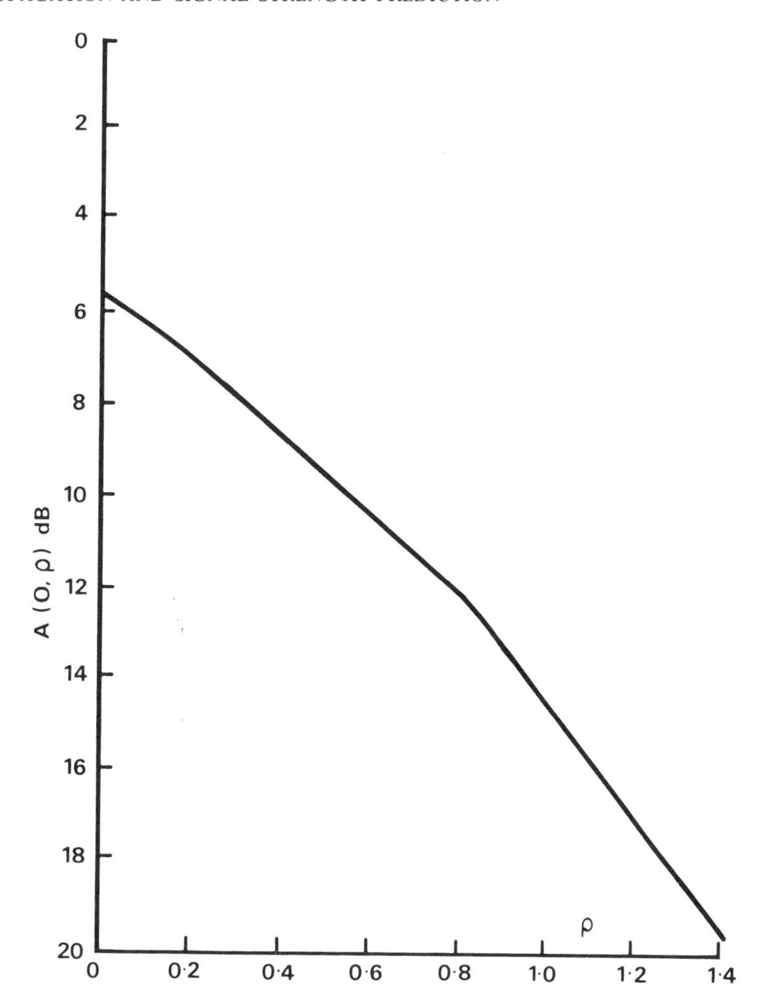

Figure 3.8 Diffraction loss over a cylinder.

where $U(v\rho)$ is a correction factor. The functions $A(v, 0)$, $A(0, \rho)$ and $U(v\rho)$ are shown in Figures 3.6, 3.8 and 3.9 respectively.

Approximations for $A(v, 0)$ have been given in section 3.3.1, and for $A(0, \rho)$ and $U(v\rho)$ are

$$
\begin{aligned}
A(0, \rho) &= 6 + 7.19\rho - 2.02\rho^2 + 3.63\rho^3 - 0.75\rho^4 && \text{for } \rho < 1.4 \\
U(v\rho) &= (43.6 + 23.5v\rho)\log_{10}(1 + v\rho) - 6 - 6.7v\rho && \text{for } v\rho < 2 \quad (3.16) \\
U(v\rho) &= 22v\rho - 20\log_{10}(v\rho) - 14.13 && \text{for } v\rho \leqslant 2
\end{aligned}
$$

The curves of Figure 3.8 and 3.9 are strictly valid for horizontal polarization only, but measurements have shown that at VHF and UHF frequencies they

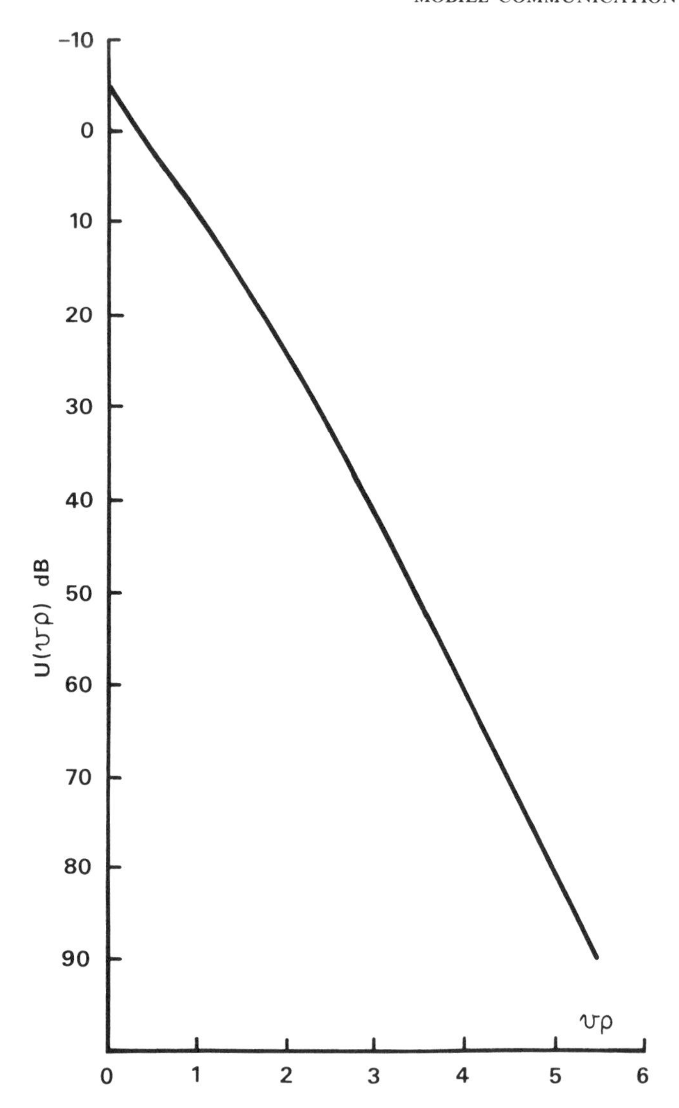

Figure 3.9 Correction factor $U(v\rho)$ for diffraction over a cylinder.

may be applied to vertical polarization with reasonable accuracy. Increasing the surface roughness of the cylinder reduces the magnitude of the surface reflections, with the result that the solution tends towards that of a perfect knife-edge.

Improvements can be made in a number of ways. There are a number of prediction methods based on empirical results, and an accurate method of predicting losses due to double knife-edge diffraction has been written up in the literature [4]. Various computer techniques for service area prediction for

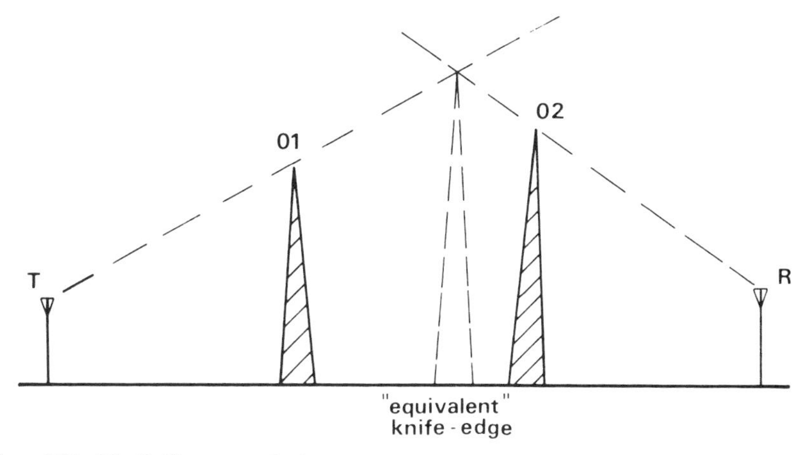

Figure 3.10 The Bullington equivalent knife-edge.

broadcast and land mobile services in the VHF and UHF bands are available, and find wide application. These techniques are capable of predicting the median field strength to within about 6 dB and this is usually considered to be satisfactory.

3.4 Multiple knife-edge diffraction

The extension of the single knife-edge diffraction theory to two or more obstacles involves considerable mathematical complexity. For two obstacles, the problem reduces to a double integral of the Fresnel form over a plane above each knife-edge and has been solved analytically [4]. However, the length and mathematical intricacy of the exact solution has focused attention on the use of approximations, especially for more than two edges.

3.4.1 The Bullington equivalent knife-edge method

An early proposal for a model which would enable the diffraction loss for a multiply-obstructed path to be evaluated was given by Bullington in 1947 [5]. This involves the replacement of the real terrain by a single equivalent knife-edge at the point of intersection of the optical paths made by each terminal and its horizon, as shown in Figure 3.1. The diffraction loss of this equivalent knife-edge is computed as representing the loss of the original real terrain situation. Unfortunately, because certain intervening obstacles are omitted from the calculation, as shown in Figure 3.11, the method seriously oversimplifies the situation, and large errors can occur.

3.4.2 The Epstein–Peterson diffraction loss method

The Epstein–Peterson diffraction loss method [6] calculates the propagation loss of a multiply-obstructed path by adding the attenuations (in dB relative to

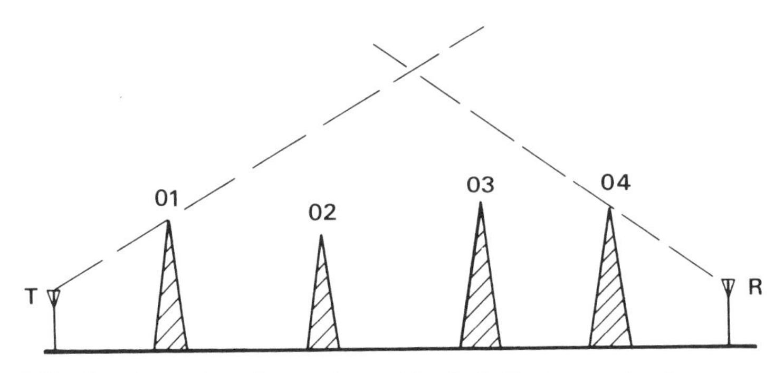

Figure 3.11 How intervening edges are ignored by the Bullington construction.

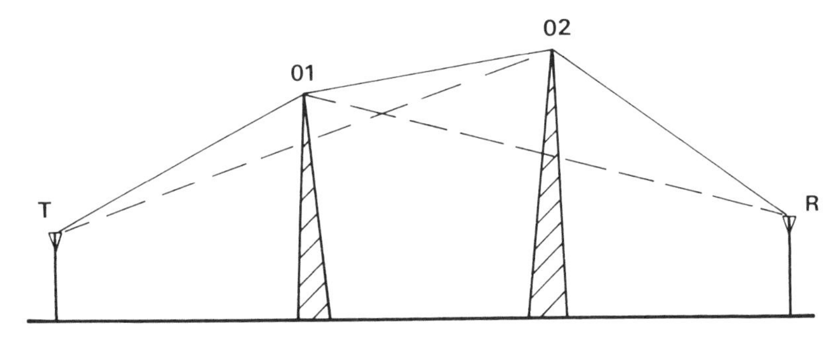

Figure 3.12 The Epstein–Peterson diffraction construction.

free space) produced by each knife-edge in turn. A two-obstacle situation, depicted in Figure 3.12, is evaluated as the sum of the diffraction losses for paths *T*-01-02 and 01-02-*R*

Comparison of this method with the rigorous solution for two knife-edges has revealed that large errors exist when the two obstacles are closely spaced.

3.4.3 The Japanese Atlas method

This method, proposed by the Japanese postal service [7], is shown in Figure 3.13 and is similar to the Epstein–Peterson method. The difference is that the diffraction loss is computed for paths *T*-01-02 and *T'*-02-*R*, where *T'* is the projection of the horizon ray 01-02 on to the plane of *T*. A correction factor can be introduced into the Epstein–Peterson method for the case of two closely-spaced edges, and it has been shown that it then becomes identical to this method. The comments on accuracy made in the previous section therefore hold.

3.4.4 Picquenard's method

Some of the limitations inherent in the methods of Bullington and Epstein and Peterson are overcome using the method proposed by Picquenard. In the example shown in Figure 3.14, the loss due to one obstruction is calculated as

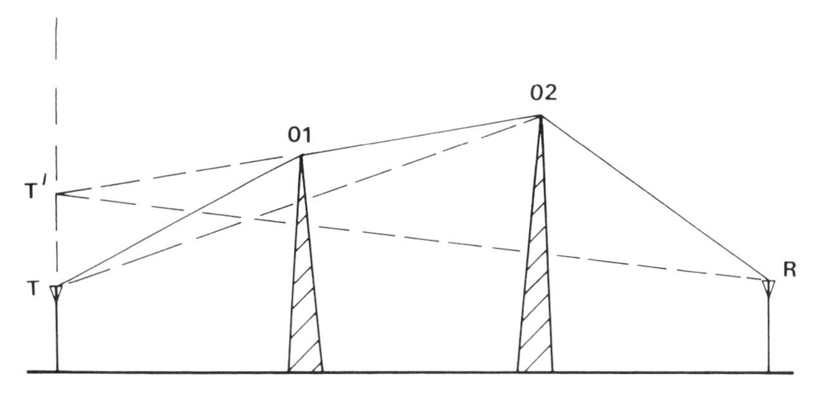

Figure 3.13 The Japanese Atlas diffraction construction.

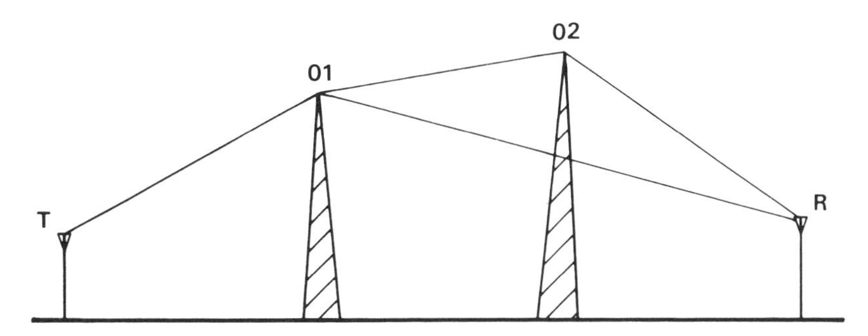

Figure 3.14 The Picquenard diffraction construction.

if the other were not present. The effective height of the second obstacle is then determined by drawing a line from the top of the first obstruction to the receiver as illustrated. The total loss is thus calculated as the sum of *T*-01-*R* and 01-02-*R*.

3.4.5 The Deygout method

The Deygout method [9] of determining the diffraction loss over obstructed paths, often called the 'main-edge' method, is best described by way of an example.

Consider the three obstacle paths of Figure 3.15. The losses due to each of the knife-edges must be evaluated in the absence of the others, i.e. paths *T*-01-*R*, *T*-02-*R* and *T*-03-*R*. The largest is termed the 'main-edge' and its diffraction loss calculated (path *T*-02-*R* in Figure 3.15). The diffraction loss over the remaining obstacles is then found with respect to the main-edge and the visible terminal (*T*-01-02 and 02-03-*R*).

For more than three knife-edges, the total loss is evaluated as the sum of all the individual losses for the edges in order of decreasing loss when the above procedure is recursively repeated. For a double knife-edge case in which one

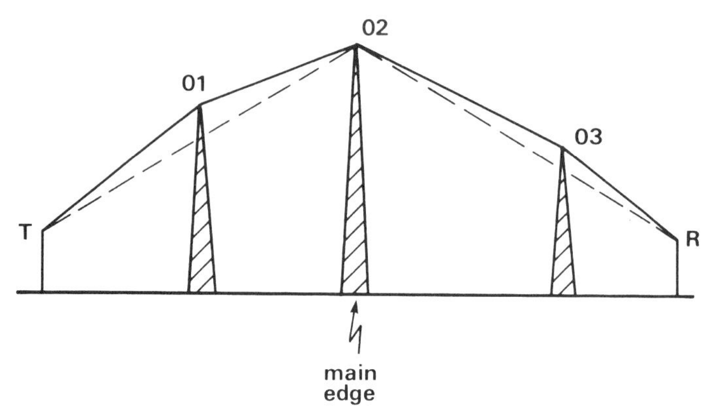

Figure 3.15 The Deygout diffraction construction.

edge is largely dominant, good agreement with the exact solution is obtained. When the magnitudes of the losses are comparable, however, the method is found to overestimate the loss.

Comparison of the predicted loss (without corrections) for three and more knife-edges with measurements at 160, 1800 and 2450 MHz has shown close agreement for highly obstructed paths. Superior accuracy to the Bullington and Epstein–Peterson methods has also been demonstrated, but the application to paths with more than four obstacles results in a complex solution which generally overestimates the losses.

3.4.6 Comparison of the diffraction models

Comparisons of the diffraction loss approximations described in the previous sections have been given in the literature [10]. Each model has been found to have a region over which it is most suitable but, as a general rule, the Epstein–Peterson construction produced consistently low errors when compared to the rigorous two-obstacle solution. Over highly irregular terrain, the Deygout method has been demonstrated to give increased accuracy. The improvement generally appears small, however, and may not warrant the additional complexity involved.

3.5 Propagation prediction models

Several methods for predicting propagation over irregular terrain have appeared in the literature [5, 11–18]. Many of the earlier techniques, such as those due to Bullington [5] and Egli [11], are based on using the theoretical equations describing propagation in free space or over a plane Earth, and then including an additional term to account for diffraction losses.

Most modern methods may be classified as terrain-based, since they are built around some form of computation for the diffraction loss based on the

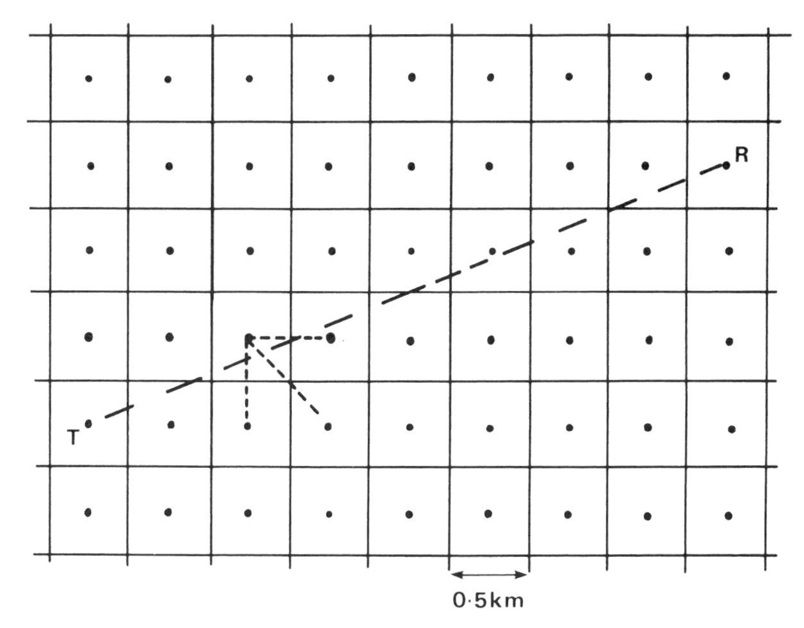

Figure 3.16 Matrix of heights, illustrating row, column and diagonal interpolation.

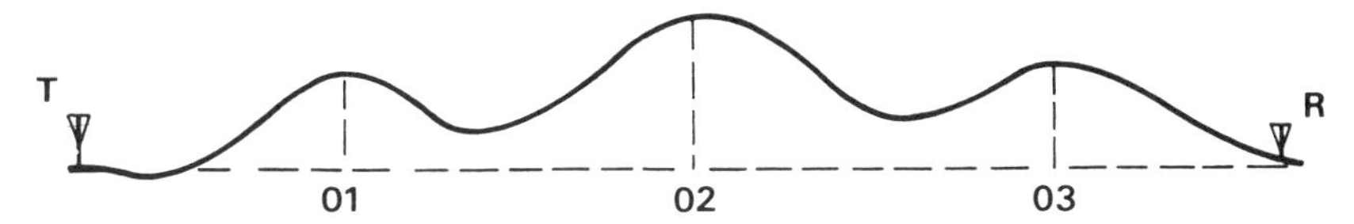

Figure 3.17 Ground profile drawn from terrain data map, showing the positions of (assumed) knife-edges.

path profile between the transmitter and the receiver. One of the most widely used in the UK is the prediction method adopted by the Joint Radio Committee (JRC) of the Nationalised Power Industries. The technique has been described at different stages of its development by Edwards and Durkin [12] and Dadson [18]. The method utilizes a computerised topographical data base, providing height reference points at 0.5 km intervals, and this matrix of terrain heights may be represented as in Figure 3.16. To calculate the received signal level, the computer first reconstructs the path profile between the transmitter and receiver. Row, column and diagonal interpolation is used to give increased accuracy, and the computer determines the height and position of obstructions along the path as shown in Figure 3.17. Having obtained a ground-path profile between the transmitter and the receiver, the computer then tests for the existence of a line-of-sight path, and whether Fresnel-zone clearance is obtained over the path. If both tests are satisfied, the free-space and plane-Earth losses are calculated and the higher value is chosen. If the tests fail, the program evaluates the loss caused by obstructions, grading them into single or multiple diffraction edges. Calculations are made for up to three diffracting edges, and any greater number of obstructions is converted into three edges using the method suggested by Bullington. The main shortcoming of the JRC method is its inability to take into account the effects of buildings, and this produces errors in built-up areas.

3.6 Signal strength prediction in urban areas

Following an extensive series of propagation measurements in and around Tokyo, Okumura *et al.* [14] produced a detailed analysis of the results, and hence arrived at an empirical method for signal strength prediction. Essentially, the method is based on determining the free-space path loss between the transmitter and the receiver, adding an urban loss, and then adding or subtracting numerous correction factors to account for the nature of the terrain, the extent of urbanization, the heights of the antennas and street orientation. The basic formulation of the Okumura technique can be expressed as

$$\text{path loss} = L_F + A_{mn} - H_{tn} - H_{tn}\,\text{dB}. \tag{3.17}$$

In this expression, L_F is the free-space path loss and A_{mn} is the median attenuation relative to L_F in urban areas over what is defined as 'quasi-smooth' terrain with a transmitter antenna height h_t of 200 m and a receiver antenna height h_r of 3 m. A_{mn} is a function of frequency and range and is expressed in graphical form by the series of curves shown in Figure 3.18. H_{tn} and H_{rn} are correction factors to account for antennas not at the reference heights of 200 m and 3 m; they are termed the height-gain factors and are also functions of frequency and range. Okumura's paper contains graphs from which the appropriate values for any specific situation can be extracted.

If the terrain cannot be treated as 'quasi-smooth' or if the environment is not urban, then further corrections have to be made. In fact, Okumura produced eight factors intended to correct for suburban and open areas, sloping terrain, hilly terrain and mixed land–sea paths, all these factors being expressed in graphical form. Street orientation in urban areas can also be taken into account.

The method is often quoted, since it is amongst the few readily available techniques that are specifically intended for use in built-up areas and which take account of variations in both terrain and urbanization. However, it is rather involved and difficult to implement directly on a computer, because the data are available only in graphical form. It therefore has to be entered point by point into a computer memory, and interpolation routines have to be devised for obtaining intermediate points. A further reservation about the method stems from the differences, topographical and otherwise, between Tokyo, where the measurements were taken, and other cities.

In order to put Okumura's technique into a form suitable for implement-

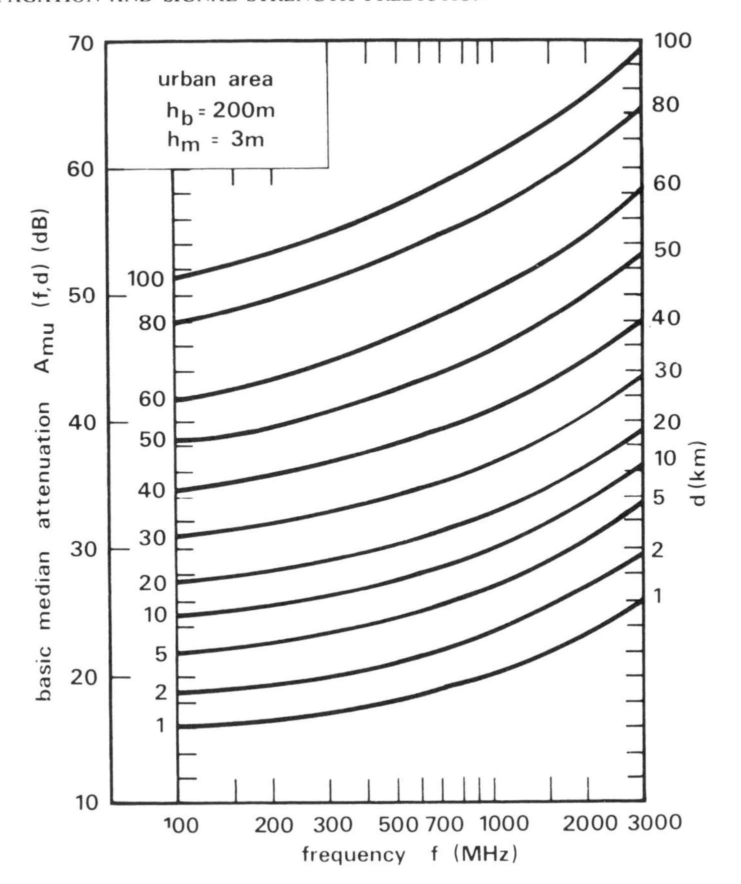

Figure 3.18 Basic median path loss relative to free space in urban areas over quasi-smooth terrain (after Okumura).

ation on a computer, Hata [19] has published an empirical formula for the propagation loss which is based on Okumura's results. The path loss expression so produced is

$$L_p = 69.55 + 26.16\log f - 13.82\log h_t - A(h_r) + (44.9 - 6.55\log h_t)\log d \text{ dB} \quad (3.18)$$

where f lies in the range 150–1500 MHz
h_t lies in the range 30–300 m
and d lies in the range 1–20 km.

$A(h_r)$ is the correction factor for the mobile antenna height and is computed as follows.

For a small or medium city

$$A(h_r) = (1.1\log f - 0.7)h_r - (1.56\log f - 0.8) \text{ dB} \quad (3.19)$$

where h_r lies in the range 1–10 m.

For a large city

$$A(h_r) = 8.29[\log(1.54h_r)]^2 - 1.1\,\text{dB} \quad (f \leqslant 200\,\text{MHz}) \tag{3.20}$$

and

$$A(h_r) = 3.2[\log(11.75h_r)]^2 - 4.97\,\text{dB} \quad (f \geqslant 400\,\text{MHz}) \tag{3.21}$$

Hata's formulae are easily computable and give predictions almost indistinguishable from those produced by the Okumura method over the restricted range of parameters for which the equations apply.

3.6.1 Allsebrook's method

Allsebrook [17] suggested a 'flat city' prediction model at VHF based on the formula

$$\text{path loss} = L_p + L_B\,\text{dB} \tag{3.22}$$

where L_p is the plane-Earth loss, and L_B is the diffraction loss incurred by the propagation of the signal over the buildings adjacent to the mobile. At UHF, he used an additional correction factor γ.

For 'hilly cities', Allsebrook proposed a modified Blomquist and Ladell model [15], given by

$$\text{path loss} = L_F + [(L_p - L_F)^2 + L_D^2]^{1/2} + \gamma\,\text{dB} \tag{3.23}$$

where L_F is the free-space path loss, and L_D the diffraction loss over terrain obstacles.

Allsebrook noted the approximate fourth-law range dependence of the path loss, which had previously been reported by others [11, 20], and termed the excess loss suffered by the signal over the plane-Earth loss the 'urban clutter factor' β.

3.6.2 Ibrahim's method

It is apparent from a study of the literature that there is no universally-accepted technique for predicting signal strength in built-up areas. Ibrahim [21] advanced some reasons for this, including the following.

(i) Although it is in built-up areas that mobile radio is widely used, most of the prediction methods fail to take the influence of buildings into account
(ii) The few prediction methods intended for urban application suffer from the inherent vagueness associated with a qualitative description of the urban environment
(iii) Little information is available about the accuracy of these models, especially when applied to different environments and at frequencies other than those used for the collection of the original data upon which a model was based.

A series of field trials was therefore conducted in London at frequencies in the range between 168 and 900 MHz. Data were collected in 500 m squares, as used by Edwards and Durkin [12], and a median path loss between isotropic antennas was extracted from the data for each square. The range from the transmitter varied from 2 to 10 km, and the various measured path losses were used to determine the effects of range, transmission frequency, terrain features and degree of urbanization.

As far as range is concerned, the fourth-power range dependence law, as in the plane-Earth equation, proved to be a good approximation at the frequencies used and for ranges up to 10 km. At all ranges and for all types of environment, the path loss was found to increase at the higher transmission frequencies and strong correlation was evident between the losses at the three frequencies used, as shown by Figure 3.19. It was concluded that the propagation mechanism at all three frequencies was essentially similar.

3.6.2.1 Classification of the urban environment. The propagation of radio waves in built-up areas is influenced considerably by the nature of the urban environment, such as the size and density of buildings, the width of roads, the existence of parks and open areas, etc. Built-up areas are usually classified into urban and suburban, the urban areas being generally defined as dominated by tall buildings, office blocks and high-rising residential towers, while a suburb is basically residential with houses and gardens, playing fields, and parks as the main features. Such qualitative definitions are not precise and can be subjective. They lead to doubts such as whether radio propagation measurements made in one city are applicable in another, or whether the suburbs of different cities are similar. There is an obvious need to describe the urban environment quantitatively, to surmount the ambiguities embodied in qualitative definitions which arise from cultural differences and subjective judgement.

An early approach to this problem was that of Kozono and Watanabe [22], who attempted to quantify the urban environment. For that purpose, they

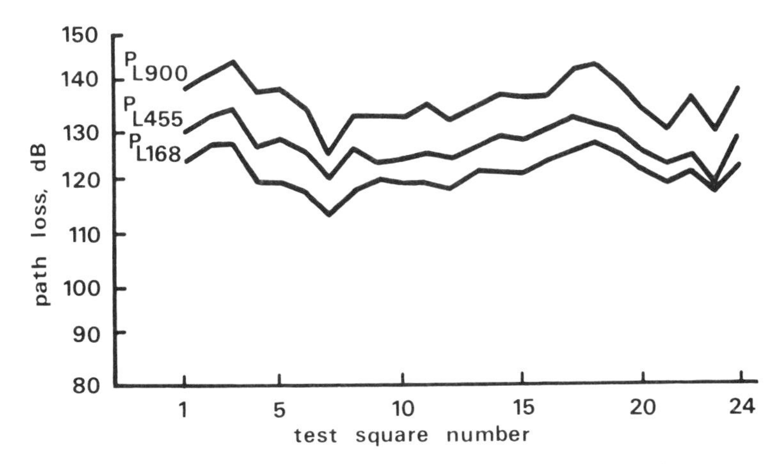

Figure 3.19 Median path loss to test squares in London at various frequencies.

proposed four parameters: area factor of occupied buildings (α); extended area factor of occupied buildings (α'); building volume over a sampled area (β); and building volume over an extended area (β'). A sampled area was defined as a 500×800 m section of the Japanese 'Community Maps'. To define an extended area, the most relevant section of the map was selected from those which overlapped a straight line connecting the base station to the sampled area. The selected section together with the sampled section constituted an extended area. They concluded that, though β' often correlated better with the median received signal, α is more suitable to use, since it is easier to extract from the Japanese 'Community Maps'.

Ibrahim used a similar approach, knowing that local authorities in the UK compile and regularly update a database of land usage, based on 500 m squares, which classifies different categories of usage. He extracted two factors, firstly a land-usage factor L, defined as the percentage of the square under consideration that is covered by buildings, and secondly, a degree of urbanization factor U, defined as the percentage of building-site area within the square which is covered by tall buildings. In this context, 'tall' means buildings with four or more floors. Analysis showed that these parameters were highly correlated with the path loss and were sensitive enough to be useful in propagation modelling.

Having extracted the relevant information, two propagation models were produced, an empirical model based on multiple-regression analysis and a semi-empirical model based on the plane-Earth propagation equation with the addition of a frequency-dependent clutter-factor to represent the excess loss. The clutter factor was calculated by regression analysis to determine the best way of including each relevant parameter. Inclusion of the factor U posed a problem, since the information necessary to compute it is available only for inner-city areas. Hence it was introduced into the models as an additional parameter to be used only in the highly urbanized parts of cities, usually the city centre. The following equation was produced as the best fit for London data:

$$\begin{aligned} L_p = & -20\log(0.7h_b) - 8\log h_m + \frac{f}{40} + 26\log\frac{f}{40} \\ & -86\log\left(\frac{f+100}{156}\right) + \left[40 + 14.15\log\left(\frac{f+100}{156}\right)\right]\log d \\ & + 0.265\,L - 0.37\,H + K \end{aligned} \tag{3.24}$$

where $K = 0.087\,U - 5.5$ for the city centre, otherwise $K = 0$

L_p is the median path loss between two isotropic antennas in dB

h_b and h_m are the effective heights of the transmitting and receiving antennas respectively in m, and $h_m \leqslant 3$ m

f is the transmission frequency in MHz

d is the range in m ($d \leqslant 10\,000$ m)

L is the land usage factor
H is the difference in height between the squares containing the transmitter and receiver
U is the degree of urbanization factor.

The semi-empirical model based on the plane-Earth equation suggests that the relationship between the received and transmitted powers is expressed by

$$\frac{P_R}{P_T} = G_b G_m \left[\frac{h_b h_m}{d^2} \right]^2 \times \text{(clutter factor)} \tag{3.25}$$

and the median path loss in dB is

$$L_p = 40 \log d - 20(h_b h_m) + \beta \tag{3.26}$$

where β, the value of the clutter factor in dB, is

$$\beta = 20 + \frac{f}{40} + 0.18L - 0.34H + K \tag{3.27}$$

with $K = 0.094U - 5.9$.

Here again, K is applicable only in the highly urbanized city centre, otherwise $K = 0$.

The rms errors produced by the two models are summarized in Table 3.1, the rms error being defined as

$$\text{rms error} = \sqrt{\frac{\sum(Y - \hat{Y})^2}{N}}$$

where $\hat{Y}$ is the measured value, Y is the predicted value, both in dB, and N is the number of values considered.

Table 3.1 The rms errors produced by the two models suggested by Ibrahim

	Frequency (MHz)		
	168	455	900
Empirical model	2.1	3.2	4.19
Semi-empirical model	2.0	3.3	5.8

3.7 Discussion

Investigations by several different research workers [11, 17, 20, 21] have revealed that at fairly short ranges, up to a few tens of km, the measured median path loss varies with the fourth power of the range from the transmitter, and this has led to several models based on the plane-Earth propagation equation [11, 17, 21]. Other empirical models exist, notably that due to Okumura *et al.*, and these are applicable over a wide range of frequencies and for distances of hundreds of km [16]. There have been only a few investigations concerned with the effect of antenna height at base and mobile, although it is apparent that a good base-station site is always desirable for good coverage, that is, the base-station antenna should be high enough to clear all the obstacles in the immediate vicinity of the site. Although it is clear that a higher antenna would provide a wider coverage, the exact relationship between antenna height and path loss requires further investigation, although some experimental evidence is available [14].

Alternatively, it is possible to employ the relationship extracted from the plane-Earth equation which relates the received signal power to the square of the antenna height. However, it is necessary to point out that, while employing high antennas increases the coverage available from a base-station site, it can also have adverse effects on channel re-use distances because of the increased possibility of causing co-channel interference. In this connection, the relationships proposed by Okumura have been applied to other measured data and found to produce quite good results [24]. In the past, mobile antenna height above local ground has not attracted much attention since most radio installations were installed in vehicles. However, the increased use of hand-portable equipment and its use inside buildings has increased the importance of studying this particular topic.

When the influence of the urban environment on the median path loss is considered, a quantitative description of the environment appears to be necessary. For that purpose, two parameters have been proposed: a land-usage factor describing the percentage of the relevant area that is occupied by buildings of any kind, and a degree-of-urbanization factor describing the percentage of the built-up area within a test area that is occupied by tall buildings. Good correlation has found to exist between the values of the measured path loss and the two urban environment factors, the path loss increasing directly with each of the two factors.

Close correlation is evident between path-loss measurements over a wide range of frequencies. This, together with the fact that the measured path loss at different frequencies responds in a similar manner to the variations in the urban parameters, suggests that the propagation mechanism is similar. Accordingly, several authors have concentrated on the formulation of path-loss prediction equations applicable over a wide frequency range.

3.8 Signal variability

Prediction of the received mobile radio signal strength is a two-stage process, involving an estimation of both the median received signal within a relatively small area, and the signal variability about the median level. The early part of this chapter has been concerned with the problem of median signal strength prediction in small areas, and we have been concerned particularly with built-up areas where there is a concentration of mobile radio services.

We now turn to the question of signal variability and address the problem of quantifying the extent to which the signal fluctuates within the area under consideration. It is apparent that there are two contributing factors. Firstly, there is the variation in the median signal itself as we move from place to place within the area under consideration. This is caused by large-scale variations in the terrain profile along the path to the transmitter, and by changes in the nature of the local topography. It is often termed 'slow fading'. Superimposed on this are the rapid and severe variations in the received signal strength caused by multipath propagation in the immediate vicinity of the receiver. This is due to scattering from man-made and natural obstacles such as buildings and trees, and is commonly called 'fast fading'.

A quantitative measure of the signal variability is essential for several reasons. It is only then, for example, that we can estimate the percentage of any given area that has an adequate signal strength, or the likelihood of interference from a distant transmitter. An estimate of the variability is no less important than a prediction of the median signal strength itself.

3.8.1 Statistical analysis of the signal

3.8.1.1 Local statistics. The scattering model proposed by Clarke [24] is based on the assumption that the received signal consists of a large number of randomly phased components, and leads to the conclusion that the probability density function of the signal envelope r follows a Rayleigh distribution given by

$$p_r(r) = \frac{r}{\sigma^2} \exp(-r^2/2\sigma^2) \tag{3.28}$$

The cumulative distribution $P_r(R) = \text{prob}\ [r \leqslant R]$ is

$$P_r(R) = \int_0^R p_r(r)\,\mathrm{d}r = 1 - \exp(-R^2/2\sigma^2) \tag{3.29}$$

This scattering model describes the local, i.e. small-area, statistics of the signal envelope in terms of only one parameter σ, the modal value. The mean of the distribution is $\sigma\sqrt{(\pi/2)}$, and this is the local mean of the signal envelope. Factors such as range, the path profile to the transmitter, the type and density of buildings near the receiver and the width and orientation of the street, combine to influence the value of the mean. It is only over distances sufficiently small to ensure that these factors are sensibly constant that the process can be considered statistically stationary.

To test whether the data collected in any small area fit this hypothesis, the cumulative distribution is often plotted on Rayleigh graph paper, on which Rayleigh-distributed data would appear as a straight line. Departures from

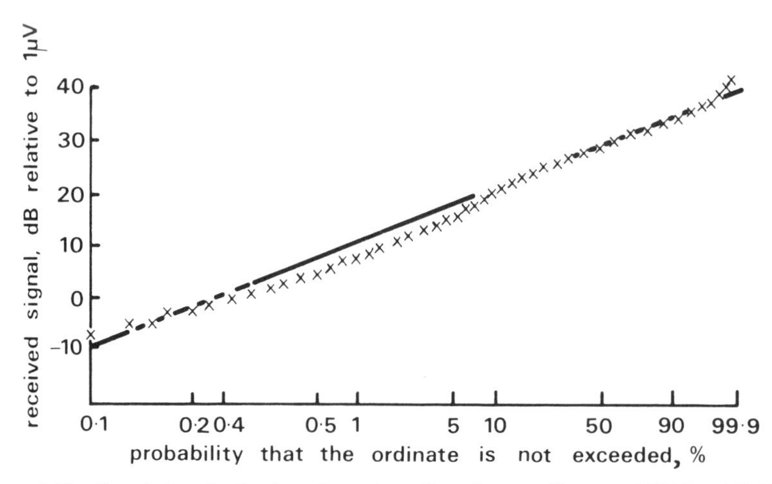

Figure 3.20 Cumulative distribution of smaples collected over a distance of 100 λ at 168 MHz.

Rayleigh are easily apparent, but since the scale is highly non-linear the tails of the distribution are over-emphasized and a quantitative judgement is difficult. An alternative [25], avoiding this drawback, is to plot $P(R)$, the cumulative distribution of the measured data, against $P_r(R)$, the cumulative distribution of a Rayleigh process having the same rms value. Departures from a straight line of unity slope indicate differences in the statistical distributions of $P(R)$ and $P_r(R)$ with no bias towards any particular range of probabilities. Figure 3.20 shows the cumulative distribution of about 6000 data samples, collected over a distance of 100λ at 168 MHz, plotted on Rayleigh paper, and Fig. 3.21 shows the same data plotted against $P_r(R)$. Comparison of the two figures shows that the data actually depart more seriously from Rayleigh near the middle of the distribution, and not at the lower tail as suggested by Figure 3.20.

A statistically valid test of this model would require a large grouping of data to be plotted, but in doing this it is almost certain that the 'small locality' assumption would be violated. Indeed, whenever a reasonably large quantity of data collected over some distance is considered, large departures from Rayleigh always appear. This is accounted for by suggesting that the process is non-stationary; over each small area the process is Rayleigh, but the mean value varies from place to place. To test whether we are dealing with a process that really is fundamentally Rayleigh, some way must be found to handle the problem of non-stationarity.

Clarke suggested the technique of normalizing the data by way of its running mean. He took a set of experimental results and divided each data point by a local mean obtained from averaging the 200 points symmetrically adjacent to it; the resulting normalized random variable was then treated in exactly the same way as the original random variable. The argument was that if a Rayleigh

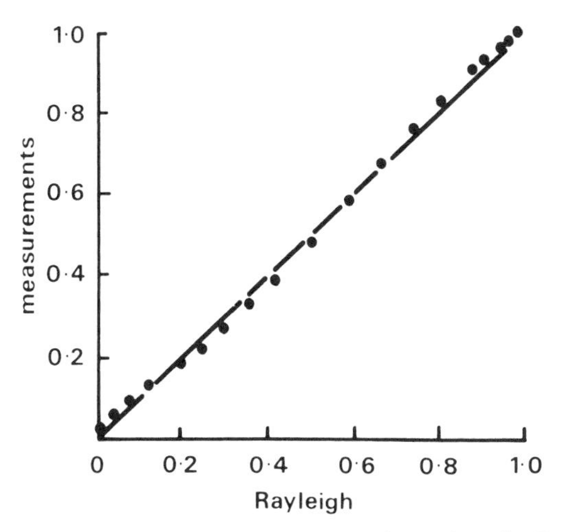

Figure 3.21 Cumulative distribution of the same samples against the Rayleigh cumulative distribution.

process is normalized to its rms value, the resultant is another Rayleigh process with an rms value of unity. If the local Rayleigh process is represented by eqn (3.28), then the rms value is $\sigma\sqrt{2}$. If we now normalize to this value, then the new variable r_n is $r_n = r/\sigma\sqrt{2}$, and since $p_n(r_n)\ \mathrm{d}r_n = p_r(r)\mathrm{d}r$, the new probability density function will be

$$p_n(r_n) = 2r_n \exp(-r_n^2) \tag{3.30}$$

which is a Rayleigh process with $\sigma^2 = 0.5$ and an rms value of unity. Normalization of a Rayleigh process therefore does not change the distribution, it only changes the rms value.

To determine a distance suitable for normalizing measured data, several investigators have experimented with windows of width between 2λ and 64λ. Davis and Bogner [26] concluded that at 455 MHz it was reasonable to regard the data as statistically stationary over a distance of 25 m, whilst Parsons and Ibrahim [27] found their data at 168 MHz to be a stationary Rayleigh process for distances up to 42 m. At 450 and 900 MHz, 20 m was a more reasonable distance. The effectiveness of normalizing by way of a running mean is illustrated in Figure 3.22, which shows a close approximation to a Rayleigh distribution. It was produced from 1.5 million samples collected in London at 168 MHz.

3.8.1.2 Statistics of the local mean. Measurements reported by Reudink [28], Black and Reudink [29] and Okumura *et al.* [14] are often quoted to suggest that the local mean of signals received at a given range and frequency, and in similar environmental areas, follows a log normal distribution. All these researchers found that, when fast fading was averaged out, the variations in the

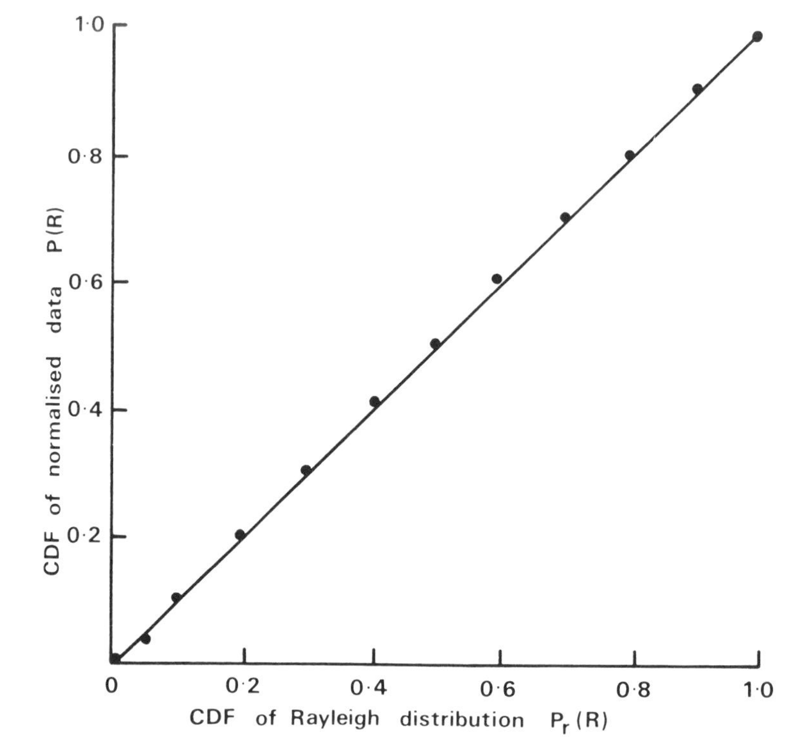

Figure 3.22 Cumulative distribution of samples collected at 168 MHz against the Rayleigh cumulative distribution.

local mean were very closely log normal. This suggests that a three-stage model might be appropriate to describe urban propagation: an inverse *n*th-power law with range from the transmitter to the area where the receiver is located (as we have seen earlier, *n* is often very close to 4); log normal variations of the local mean within that area; and superimposed fast fading which follows a Rayleigh distribution.

Okumura's experiments in Tokyo showed that when the median signal strength was computed over 20 m sectors and the standard deviation determined over areas of diameter 1–1.5 km, the values all lay in the range 3–7 dB. They increased slightly with frequency, but appeared insensitive to range. Similar values had been measured in London at frequencies between 75 and 400 MHz [27]. Analysis of Ibrahim's data confirmed that the standard deviation increased slightly with frequency, from 5.4 dB at 168 MHz to 6.4 dB at 900 MHz, and also showed that the values measured at short ranges were lower than those at longer ranges. There are two possible explanations for this. Firstly, the standard deviation might be range-dependent, but if it is, this would contrast with Okumura's findings which did not indicate such a trend. A more likely explanation is that the value is influenced by the degree of urbanization. At 2 km range, the test area used by Ibrahim was dominated by high-rise buildings, whereas at 9 km the area was mainly residential, houses and gardens being the main features. Reudink [28] faced by apparently contradicting results from tests at 800 MHz and 11.2 GHz in Philadelphia and New York, where the standard deviation decreased with range in Philadelphia but increased in New York, also suggested that the local environment exerted a much stronger influence than did the range from the transmitter.

3.9 Large-area statistics

We have seen in the previous section that over a relatively small distance (a few tens of wavelengths) the signal is well described by Rayleigh statistics, with the local mean over a somewhat larger area (with homogeneous environmental characteristics) being log normally distributed. It is of interest, therefore, to examine the overall distribution of the received signal in these larger areas. We might reasonably expect it to be a mixture of Rayleigh and log normal, and investigations have indeed shown this to be the case. Parsons and Ibrahim [27] examined the Nakagami-*m* and Weibull distributions, which both contain the Rayleigh distribution as a special case, but came to the conclusion, suggested earlier by Suzuki [30], Hansen and Meno [31] and Lorenz [32], that the statistics of the mobile radio signal can be represented by a mixture of Rayleigh and log normal statistics in the form of a Rayleigh distribution with a log-normally varying mean.

The expression suggested by Suzuki was

$$p_s(x) = \int_0^\infty \frac{x}{\sigma^2}\exp\left(-\frac{x^2}{2\sigma^2}\right)\frac{1}{\sigma\lambda\sqrt{2\pi}}\exp\left[-\frac{(\log\sigma-\mu)^2}{2\lambda^2}\right]\mathrm{d}\sigma \qquad (3.31)$$

that is, the integral of the Rayleigh distribution over all possible values of σ, weighted by the PDF of σ, and this attempts to provide a transition from local to global statistics. However, although Suzuki compared the fit of four distributions, the Rice, Nakagami, Rayleigh and log normal to experimental data, he did not use this mixture distribution, probably because the PDF exists in integral form and presents computational difficulties. It was left to Lorenz to evaluate the expression, and it was he who termed this mixture the 'Suzuki distribution'.

In comparing the possible statistical distributions that can be used to describe the signal strength, it is important to establish a suitable criterion as the basis for comparison. Interest in the signal statistics stems mainly from the fact that without a reliable statistical model, prediction of the median signal strength is of little help either to the systems engineer or to frequency-management bodies. What they really need is an accurate prediction of values near the tails of the distribution, which are vital for coverage estimation and frequency re-use planning. Thus, the most suitable model is one which, given

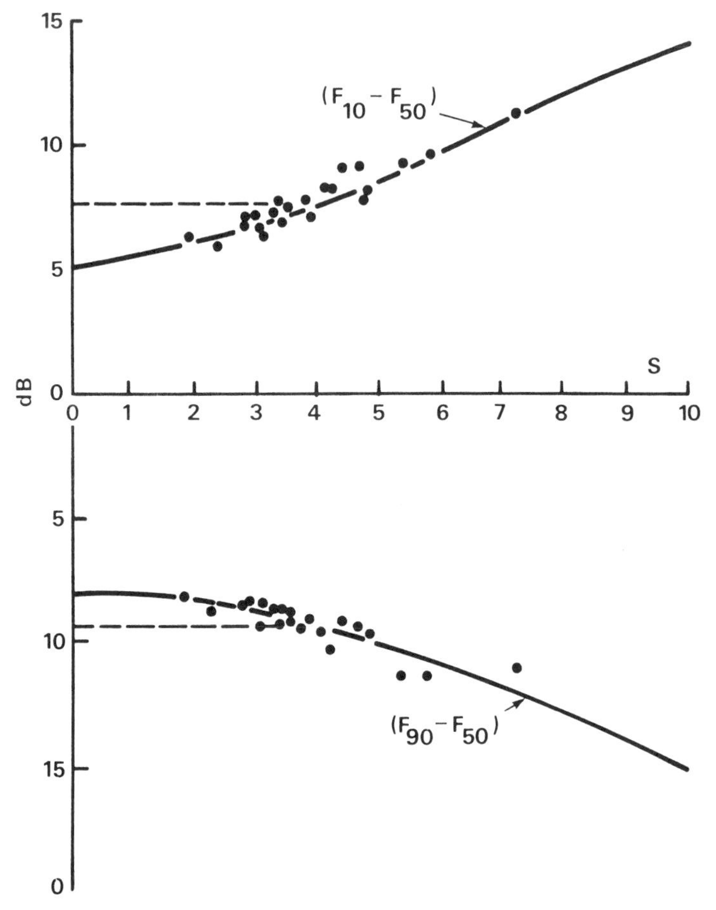

Figure 3.23 Theoretical and experimental values of $(F_{10} - F_{50})$ and $(F_{90} - F_{50})$ for data at 455 MHz plotted against s, the parameter of the Suzuki distribution.

the median value, will predict with least error the values between, say, 1% and 20% at one end and between 80% and 99% at the other. It is on this basis that the Suzuki distribution proves best.

Quantiles related to the median value, i.e. $(F_q - F_{50})$ are often of practical interest, where F_q is the value exceeded at q% of locations. Figure 3.23 shows a typical plot of experimental data for values of $q = 10$ and 90 (%) as a function of the parameters of the Suzuki distribution. Lorenz has shown that for a given set of experimental points, drawn from an (assumed) Suzuki distribution, the value of s is given by $s = \sqrt{(\mu - 31)}$.

Figure 3.23 shows that the values of F_{90} and F_{10} can be estimated, given the median value F_{50}, by subtracting 9.5 dB and adding 7.5 dB respectively. Thus, it is possible to use methods discussed earlier in the chapter not only for median signal strength prediction but also for area coverage estimation and frequency re-use planning.

3.10 Building penetration losses

The use of hand-portable equipment in cellular radio systems, and the increased availability of cordless telephones, have made it essential to study radio propagation within and into buildings. There have been several surveys at various frequencies [33–35].

A number of authors have studied the problem of building penetration loss, and model it as the distance-dependency of the path loss when the mobile is outside a building, plus a building loss factor. The building loss factor is included in the model to account for the increase in attenuation of the received signal observed when the mobile is moved from outside a building to inside. This model was first proposed by Rice [36] in 1959, and has been used in all subsequent investigations. The model assumes that the building loss is independent of the distance between base and mobile, although no verification of this assumption has been reported. Furthermore, a building has an internal distance through which the signal has to propagate, therefore the building loss will include any spatial variation of the signal within a building.

The determination of the distance dependency of the path loss between a base station and mobile, when both are situated outside a building, has been covered earlier in this chapter. The model of path loss distance dependency is either a free-space or plane-Earth theoretical loss, plus a clutter factor. The clutter factor accounts for a consistent increase in attenuation noticed between the theoretical and observed path losses. Given that the outside distance dependency has been described in existing propagation studies by the above model, the thrust of building propagation studies has been to determine the building loss and its variation in a form that can easily be combined with vehicular mobile studies to give coverage into buildings. Therefore the building loss used in the literature is defined, initially by Rice, as the difference between the median field intensity in the adjacent streets and the field intensity at a location on the main floor of a building. If the median field intensity in the streets adjacent to a large number of buildings is measured, then the resulting distribution will approach the distribution of the median field intensity in the streets, that is, the distribution of the clutter factor. To reduce the need for calculating distributional effects of the building loss and clutter factor separately, some authors also present a combined clutter factor–building loss distribution for their particular situation. The path loss from base to mobile is then calculated from the sum of the theoretical path loss, median clutter factor, median building loss and a distributional factor. The grade of service will determine the distributional factor which may be calculated from the combined clutter factor–building loss distribution.

In general, measurements have been made inside buildings using portable signal-strength measuring equipment, and the signal statistics have been modelled as superimposed Rayleigh and log normal. It is often necessary to proceed with caution in interpreting the results of propagation measurements into buildings since, in many cases, although there is no line-of-sight path to the lower floors, such a path may exist as the receiver is moved to higher floors within the same building. However, in general, the small-scale signal variations follow a Rayleigh distribution, and the large-scale variations are log normally distributed, with a standard deviation related to the frequency and the transmission conditions relevant to the area under investigation.

For areas where there is no line of sight, scattering is the dominant propagation mechanism and the standard deviation is typically 4 dB; where a partial line of sight exists, the standard deviation increases to 6–9 dB. In general, it has been found that the loss decreases at a rate of about 2 dB per floor [35, 37], although it has sometimes been observed that at higher levels within buildings the loss increases again [38]. This has been attributed to the shadowing effects of adjacent buildings. However, as the transmission frequency rises, the building penetration loss is reduced.

3.10.1 Interference effects

To conclude this chapter we cover, very briefly, the problem of interference, which is of growing importance in high-capacity mobile radio systems. The effects of co-channel and adjacent channel interference in cellular systems will be dealt with in later chapters.

If a certain band of frequencies is allocated for a mobile radio system, a fundamental requirement in the efficient use of this band is to re-use frequencies at as small a geographical separation as possible. Cellular systems, for example, rely on frequency re-use to achieve high spectrum efficiency. Wherever a band of frequencies is used, however, interference effects have to be taken into account. These can be mainly classified as 'co-channel' or 'adjacent-channel'. Co-channel interference lies within the bandwidth of the victim receiver, and arises principally from other transmitters using the same carrier frequency or from intermodulation between other transmitters in the same band. Adjacent channel interference arises from the same sources, and causes problems because the receiver filters do not have perfect selectivity. In general, adjacent channel interferers have to be at a much higher power level than co-channel interferers, to cause similar problems.

To achieve reliable communication between a base station and a mobile within the desired service area, we require, firstly, an acceptable signal level, and the probability of satisfactory reception using this criterion alone has been treated in the literature [39, 40]. Use of a signal/interference ratio alone has also been considered [41], and leads to different results. To be realistic, however, we actually need an acceptable signal and an acceptable signal/interference ratio simultaneously. This problem has received some attention; the case where reception is subject to shadowing, i.e. the received signal strength from both the wanted and unwanted transmitters follows a log normal distribution, was reported some years ago [42]. More recently the 'fading plus shadowing' case has been considered. This is more complicated, since the signal statistics follow a superimposed Rayleigh and log normal distribution. Interesting results have been reported [43] for the case of a single interfering transmitter. It has been shown that close to the wanted transmitter, where the signal strength is high, interference conditions predominate, whereas further away, coverage conditions are more important.

For multiple interfering signals, the problem of predicting the probability of interference becomes much more involved. Independent interfering signals add together on a linear power basis, so to predict the probability of interference it is necessary to find an equivalent probability density function for the sum of the interfering signals. Approaches to the problem have been made for the case when Rayleigh fading exists [44], but the general problem of multiple interferers in the presence of fading and shadowing (Rayleigh plus log normal) is still the subject of research.

References

1. Jordan, E.C. and Balmain, K.G. *Electromagnetic Waves and Radiating Systems.* Prentice-Hall, New York (1968).
2. Griffiths, J. *Radio Wave Propagation and Antennas: an Introduction.* Prentice-Hall International, London (1987).
3. Dougherty, H.T. and Maloney, L.J. Application of diffractions by convex surfaces to irregular terrain situations. *Radio Science* **68D** (2) February 1964.
4. Millington, G., Hewitt, R. and Immirzi, F.S. Double knife-edge diffraction in field-strength predictions. *Proc. IEE* **109-C** (1962) 419–429.
5. Bullington, K. Radio Propagation at frequencies above 30 Mc/s. *Proc. IRE* **35** (1947) 1122–1136.
6. Epstein, J. and Peterson, D.W. An experimental study of wave propagation at 850 Mc. *Proc. IRE* **41** (5) (1953) 595–611.
7. Atlas of radio wave propagation curves for frequencies between 30 and 10,000 Mc/s. Radio Research Laboratory, Ministry of Postal Services, Tokyo, Japan (1957) 172–179.
8. Picquenard, A. *Radio Wave Propagation.* Wiley, New York (1974).
9. Deygout, J. Multiple knife-edge diffraction of microwaves, *IEEE Trans.* **AP-14** (4) (1966) 480–489.
10. Wilkerson, R.E. Approximation of the double knife-edge attenuation coefficient. *Radio Science* **1** (12) (1966) 1439–1443.
11. Egli, J.J. Radio propagation above 40 Mc over irregular terrain. *Proc. IRE* **45** (1957) 1383–1391.
12. Edwards, R. and Durkin, J. Computer prediction of service area for VHF mobile radio networks. *Proc. IRE* **116** (9) (1969) 1493–1500.
13. Causebrook, J.H. Computer prediction of UHF broadcast service areas. *BBC Research Rept.* **RD 1974 4** (1974).
14. Okumura, Y., Ohmori, E., Kawano, T. and Fukuda, K. Field strength and its variability in VHF and UHF land mobile service. *Rev. Electr., Commun. Lab.* **16** (1968) 825–873.
15. Blomquist, A. and Ladell, L. Prediction and calculation of transmission loss in different types of terrain. *NATO-AGARD Conf.*, Publication CP–144 (1974).
16. Longley, A.G. and Rice, P.L. Prediction of tropospheric transmission loss over irregular terrain—a computer method. *ESSA Technical Rept* **ERL 79IT567** (1968).
17. Allsebrook, K. and Parsons, J.D. Mobile radio propagation in British cities at frequencies in the VHF and UHF bands. *Proc. IEE* **124** (2) (1977) 95–102.

18. Dadson, C.E. Radio network and radio link surveys derived by computer from a terrain data base. *NATO-AGARD Conf.*, Publication CPP–269 (1970).
19. Hata, M. Empirical formula for propagation loss in land-mobile radio services. *IEEE Trans.* **VT-29** (1980) 317–325.
20. Young, W.R. Comparison of mobile radio transmission at 150, 450, 900 and 3700 MHz. *Bell Syst. Tech. J.* **31** (1952) 1068–1085.
21. Ibrahim, M.F. and Parsons, J.D. Signal strength prediction in built-up areas, Part 1: Median signal strength. *Proc. IEE Part F*, **130** (5) (1983) 377–384.
22. Kozono, S. and Watanabe, K. Influence of environmental buildings on UHF land mobile radio propagation. *IEEE Trans.* **COM-25** (1977) 1133–1143.
23. Atefi, A. and Parsons, J.D. Urban radio propagation in mobile radio frequency bands. *Communications 86 Conf.*, Birmingham, UK, (*IEE Conf. Publ.* **262**, 13–18).
24. Clarke, R.H. A statistical theory of mobile radio reception. *Bell Syst. Tech. J.* **47** (1968) 957–1000.
25. Wilk, M.B. and Ganadesikan, R. Probability plotting methods for the analysis of data. *Biometrika* **55** (1968) 1–7.
26. Davis, B.R. and Bogner, R.E. Propagation at 500 MHz for mobile radio. *Proc. IEE Part F*, **132** (5) (1985) 307–320.
27. Parsons, J.D. and Ibrahim, M.F. Signal strength prediction in built-up areas, Part 2: Signal variability. *Proc. IEE, Part F*, **130** (5) (1983) 385–391.
28. Reudink, D.O. Comparison of radio transmission at X-band frequencies in suburban and urban areas. *IEEE Trans.* **AP-20** (1972) 470.
29. Black, D.M. and Reudink, D.O. Some characteristics of mobile radio propagation at 836 MHz in the Philadelphia area. *IEEE Trans.* **VT-21** (1972) 45–51.
30. Suzuki, H. A statistical model for urban radio propagation. *IEEE Trans.* **COM-25** (1977) 673–680.
31. Hansen, F. and Meno, F.I. Mobile fading—Rayleigh and lognormal superimposed. *IEEE Trans.* **VT-26** (1977) 332–335.
32. Lorenz, R.W. Theoretical distribution functions of multipath fading processes in mobile radio and determination of their parameters by measurement. *Tech. Ber.* **455 TBr 66**, Forschungsinstitut der Deutschen Bundespost (in German).
33. Alexander, S.E. Radio propagation within buildings at 900 MHz. *Electronics Lett.* **18** (21) (1982) 913–914.
34. Durante, J.M. Building penetration loss at 900 MHz. *Proc. Conf. IEEE Vehicular Technology Group* (1973) 1–7.
35. Turkmani, A.M.D., Parsons, J.D. and Lewis, D.G. Radio propagation into buildings at 441, 900 and 1400 MHz. *Proc. 4th Int. Conf. on Land Mobile Radio*, Warwick (*IERE Conf. Publ.*, 1987).
36. Rice, L.P. Radio transmission into buildings at 35 and 150 Mc. *Bell Syst. Tech. J.* **38** (1959) 197–210.
37. Cox, D.C. Murray, R.R. and Norris, A.W. Measurement of 800 MHz radio transmission into buildings with metallic walls. *Bell Syst. Tech. J.* **62** (9) (1983) 2695–2717.
38. Walker, E.H. Penetration of signals into buildings in the cellular radio environment. *Bell Syst. Tech. J.* **62** (9) (1983) 2719–2734.
39. Jakes, W.C. Jr. (ed.) *Microwave Mobile Communications.* Wiley-Interscience, New York (1974).
40. Parsons, J.D. Propagation and interference in cellular radio systems. *IEE Conf. on Mobile Radio Systems and Techniques.* York (*Conf. Publ.* **238** (1984) 71–75).
41. French, R.C. The effect of fading and shadowing on channel reuse in mobile radio. *IEEE Trans.* **VT-28** (1979) 171–181.
42. Sachs, H.M. A realistic approach to defining the probability of meeting acceptable receiver performance criteria. *IEEE Trans.* **EMC-13**, (1971) 3–6 (see also **EMC-14** (1972) 74–78).
43. Williamson, A.G. and Parsons, J.D. Outage probability in a mobile radio system subject to fading and shadowing. *Electronics Lett.* **21** (14) (July 1985) 622–623.
44. Sowerby, K.W. and Williamson, A.G. Outage probability calculations for a mobile radio system having multiple Rayleigh interferers. *Electronics Lett.* **23** (1987) 600–601.

4 Modulation techniques

4.1 Introduction

From the discussion in the earlier chapters it has become clear that, for mobile radio systems, frequencies in the VHF and UHF bands are most suitable. The signals that need to be transmitted over mobile radio channels are principally analogue speech and data, although digitized speech will be used in the next generation of cellular radio systems, and in this chapter we briefly review some of the techniques used to modulate this kind of information on to a VHF or UHF carrier. Analogue modulation techniques are well described in many textbooks [e.g. 1, 2, 3] and so we deal with the principles only very briefly. The purpose of the early part of the chapter is to highlight those techniques that have particular application in the mobile radio field and to discuss the effects that phenomena such as noise and fading have on the transmitted signal. The later part of the chapter deals with digital modulation techniques, again concentrating only on applications in mobile radio. More general treatments can be found elsewhere [4, 5].

4.2 Amplitude modulation

If two signals, $v_c(t)$ and $v_m(t)$ are applied to a multiplier, as in Figure 4.1*a*, then if the modulating signal $v_m(t) = V_m \cos \omega_m t$ and the carrier $v_c(t) = V_c \cos \omega_c t$,

$$v(t) = \tfrac{1}{2} k V_m V_c \{\cos(\omega_c + \omega_m)t + \cos(\omega_c - \omega_m)t\} \tag{4.1}$$

where k is the constant of the multiplier. The output of the modulator therefore contains two components, one above and one below the carrier. Each is separated from the carrier by an amount $\omega_m (= 2\pi f_m)$, so that the bandwidth occupied by the signal is $\pm f_m (= 2f_m)$. In this case there is no component at f_c, i.e. this is a double-sideband suppressed-carrier (DSBSC) signal.

In practice the modulating signal will occupy a range of frequencies, and the result of the modulation process will then be to produce a band of frequencies above the carrier (known as the upper sideband) and another band below the

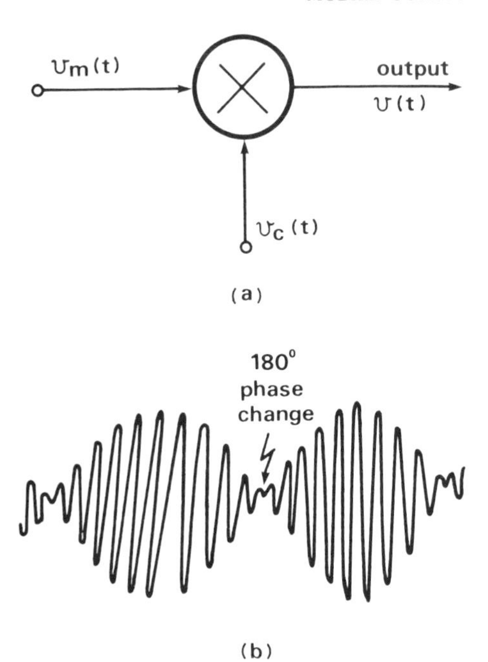

Figure 4.1 (*a*) Low-level generation of amplitude-modulated signals; (*b*) time waveform of a DSBSC signal.

carrier (known as the lower sideband). It is important to realize that the lower sideband is inverted, in the sense that the lowest frequency in the lower sideband is the highest baseband frequency, and vice versa.

It is obvious from Figure 4.1*b* that the envelope of the DSBSC signal is not the same shape as the original modulating sinusoid, and this is because no carrier is present. However, demodulation of the signal is simplified if a carrier component is present, and one way to introduce this is to add a DC component to the modulating signal.

The output signal becomes

$$v(t) = k_1 V_c \cos \omega_c t + k V_m V_c \cos \omega_m t \cos \omega_c t \tag{4.2}$$

$$= k_1 V_c (1 + m \cos \omega_m t) \cos \omega_c t \tag{4.3}$$

where $m = (k/k_1) V_m$, and is called the modulation index. If $m < 1$, the signal is termed 'full-AM'.

In the limiting case for which $m = 1$ (this is often termed 100% modulation, since m is commonly expressed as a percentage), only 33.3% of the transmitted power is in the sidebands, whereas it is apparent that *all* the information content is in the sidebands. In the DSBSC case there is no carrier component, so power is only transmitted when information is being sent.

4.2.1 Reduced carrier modulation

The case of 100% modulation discussed above is such that the sum of the sideband amplitudes is equal to that of the carrier i.e. $2(m/2) = m = 1$, so that the minimum signal amplitude is just zero and no phase reversals occur. However, it is possible for a reduced carrier component to be used with a value less than the sum of the sideband amplitudes. Such a signal is termed double-sideband reduced carrier (DSBRC) or double-sideband diminished carrier (DSBDC). For such signals $m > 1$, and the signal is said to be over-modulated.

There have been proposals [6, 7] for the use of both DSBDC and DSBSC in mobile radio systems. The apparent advantages are a reduction in carrier fading and increased transmitter efficiency, since no energy is radiated when no information is being sent. There are some points in favour of reduced carrier, since the presence of a small carrier component aids automatic frequency control (AFC) and automatic gain control (AGC), but there is no saving in bandwidth, and both the transmitter and receiver become more complex, so that the proposals have never been implemented except in experimental systems.

4.2.2 Single-sideband modulation

Although the AM systems described in the previous section differ from the point of view of transmitter efficiency, they all use the same amount of bandwidth to transmit the information, i.e. their spectral efficiency is the same. However, both sidebands carry identical information, so there are obvious advantages in terms of power and spectral efficiency if one of them can be eliminated. The signal so formed is termed single-sideband (SSB), and such signals have been used for many years over HF links where spectral efficiency is of paramount importance and skilled radio operators are normally employed.

The attractions of SSB for mobile radio use are also obvious. It appears that the already-congested mobile radio bands could accommodate twice as many users if a modulation system requiring half the bandwidth could be employed, and, moreover, transmitter efficiency would be maximized. However, this simple argument does not present the true picture, as will be seen later.

4.3 Angle modulation

The principle of angle modulation is to cause either the frequency or phase of a carrier signal to vary in sympathy with the instantaneous amplitude of a modulating signal $v_m(t)$, such that either

$$f = f_c + K_f v_m(t)$$

or

$$\phi = \phi_c + K_p v_m(t)$$

where K_f and K_p are arbitrarily chosen. An FM signal can therefore be expressed as

$$v_f(t) = V_c \cos\left[2\pi f_c t + 2\pi K_f \int_0^t v_m(t)\,dt\right] \quad (4.4)$$

and a PM signal as

$$v_p(t) = V_c \cos[2\pi f_c t + \phi_c + K_p v_m(t)]$$

The phase of the unmodulated carrier ϕ_c is constant and can be arbitrarily put equal to zero, giving

$$v_p(t) = V_c \cos[2\pi f_c t + K_p v_m(t)] \quad (4.5)$$

Comparison of these equations, which are of the same general form, shows that PM is angle modulation with ϕ proportional to $v_m(t)$, whereas in FM, ϕ is proportional to $\int v_m(t)$.

4.3.1 Bandwidth of angle-modulated signals

For AM systems, determining the bandwidth is fairly straightforward, since each component in the modulating waveform gives rise to only two side-frequencies whose amplitude and frequency can be easily deduced from elementary trigonometry. Angle modulation is more complicated, since putting $v_m(t) = V_m \cos\omega_m t$ in eqn (4.4), gives rise to an equation containing the functions $\cos(\beta \sin\omega_m t)$ and $\sin(\beta \sin\omega_m t)$. These represent a power series of a power series, and are known as Bessel functions. Using the definitions of Bessel functions, it can be shown [3] that the FM equation can be written as

$$\begin{aligned} v_f(t) = V_c\{ & J_0(\beta)\cos\omega_c t \\ & + J_1(\beta)[\cos(\omega_c + \omega_m)t - \cos(\omega_c - \omega_m)t] \\ & + J_2(\beta)[\cos(\omega_c + 2\omega_m)t + \cos(\omega_c - 2\omega_m)t] \\ & + J_3(\beta)[\cos(\omega_c + 2\omega_m)t - \cos(\omega_c - 3\omega_m)t] \\ & \quad \vdots \ \} \end{aligned} \quad (4.6)$$

This shows the existence of an infinite number of sidebands, all harmonically related to the modulating frequency f_m and implying an infinite bandwidth.

In general, however, the higher-order sidebands are of smaller amplitude, and since the value of β is under the designer's control (because k_f is a design parameter) the designer can thereby determine the number of pairs of significant sidebands, in other words can choose the effective bandwidth. Furthermore, the amplitudes of the carrier and sideband components vary with β, and although they can all reverse phase with various values of β, the overall magnitude of the FM signal remains constant.

4.3.2 Criteria for determining bandwidth

It has been implied in the previous section that bandwidth is under the control of the designer through a choice of the number of significant sidebands. The question therefore arises as to what constitutes significance? In many practical cases it is accepted that the nth pair of sidebands is insignificant if their magnitude is less than 1% of the unmodulated carrier amplitude, i.e. if

$$J_n(\beta) < 0.01\,J_0(0)$$

However, $J_0(0) = 1$, so the criterion becomes

$$J_n(\beta) < 0.01 \tag{4.7}$$

It is worth emphasizing that this is to some extent an artificial criterion, since any other value could just as easily have been chosen. It is however in widespread use, and it means that components more than 40 dB below unmodulated carrier level are considered negligible.

In practice, the bandwidth available for any channel is limited, so a compromise must be reached in choosing β. On one hand, β should be large for maximum enhancement of the information; on the other hand, it should be small enough to avoid generating sidebands that would interfere with other users. Finally, although practical baseband signals occupy a spectrum of frequencies, as far as bandwidth is concerned it is usually sufficient to base calculations on the highest frequency in the modulating signal.

4.3.3 Carson's rule

An alternative rule for determining the bandwidth of an FM signal due to J.R. Carson [8] is available. If we consider the significant sideband criterion and examine tables of Bessel functions, it becomes apparent that $J_n(\beta)$ diminishes very rapidly if $n > \beta$, particularly if β is large. Indeed, the ratio n/β approaches unity as β increases, and we can obtain an approximation to the bandwidth in this case by assuming that the last significant sideband occurs at $n = \beta$. So

$$\text{bandwidth} = 2nf_\mathrm{m} = 2\beta f_\mathrm{m} = \frac{2\Delta f}{f_\mathrm{m}} \cdot f_\mathrm{m} = 2\Delta f$$

On the other hand, for very small values of β, only $J_0(\beta)$ and $J_1(\beta)$ are significant, so we can write

$$\text{bandwidth} = 2f_\mathrm{m}$$

These are the two limiting cases which establish the bounds. Carson suggested that a convenient rule which covers the intermediate cases in a continuous manner would be

$$\text{bandwidth} = 2(\Delta f + f_\mathrm{m}) = 2f_\mathrm{m}(1 + \beta) \tag{4.8}$$

This rule approaches the correct limits as β becomes very large and very small. It is widely used in practice, even for non-sinusoidal modulation, using the highest frequency in the modulating waveform. Agreement between Carson's rule and the significant sideband criterion is always reasonable, although the bandwidth as calculated using the latter is always greater.

4.3.4 Narrowband modulation

A special and interesting case of angle modulation arises when the modulation index $\beta \ll 1$. To distinguish this from the general case considered previously when β could take any value, we refer to the case now under consideration as narrowband.

Narrowband FM (or PM) is used in mobile radio where the channel spacing is fairly small (12.5 kHz or 25 kHz). Under conditions where β is small,

$$\cos(\beta \sin \omega_m t) \simeq 1$$

$$\sin(\beta \sin \omega_m t) \simeq \beta \sin \omega_m t$$

so that

$$v_f(t) \simeq V_c[\cos \omega_c t - \beta \sin \omega_m t \sin \omega_c t]$$

$$= V_c[\cos \omega_c t + \beta/2 \sin (\omega_c + \omega_m)t - \beta/2 \sin (\omega_c - \omega_m)t]$$

which can be compared with eqn (4.1) for AM. Both equations show that only one sideband-pair exists, the amplitude being $m/2$ in the AM case and $\beta/2$ in the narrowband FM (NBFM) case.

In an AM signal, the sidebands are symmetrically placed about the carrier, so their sum is always in phase with the carrier, and the total signal is therefore one which varies in amplitude but not in phase. For NBFM, the sidebands are symmetrically placed about a line at 90° to the carrier which gives rise to a signal with a varying frequency but (almost) constant amplitude. For NBFM, it is easy to show that the signal is represented by

$$v_p(t) = E_c[\cos \omega_c t - \beta/2 \sin(\omega_c + \omega_m)t - \beta/2 \sin(\omega_c - \omega_m)t]$$

Theoretically, of course, the amplitude of an angle-modulated signal is only constant if there is an infinite number of sidebands. The assumptions made for narrowband modulation has restricted the number of sidebands to the first pair, so it is not surprising that in these signals there is a small amount of incidental AM.

4.4 Implementation of AM systems

A full-AM signal as described in section 4.2 can be generated at a high power level because no information is contained in the phase of the signal. A typical and very efficient way to generate such a signal, widely used in mobile and fixed transmitters, is to use a class-C RF amplifier, the modulating signal being used to vary the supply voltage to the amplifier between zero and twice the nominal supply voltage. This causes between 0 and 100% modulation. The peak envelope power is four times the unmodulated carrier power, and this means that the transmitter output stage must be rated at four times the carrier power. Efficiency can therefore reach 80% because the amplifier is operated in class C. The modulator is effectively acting as a multiplier, and because the modulating signal is added to a DC voltage a carrier component is present, as shown in section 4.2.

Alternatively, generation at a low power level is possible by using the modulating signal to vary the bias on the base of a bipolar transistor or the gate of a FET, and hence to alter the RF conduction angle of the amplifier which is normally operated in class B or class C. Changing the bias also alters the efficiency of the stage, leading to a lower efficiency than in high-level class-C amplifiers. Generally, if a high modulation index is required, some linearization technique must be used at the modulation input point. Finally, generation at a low power level is possible, using a balanced mixer or multiplier as in Figure 4.1, followed by a linear power amplifier. Indeed, the widespread introduction of semiconductor devices as RF amplifiers in radio transmitters has increased the popularity of low-level modulation. There is a large change in junction capacitance as the voltage across a semiconductor device is varied, and this leads to detuning and the generation of spurious phase modulation if high-level modulation is attempted. For suppressed-carrier signals low-level generation is essential.

4.5 Single-sideband implementation

4.5.1 Filter method

In common with double sideband suppressed carrier systems, the generation of SSB signals must be done at low power levels, and an immediately attractive method uses the combination of a mixer and filter shown in Figure 4.2. This method effectively uses a multiplier to generate a double sideband suppressed carrier signal as in Figure 4.1, relying on a filter to remove one of the sidebands and any residual carrier resulting from imbalance in the multiplier. It is apparent that the filter must have very sharp cut-off characteristics, and the higher the frequency at which the signal is generated, the more difficult this becomes. Special designs have evolved, using either electromechanical resonators in the range 50–500 KHz or crystal lattice filters at up to 10 MHz.

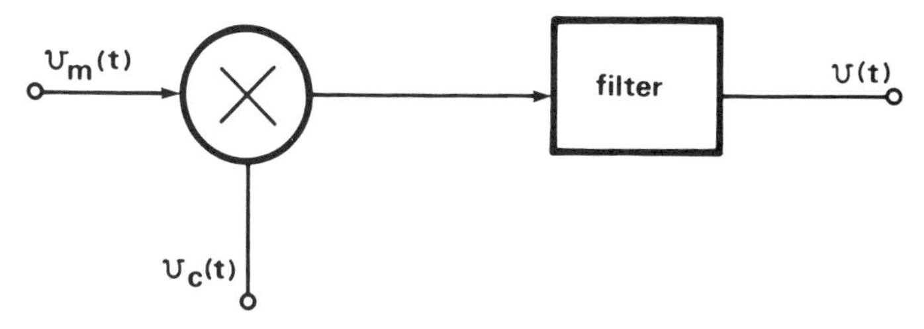

Figure 4.2 Generation of a single-sideband (SSB) signal using the filter method.

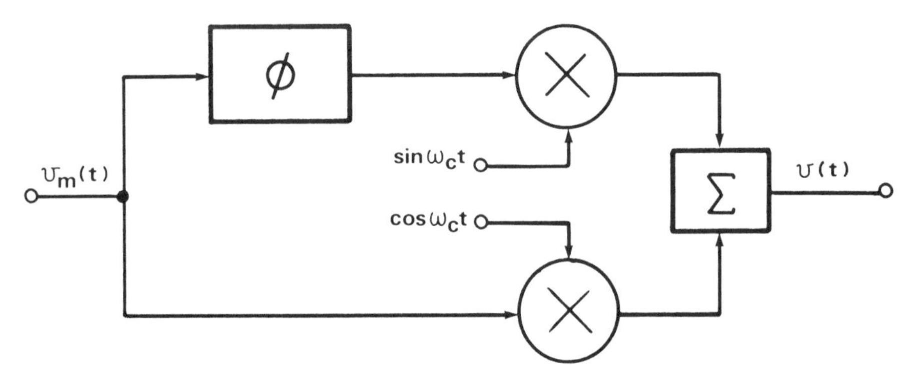

Figure 4.3 Generation of a single-sideband (SSB) signal using the phase-shift (Hartley) method.

4.5.2 Outphasing methods

An alternative is possible, the phase shift (or Hartley) method illustrated in Figure 4.3. In this method the modulating signal is processed in two parallel paths, one of which contains a 90° phase shifter. Unfortunately any imperfections, such as occur if the two carrier oscillators are not exactly 90° out of phase or the phase shift network does not produce exactly 90° phase shift over the whole bandwidth of the modulation, lead to some of the unwanted sideband being generated, and this causes interference to other radio users. In practice, for modulating signals which occupy a relatively wide bandwidth, it is extremely difficult to design a phase shifter with acceptable characteristics. Some improvement can be obtained by using two phase shifters, one in each branch, with the differential phase shift between them maintained at 90°. However, it is necessary to maintain the differential phase shift within a few degrees if suppression of the unwanted sideband is to exceed about 40 dB.

The problem of maintaining the 90° phase shift over the full bandwidth of the baseband signal can be overcome by using two stages of quadrature modulation. This technique was first described by Weaver [9], and is illustrated in Figure 4.4. It is sometimes referred to as the 'Third Method', and is attractive because there is now only a requirement for 90° phase shifts at single frequencies.

Referring to Figure 4.4, the outputs after the first stage of quadrature modulation are:

$$e_1(t) = \tfrac{1}{2} \sum_{n=1}^{N} A_n[\cos(\omega_n - \omega_0)t + \cos(\omega_n + \omega_0)t]$$

$$e_2(t) = \tfrac{1}{2} \sum_{n=1}^{N} A_n[-\sin(\omega_n - \omega_0)t + \sin(\omega_n + \omega_0)t]$$

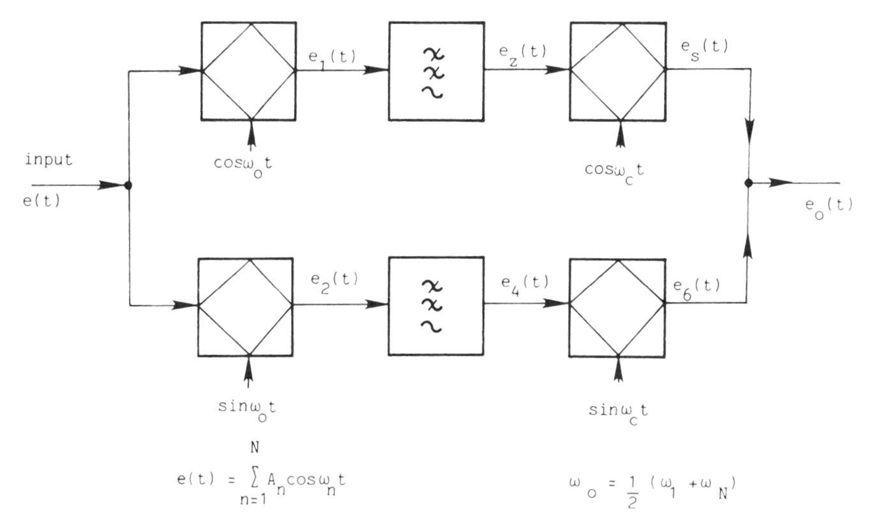

Figure 4.4 Weaver's method of SSB generation.

After low-pass filtering

$$e_3(t) = \tfrac{1}{2} \sum_{n=1}^{N} A_n \cos(\omega_n - \omega_0)t$$

$$e_4(t) = \tfrac{1}{2} \sum_{n=1}^{N} - A_n \sin(\omega_n - \omega_0)t$$

After the second stage of modulation

$$e_5(t) = \tfrac{1}{4} \sum_{n=1}^{N} A_n[\cos(\omega_c - \omega_n + \omega_0)t + \cos(\omega_c + \omega_n - \omega_0)t]$$

$$e_6(t) = \tfrac{1}{4} \sum_{n=1}^{N} A_n[-\cos(\omega_c - \omega_n + \omega_0)t + \cos(\omega_c + \omega_n - \omega_0)t].$$

Adding $e_5(t)$ and $e_6(t)$ gives

$$e_0(t) = \tfrac{1}{2} \sum_{n=1}^{N} A_n \cos(\omega_c + \omega_n - \omega_0)t \tag{4.9}$$

which is an upper sideband associated with the effective carrier frequency $(\omega_c - \omega_0)$.

Careful balance in the networks is still necessary, however, since the first stage of quadrature modulation inserts a carrier in the centre of the baseband signal, and imbalance in either of the first stage modulators results in the appearance of an unwanted inband tone at the modulator output. Nevertheless, the Weaver SSB generator is in principle realizable in VLSI and continues to attract interest.

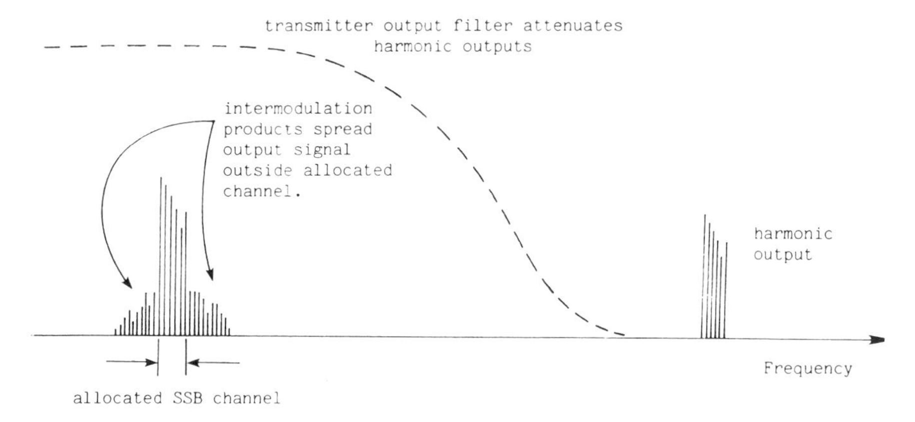

Figure 4.5 Effect of intermodulation on SSB transmitter output.

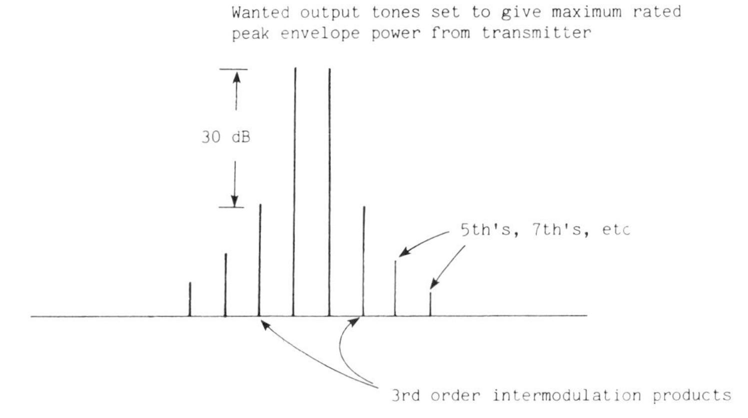

Figure 4.6 Typical linearity requirement for marine HF SSB transmitter.

4.5.3 Problems in SSB implementation

The generation methods described above all require the linear amplification of the single sideband signal, usually after a further stage of frequency shifting to the required output frequency. After RF gain, filtering can be again used to reduce output components at harmonic frequencies, but close-in filtering of output components is not realistically feasible. Figure 4.5 illustrates the problem which arises when the single sideband signal is amplified by a practical RF device. Intermodulation among the components of the SSB signal results in the appearance of spectral components outside the band occupied by the original SSB signal. In terms of the operation of SSB equipments in isolation, these effects can be tolerated—the marine applications of SSB in the HF bands make only modest demands on RF amplifier linearity, as suggested in Figure 4.6, where third-order products are typically 30 dB below the principal outputs in a two-tone test. However, in the land

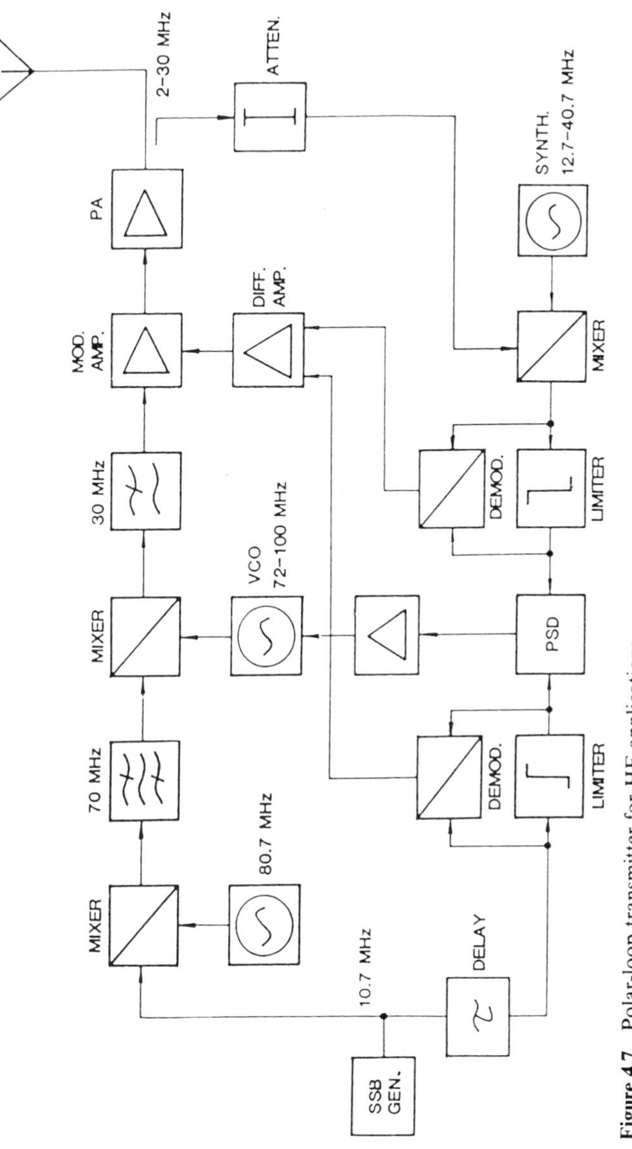

Figure 4.7 Polar-loop transmitter for HF applications.

mobile environment with a large number of transmitters and receivers operating in the same geographical areas, the general level of noise produced by transmitter intermodulation must be reduced very substantially if the potential improvements in spectral utilization obtainable in principle with SSB are to be achieved in practice. This problem has been addressed by Petrovic [10], who has developed a number of related techniques for improving the linearity of SSB transmitters and applied these methods to transmitters operating at frequencies from HF to UHF.

Figure 4.7 shows a block diagram of the polar-loop transmitter. The operating mechanism of this system is readily perceived—the non-linear power amplifier (PA) introduces amplitude (envelope) non-linearity and amplitude-to-phase-shift conversion (AM-PM). The polar loop arrangement continuously samples the PA output and compares it with the original signal from the SSB generator; the VCO loop corrects for phase variations and the differential amplifier corrects for envelope variations. Improvements in third-order intermodulation product levels of 30–40 dB are obtainable.

A further feature of SSB speech transmissions is their very wide dynamic range (peak amplitudes in speech are typically four times greater than the average power level) and this accentuates the linearity problem. Modern SSB systems employ amplitude companding to ease the demands on linearity for speech transmissions.

4.6 Demodulation

In the previous sections we have discussed the methods of generating AM and AM-derived signals. We now take a brief look at the demodulation process required at the receiving end of the link to recover the original baseband signal. In general terms, the available techniques can be divided into two categories, coherent (or synchronous) demodulator systems and non-coherent (asynchronous) systems.

4.6.1 Coherent demodulation

The process of coherent demodulation can be used to recover the baseband signal from any AM type of signal. The term coherent arises from the fact that the demodulator uses a carrier signal identical with that at the transmitter to effectively reverse the modulation process. This carrier can either be obtained from the received signal or, if that is not possible, generated locally. DSBSC and SSB are linear modulation schemes, and coherent demodulation must be employed at the receiver.

The coherent demodulator shown in Figure 4.8 consists of a multiplier followed by a low-pass filter. The receiver has no direct knowledge of the frequency or phase of the carrier signal used at the transmitter so in general the requirements for oscillator stability are very demanding. If we assume that at the receiver a locally-generated signal $\cos(\omega_c t + \theta)$ is used, then the output

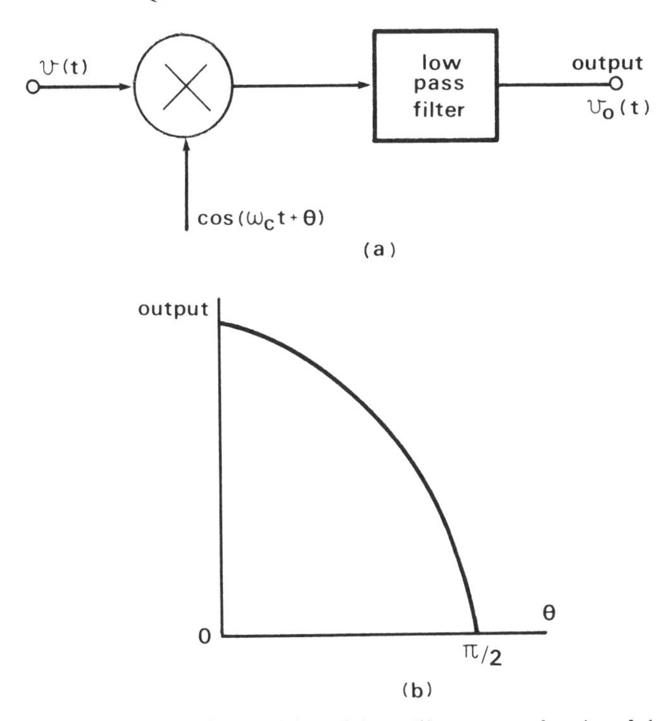

Figure 4.8 (*a*) Coherent (or synchronous) demodulator; (*b*) output as a function of phase angle θ.

following the low-pass filter, when a DSBSC signal is received, is

$$v_0(t) = \tfrac{1}{2}\cos(\omega_m t - \theta) + \tfrac{1}{2}\cos(\omega_m t + \theta) \tag{4.10}$$

$$= \tfrac{1}{2}[\cos\theta\cos\omega_m t + \sin\theta\sin\omega_m t + \cos\theta\cos\omega_m t - \sin\theta\sin\omega_m t]$$

$$= \cos\phi\cos\omega_m t \tag{4.11}$$

In this equation, $\cos\phi$ is a constant, representing the amplitude of the recovered baseband signal, and it is now clear that the phase of the locally-generated carrier is very important. If $\theta = 0$, i.e. the carrier has the same phase as used at the transmitter, then the baseband signal is recovered with unity amplitude. If $\theta = \pm 90°$, however, the output is zero. An interesting situation exists if the phase of the locally-generated carrier is exactly 180° out of phase with the transmitter carrier, because then all modulation components are recovered with unity amplitude but with their original phase reversed. This is not very important if speech is being used, because the human ear is relatively insensitive to phase, but if a data signal is recovered with reverse phase all 1s are recovered as 0s and vice-versa, which is disastrous!

It is clear that the coherent demodulator is equally effective with SSB and full-AM signals. As far as SSB is concerned, if the lower sideband is transmitted, then eqn (4.10) shows that after the filter we are left with only

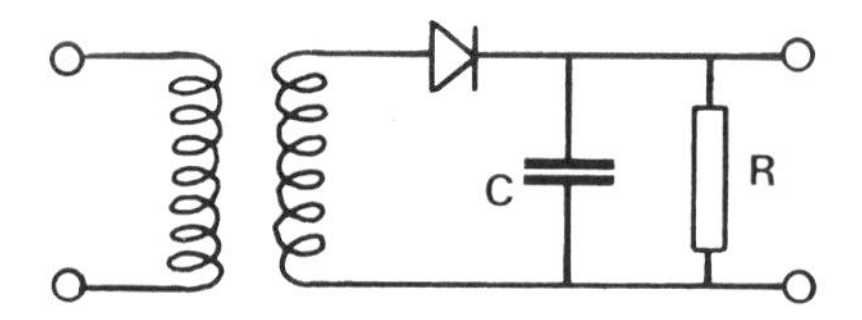

Figure 4.9 Simple diode detector (envelope detector).

$\cos(\omega_m t + \theta)$. It is apparent, therefore, that with SSB each component of the baseband signal is recovered with full amplitude, but with a phase shifted by an amount θ. In this case there is no problem of reduced amplitude due to the presence of two demodulated components, as with DSBSC. The phase shift is not important with speech, as mentioned above, but again with data it can cause severe problems. Indeed, as far as speech is concerned, small offsets in the frequency of the locally-generated carrier can be tolerated. The pitch of the recovered speech may be changed slightly, but provided the difference is less than about 20 Hz the intelligibility is unaffected.

A second feature of linear modulation schemes relates to their inability in general to suppress co-channel interference. Non-linear schemes such as FM provide for receiver discrimination in favour of strong received signals, giving reduced output from lower level interferers. This aspect of performance will be described further in Chapter 8 in the context of cellular radio systems, where C/I performance of receivers is very significant in determining user capacity.

4.6.2 Non-coherent (envelope) detection

In the discussion on the generation of full-AM signals, some stress was laid on the fact that the signal envelope follows the same shape as the modulating waveform. The reason for this is that the envelope can easily be recovered using a simple rectifier circuit as shown in Figure 4.9. This is much easier to build than a coherent detector—it requires no locally-generated carrier and is very cheap. Indeed, the popularity of AM for medium-wave radio broadcasting is almost entirely due to the fact that such signals can be non-coherently demodulated so that low-cost receivers can be mass-produced. The fact that AM is wasteful in terms of transmitted power is relatively insignificant in the broadcasting context.

As far as the circuit itself is concerned, some care needs to be taken in choosing the values of capacitance and resistance to avoid distortion at high values of modulation index, but several textbooks [e.g. 2] give the necessary design procedures.

4.7 Generation of FM signals

For the generation of FM signals, we require a circuit that produces a frequency change proportional to the input voltage, i.e. a voltage-to-frequency

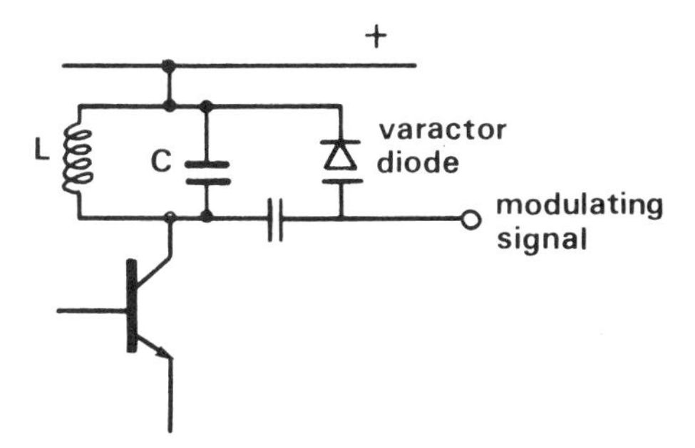

Figure 4.10 Principle of direct frequency modulation.

converter. Methods of generating FM fall into two general categories, direct and indirect.

4.7.1 Direct methods

These usually rely on the use of a solid-state device known as a varactor diode. When reverse-biased, these devices exhibit a capacitance inversely proportional to the applied voltage. Typically, the capacitance varies from 10 to 50 pF if the applied voltage is changed by 20 V. Figure 4.10 shows in principle how a varactor diode is used to change the frequency of a simple oscillator and hence to produce an FM signal. It is relatively simple to show that the change of frequency Δf produced by a change of capacitance ΔC is

$$\Delta f = f_0 \frac{\Delta C}{2C} \tag{4.12}$$

Quite small changes in capacitance can cause quite large changes in frequency if f_0 is high. This makes the circuit difficult to control, and the solution usually adopted is to generate the FM signal at a lower frequency where $\Delta f / f$ is larger. It can then be translated to the required frequency range.

At frequencies in the microwave range, it is possible to produce FM directly, by using semiconductor devices such as Gunn diodes. These produce oscillations at a frequency determined by an applied voltage, thus producing FM directly. Typically they operate at 10 GHz and provide a frequency deviation of about 30 MHz/V.

4.7.2 Indirect methods

In the discussion on narrowband FM in section 4.3.4 it became clear that the only difference between AM and narrowband FM or PM is in the phase of the carrier with respect to the sidebands. This points the way to an interesting method of generating FM indirectly, first proposed by Armstrong [11]. Equation (4.4) shows that to generate an FM signal the modulation waveform must first be integrated, so when this and a carrier signal are applied to the multiplier, a DSBSC signal results. It is now necessary to add a carrier

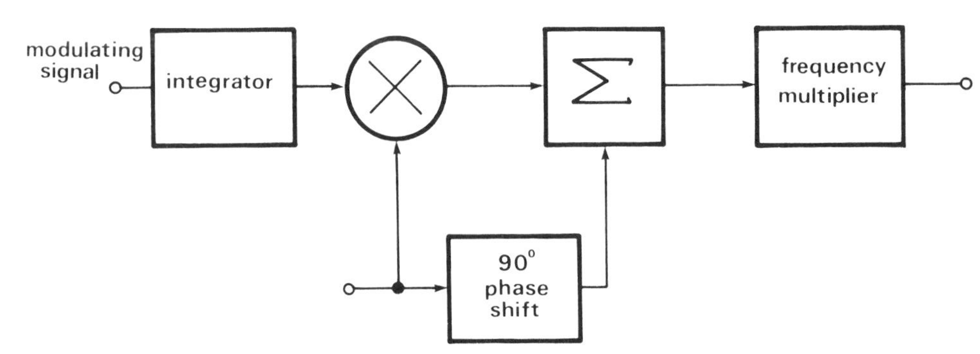

Figure 4.11 Indirect frequency modulation (Armstrong method).

component at 90°, and Figure 4.11 shows show this is done. The output of the circuit is a NBFM signal, or, if the integrator is omitted, NBPM. For some applications the narrowband signal is used as generated, but in the majority of cases further processing is necessary to generate a signal with the required carrier frequency and bandwidth.

The carrier frequency and bandwidth can be increased by a process of frequency multiplication. This differs fundamentally from frequency changing, which merely translates a signal from one part of the spectrum to another, without changing its characteristics in any way. A frequency multiplier is a circuit which, in effect, multiplies all frequencies at its input by a constant number and hence increases the centre frequency and bandwidth of a signal. If frequency multiplication is used as shown in Figure 4.11, then the signal applied to the multiplier would not be at the final carrier frequency but at an appropriate sub-multiple.

A detailed discussion of frequency multipliers is beyond the scope of this book, although information is readily available elsewhere [e.g. 1]. One simple method, however, is to drive an amplifier slightly into saturation and then select a required harmonic by means of a tuned circuit. Low-order multipliers (doublers and triplers) often use this technique. Alternatively, phase-locked loops are now readily available in integrated-circuit form. They are ideally suited to the frequency-multiplication function and are extensively used.

4.8 FM demodulators

Recovery of the modulating waveform from an FM signal requires a circuit that provides a frequency-to-voltage conversion. There are several ways of doing this; we will restrict attention to a brief description of the more popular techniques.

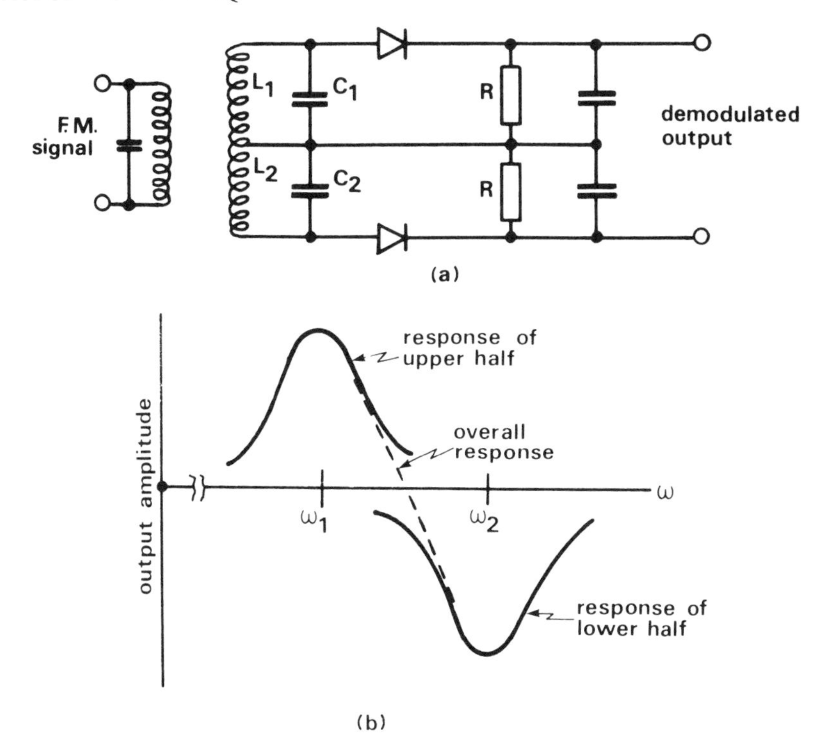

Figure 4.12 Slope detector (discriminator). $\omega_1 = \dfrac{1}{\sqrt{L_1 C_1}}$, $\omega_2 = \dfrac{1}{\sqrt{L_2 C_2}}$

4.8.1 Frequency discriminators

If we differentiate eqn (4.4), which describes an FM signal, then we obtain

$$\frac{\mathrm{d}}{\mathrm{d}t} v_{\mathrm{f}}(t) = -V_{\mathrm{c}}[2\pi f_{\mathrm{c}} + k_{\mathrm{f}} v_{\mathrm{m}}(t)] \sin\left[2\pi f_{\mathrm{c}} t + k_{\mathrm{f}} \int_0^t v_{\mathrm{m}}(t)\mathrm{d}t\right] \qquad (4.13)$$

It can be seen that if $k_{\mathrm{f}} v_{\mathrm{m}}(t) \ll 2\pi f_{\mathrm{c}}$, then an AM signal is observed at the output of the differentiator, and this can be non-coherently demodulated using an envelope detector as in Figure 4.9.

A convenient way to obtain a voltage proportional to frequency is through the use of a tuned circuit, and a practical demodulator which uses two tuned circuits to produce a linear output is shown in Figure 4.12. Two slightly different realizations of this kind of circuit find widespread use, the Foster–Seeley discriminator and the ratio detector. It is very important to remove any amplitude variations in the received FM signal (which could, for example, be caused by signal fading) before the discriminator. Otherwise, after the differentiator it will be impossible to distinguish the unwanted amplitude fluctuations from those deliberately generated by the FM-to-AM conversion process. For this reason a limiter is invariably used preceding the actual discriminator circuit.

4.8.2 Pulse-counting discriminators

It is apparent that the amplitude of an FM signal carries no information; furthermore, the exact shape of the FM signal is largely irrelevant as far as information is concerned, since this is contained in the zero-crossings. If the receiver uses a fairly low IF, then it is possible to use a pulse-counting discriminator as in Figure 4.13. The FM signal is first passed to a limiter and differentiator, the differentiator output being used to trigger a monostable circuit which produces a train of constant-width pulses with variable spacing. Finally, a low-pass filter is used to give an output proportional to the short-term average of its input. It is apparent that, when the frequency is high, the

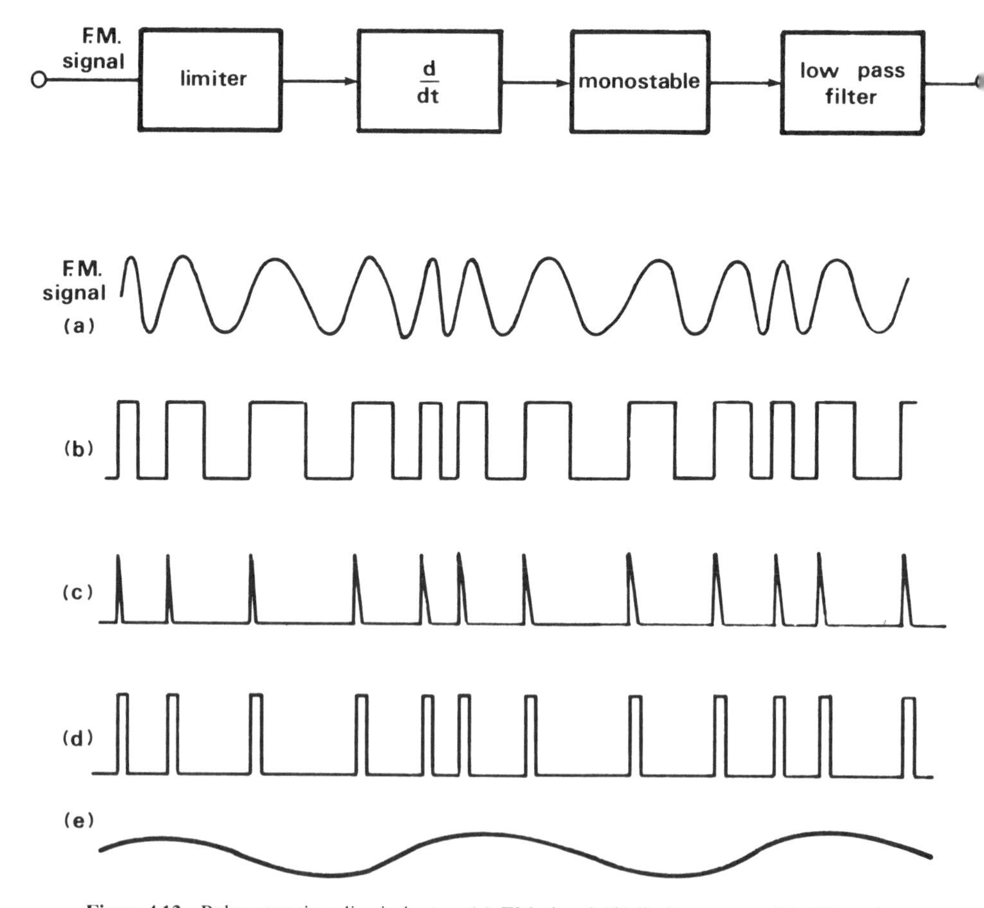

Figure 4.13 Pulse-counting discriminator. (*a*) FM signal; (*b*) limiter output; (*c*) differentiator output; (*d*) monostable output; (*e*) recovered modulation.

monostable produces more pulses in a given time than when the frequency is low—the modulating waveform is thereby recovered at the output.

Pulse-counting discriminators have to be designed so that the period of the monostable pulse is shorter than that of the highest instantaneous frequency reached by the carrier. This is essential to avoid overlapping, but otherwise the circuit does not have to be tuned and this gives it an advantage over the 'slope' discriminator discussed previously.

4.8.3 Phase-locked loops

A different approach to FM demodulation is to place a frequency modulator in the return path of a feedback system. Since a high-gain feedback system performs in its forward path the inverse operation to that in its return path, the forward path demodulates the FM signal. The phase-locked loop (PLL) is one example of such a system. A detailed study of the behaviour of the PLL is quite complicated and well beyond the scope of this book. We shall therefore restrict attention to a brief description of the steady-state behaviour of a simple loop to illustrate its application as an FM demodulator. The simple diagram in Figure 4.14 shows a phase-comparator, a filter, and a voltage-controlled oscillator (VCO), which is an FM modulator. The phase-comparator produces at its output a voltage that is proportional to the phase difference between the input signal and the output of the VCO; this voltage is filtered and is used to control the frequency of the VCO.

If we assume initially that an unmodulated carrier with phase $\phi(t) = \omega_c t + \phi_0$ is present at the input, then if the VCO output phase $\theta(t)$ has the same value, the output of the phase detector is zero. The loop is then said to be locked. If for any reason $\theta(t)$ drifts by a small value $\delta\theta$, then the phase detector produces a voltage proportional to $-\delta\theta$ which, when amplified and applied to the VCO input, acts to reduce $\delta\theta$ to nearly zero. Thus without modulation the loop stays locked to the exact carrier frequency and the approximate carrier phase. If frequency modulation is present, the VCO phase $\theta(t)$ will follow the instantaneous phase of the input signal with only a small error. The amplified error voltage appearing at the VCO output will be proportional to the

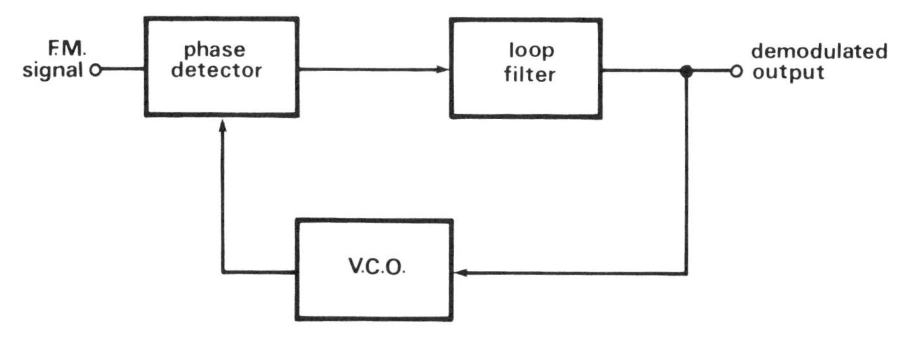

Figure 4.14 Simple phase-locked loop.

instantaneous frequency $f(t)$, since the VCO acts as an integrator. This can also be seen by remembering that the phase variations in $\phi(t)$ are proportional to $\int f(t)\,dt$ and $\theta(t) \simeq \phi(t)$. Thus $\theta(t)$ will have phase variations which are approximately proportional to $\int f(t)\,dt$. However, the output frequency of the VCO is proportional to the input voltage, so that to produce the required phase variations in $\theta(t)$, the VCO input voltage must be proportional to $f(t)$, since phase is the integral of frequency.

The PLL demodulator is particularly advantageous at low signal levels, and its linearity as a demodulator is directly related to the linearity of the VCO characteristic. The fact that highly linear PLL circuits are available in integrated circuit form has increased the popularity of this type of demodulator, which now finds widespread application in FM mobile radio receivers.

4.8.4 Quadrature detection

Consider the circuit shown in Figure 4.15*a* in which an FM signal $v_f(t)$ is multiplied by a phase-shifted version $v_q(t)$. We will show that if the network $\phi(\omega)$ has a constant amplitude response and a linear phase response (i.e. phase shift is directly proportional to change of frequency) over the whole spectrum occupied by the signal, and if the phase shift is nominally $\pi/2$ at the carrier frequency, then the circuit functions as an FM demodulator.

To achieve the required characteristics, the network $\phi(\omega)$ must produce a phase shift linearly proportional to the instantaneous frequency f_i of the FM signal, i.e.

$$v_q(t) = \rho V_c \cos\left[\omega_c t + k_f \int v_m(t)\,dt + \phi(\omega_i)\right] \tag{4.14}$$

where $\omega_i = (d/dt)\phi(t) = \omega_c + k_f v_m(t)$ and ρ is a constant.

If the phase shift at the carrier frequency is $\pi/2$, we can write

$$\phi(\omega_i) = \pi/2 + k\,\Delta\omega \quad \text{where } \Delta\omega = \omega_i - \omega_c$$

Figure 4.15*b* shows the phasor diagram when $\omega_f(t)$ and $v_q(t)$ are multiplied. The output of the multiplier is proportional to $\cos\theta$, so that

$$v_0(t) = \rho^2 V_c^2 \cos\theta = \rho^2 V_c^2 \sin\xi$$

since

$$\xi = k\,\Delta\omega = k(\omega_i - \omega_c)$$

$$= k[\omega_c + k_f v_m(t) - \omega_c] = k k_f v_m(t)$$

Hence

$$v_0(t) = \rho^2 V_c^2 \sin[k k_f v_m(t)] \tag{4.15}$$

If the phase shift varies over a small angle only we obtain

$$v_0(t) = \rho^2 V_c^2 k k_f v_m(t) \tag{4.16}$$

which is the modulating signal multiplied by constant factors.

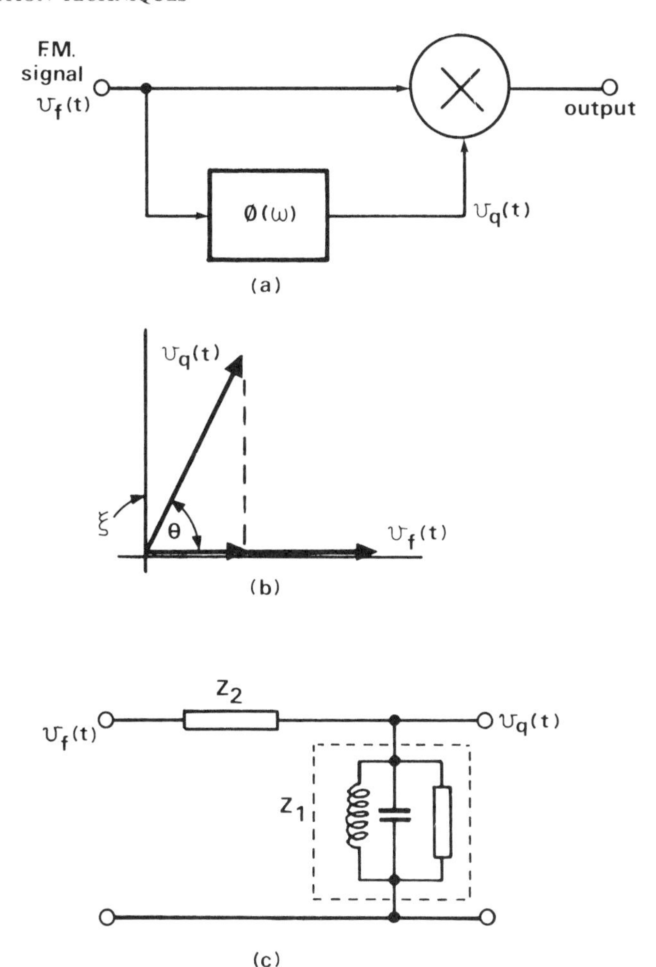

Figure 4.15 Quadrature demodulator.

In practice, the network shown in Figure 4.15*c* is usually used to realize the linear phase network $\theta(\omega)$. Z_1 is the well-known parallel RLC circuit with a low Q-factor, and Z_2 is either an inductor or capacitor whose function is to produce the $\pi/2$ phase shift at the carrier frequency. The advantage of the quadrature detector compared with a conventional discriminator is that it avoids the use of a transformer and by using an integrated circuit multiplier, can be made quite cheaply.

4.9 The effect of noise on AM systems

In addition to the transmission impairments such as fading discussed in Chapter 3, all communication systems are subject to noise which also degrades the overall performance. Noise generated within the receiving system is generally Gaussian in nature, but external noise may be impulsive. The latter topic will be discussed in Chapter 5; for the present, we concern ourselves only with the effects of Gaussian noise on AM and FM systems.

In general, additive noise affects both the amplitude and phase of a signal; a noisy sinusoidal waveform has a varying amplitude and zero-crossings which are not uniformly spaced in time. The amplitude variations will be detected by an AM demodulator, together with the intended modulation, and similarly an FM demodulator will sense the phase variations. In both cases, signal quality will be affected, and in high-noise conditions, intelligibility will be severely degraded. As far as analogue communications are concerned, the effect of noise can be quantified by comparing the signal-to-noise power ratio (SNR) at the input of the demodulator with that at the output; in other words, we compare the predetection and postdetection SNR. In order to do this we need to know something of the signal characteristics, the noise characteristics and the properties of the demodulator.

4.9.1 Noise performance of DSBSC and SSB

Coherent demodulation is necessary for these systems and is commonly accomplished using a circuit of the type shown in Figure 4.8. Several textbooks [e.g. 1, 2] treat the case when the signal is corrupted by additive Gaussian noise and show that for a DSBSC signal the output signal-to-noise ratio SNR_0 is twice the input ratio SNR_i:

$$\mathrm{SNR}_0 = 2\mathrm{SNR}_i \tag{4.17}$$

Coherent detection of DSBSC therefore provides a 3 dB improvement in SNR, i.e. the post-detection SNR exceeds the pre-detection SNR by 3 dB. The improvement can be considered as being due to the rejection of the quadrature noise component. Alternatively, we can say that the two demodulated sidebands (see eqn 4.10) add in-phase on a voltage basis, whereas the noise components add on a power basis, giving a net 3 dB improvement in SNR at the output.

As far as SSB is concerned, there is only one demodulated sideband and likewise the noise. Since we do not have two sidebands that can be coherently added, the demodulator gives no SNR advantage, and for SSB

$$\mathrm{SNR}_0 = \mathrm{SNR}_i \tag{4.18}$$

4.9.2 Noise performance of AM

As far as AM is concerned it can either be demodulated coherently as for suppressed-carrier systems, or a simple non-coherent detector can be used.

For coherent demodulation the input and output signal-to-noise ratios are related through the modulation index m by

$$\mathrm{SNR_0} = \left[\frac{2m^2}{2+m^2}\right]\mathrm{SNR_i} \tag{4.19}$$

and for the limiting case of $m = 1$, the maximum value of $\mathrm{SNR_0}$ is

$$\mathrm{SNR_0} = \tfrac{2}{3}\mathrm{SNR_i} \tag{4.20}$$

The envelope detector is a non-linear circuit and an exact analysis is rather lengthy. For high $\mathrm{SNR_i}$, however, it can be shown that the performance is identical with that for coherent demodulation. If $\mathrm{SNR_i}$ is low, the message waveform is hopelessly mutilated and is not recoverable in its original form. The noise completely swamps the signal and the envelope then bears little or no resemblance to the original modulating waveform.

4.9.3 Comparison of the modulation methods

Equations (4.17)–(4.20) do not necessarily give a realistic comparison of the different modulation methods. Generally, transmitters are power-limited, and is clear that if we consider three transmitters of equal output power P_T, then for an AM transmitter the total output power consists of carrier power P_c plus the power of two sidebands P_{SB}. For DSBSC and SSB, there is no carrier, so $P_{SB} = P_T$.

As far as noise is concerned, we have to consider the bandwidth of the signals. Both AM and DSBSC occupy the same bandwidth, but SSB needs only half that of these systems. When all relevant factors have been taken into account, the result obtained is

$$(\mathrm{SNR_0})_{\mathrm{SSB}} = (\mathrm{SNR_i})_{\mathrm{SSB}} = 2(\mathrm{SNR_i})_{\mathrm{DSB}}$$

So

$$(\mathrm{SNR_0})_{\mathrm{SSB}} = (\mathrm{SNR_0})_{\mathrm{DSB}} \tag{4.21}$$

In other words, although the DSB system requires twice as much bandwidth, coherent addition of the two sidebands compensates for the 3 dB loss in predetection SNR.

Comparison of the modulation systems is illustrated in Figure 4.16, which shows the $\mathrm{SNR_0}$ of SSB and AM compared with that of DSBSC. Note that there is a threshold effect at low levels for AM, corresponding to the point at which the noise is large enough to significantly distort the detected envelope waveform.

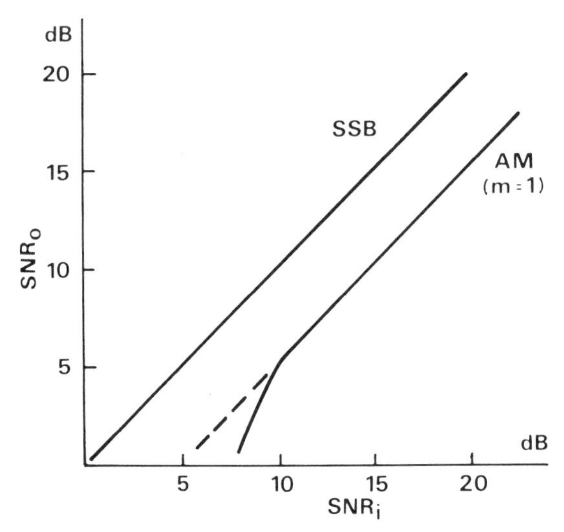

Figure 4.16 Output and input SNR for amplitude modulation systems in comparison with DSBSC.

4.10 The effect of noise on FM systems

The effect of noise on FM communication systems is slightly more complicated than on AM systems. The receiver block diagram is as in Figure 4.17, and we define the output SNR as

$$\mathrm{SNR}_0 = \frac{\text{mean signal power}}{\text{mean noise power with unmodulated carrier}}$$

Because of the inclusion of a limiter, we are now concerned only with the effect of noise on the phase of the signal.

The predetection signal-to-noise ratio SNR_i is often called the carrier-to-noise ratio CNR, and in the simple case of single-tone modulation

$$\mathrm{SNR}_0 = 3(\beta + 1)\beta^2\mathrm{SNR}_i \simeq 3\beta^3\mathrm{SNR}_i \tag{4.22}$$

$3\beta^3$ is therefore the SNR improvement factor.

4.10.1 The SNR-bandwidth interchange and FM threshold

If we re-examine eqn (4.22) and use Carson's rule to write, for large values of β, the relationship $\beta \simeq B/2\omega_m$, we obtain

$$\mathrm{SNR}_0 \simeq 3\left\{\frac{B}{2\omega_m}\right\}^3 \mathrm{SNR}_i$$

This equation reveals an important fact: in FM systems we can trade off bandwidth and signal-to-noise ratio, whereas this is not possible in AM systems. In FM systems, however, SNR_0 can be increased by increasing the

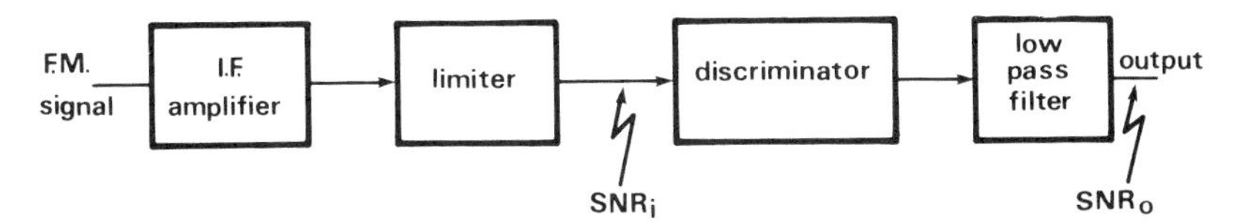

Figure 4.17 FM receiver structure.

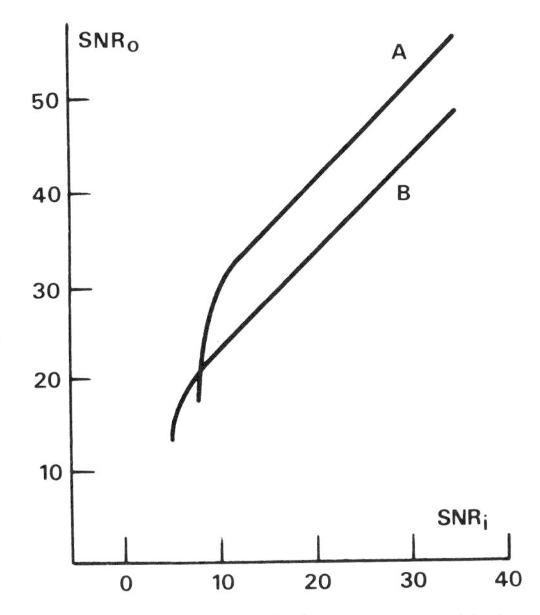

Figure 4.18 Typical demodulation characteristics for FM systems with sinusoidal modulation. (*A*) Typical wideband system; (*B*) Typical narrowband system.

modulator constant k_f without having to increase the transmitter power. Of course, this interchange cannot continue indefinitely since a point will be reached where SNR_i is no longer large enough for the above analysis to be valid. This point is called the FM threshold.

As the noise level increases with respect to the signal, the phase variations in the noise become dominant. Changes in noise can then change the carrier phase by 2π radians, and these rapid changes appear as large-amplitude spikes at the discriminator output. These spikes produce an audible 'click' on a loudspeaker and are known as click noise. Its presence indicates the onset of the FM threshold and it becomes a crackle as the SNR deteriorates. Figure 4.18 shows the output SNR as a function of the input SNR for sinusoidal frequency modulation, from which the threshold is easily observable.

4.10.2 Pre-emphasis and de-emphasis

A detailed analysis of the noise response of the FM discriminator produces an expression for the noise output power spectrum which contains a factor ω^2, a factor which is not present in the expression for the corresponding output signal power spectrum. There is therefore an emphasis on high-frequency output noise, and a de-emphasis filter having a frequency roll-off proportional to $1/\omega^2$, placed in the output, will reduce the output noise. This is desirable, but if nothing else were done this de-emphasis filter would distort the signal message severely. To counteract this effect, a pre-emphasis filter is used in the transmitter to pre-distort the modulating signal so that the overall effect on the signal is zero. The transmitter filter necessarily has a transfer function which is the inverse of the receiver filter transfer function:

$$H_p(\omega) = \frac{1}{H_d(\omega)} \tag{4.23}$$

Overall, if the signal is unaffected by the pre-emphasis and de-emphasis filters, the gain in output SNR will be simply the ratio of noise power with emphasis to that without emphasis. In a typical FM system this can easily be 10 dB, so the inclusion of the filters is well worthwhile. It is also worth pointing out that the $1/\omega^2$ characteristic mentioned above is not vital, although to avoid message distortion eqn (4.23) should always apply.

4.11 The effects of multipath propagation

We have seen in Chapter 2 that multipath propagation has significant effects on the received signal. In narrowband transmissions the envelope and phase are subjected to a random variation which is superimposed on the deterministic variations that result from the modulation process. It is clear that if these random variations are small compared with the deterministic variations then the message will be almost undistorted, but if this is not the case then intelligibility will be reduced and severe problems will arise. In wideband digital transmissions, delayed versions of the signal resulting from multipath propagation will cause intersymbol interference and hence produce a higher probability of error. Provided the bandwidth of the transmitted signal is less than the coherence bandwidth of the channel, the channel transfer function is largely independent of frequency, or perhaps, more accurately, it is independent of frequency separation within the channel. All frequency components within the message will then fade together, and this is termed 'flat fading'. However, if the signal bandwidth exceeds the coherence bandwidth, components at different frequencies will fade in an uncorrelated manner and this is termed 'frequency selective fading'.

We have also seen in Chapter 2 that the rate at which the signal envelope fades depends on the speed at which the receiver moves through the multipath field. The random FM (Doppler shift) is also a function of speed. It is worth emphasizing and clarifying, however, that the rate of fading has nothing to do with frequency selectivity. The selectivity is governed by the ratio between the signal bandwidth and the coherence bandwidth, whereas the fading rate is determined by the rate at which the observed effects change, because of receiver motion and consequent dynamic changes in the multipath structure. Most calculations of fading, particularly with regard to its effect on narrowband transmissions, are based on the assumption of flat (non-selective) fading where the fading can be represented as multiplicative noise. Separately, and additionally, most analyses are based on an assumption of slow fading; that is, effects due to the fading are slow compared with changes in the message. These are independent assumptions. Consequently, even where selective fading is concerned, slow fading is often assumed.

In practice, envelope fading means that the received SNR is continually changing, and determination of the average SNR therefore involves an integration over the envelope characteristics (usually assumed Rayleigh). In some cases it is straightforward to obtain a quantitative relationship between the average SNR and a suitable criterion of system performance, for example BER in data systems is readily related to average SNR (as will be seen in section 4.12). In other cases the relationship is not so easily obtained, and its interpretation is more difficult, for example, the relationship between average SNR and the perceived intelligibility of a voice message has to be determined by subjective testing. As far as multipath effects are concerned, much more attention has been devoted to the effects of envelope fading than to the effects of random FM. However, FM demodulators will respond to the random phase variations and this can also cause distortion in the message.

As far as mobile radio is concerned, it is clear that in AM and AM-related systems, the fading and the modulation affect the same parameter of the signal, namely its amplitude. Generally, the spectrum of the fading will lie below the information spectrum, but at high carrier frequencies and vehicle speeds the spectra may overlap. An AM detector will respond to both, and loss of signal quality may result. The subjective effects of fading at the mobile receiver therefore depend on the speed of the vehicle, since the signal suffers multiplicative modulation by the fading waveform. In practice, this has steep 'dropout' spikes, characteristic of Rayleigh fading, and is rich in harmonics. The net effect is a complex distortion of the signal spectrum, with the fading modulation index decreasing with increasing harmonics of the fundamental fading frequency.

Fast fading (hundred of fades per second upwards) can be very objectionable even if the signal does not drop below the sensitivity level of the receiver. In this case, the receiver AGC time-constant is much longer than the fading period, and large variations in the mean amplitude of the speech signal can occur. The SNR required in a vehicle is somewhat higher than for static communication links, approximately 25–30 dB. This is because the user is in an acoustically noisy environment, is often occupied in another role, and uses the radio as an aid only. Fading at a rate commensurate with the upper part of the speech band (above 1.5 kHz) is not particularly detrimental to intelligibility, and the effect is something akin to tone interference at the fading frequency. As the fading rate falls into the 300 Hz–1 kHz range (equivalent to the lower part of the speech band), severe distortion of the speech results, and if the fading depth is large, renders the information unintelligible. Sub-audio fading above about 50 Hz is moderately well tolerated by the human ear, as the variation of the speech spectrum becomes relatively small compared with its bandwidth. At slower fading rates, however, a secondary effect becomes dominant. In this range (a few fades per second), fading is more easily visualized as a 'keying' effect, and indeed sounds like it. At these rates the receiver AGC system will respond, and for mild fading will attempt to provide a constant output. However, when the fading signal falls to the receiver noise level, for example in low signal areas, the AGC is unable to cope, and fading at word and syllabic rate (2–8 fades per second) occurs, giving poor overall intelligibility. Generally the redundancy inherent in the language becomes no longer sufficient to overcome the loss of information in the fade periods, and this is of particular concern in police and other public safety applications where low-redundancy voice codes are frequently employed. Fading rates of about 10 Hz are particularly bad, as the ear has an apparent capture effect for the fading frequency, the speech being totally unintelligible.

At a carrier frequency of 100 MHz the fading distance is approximately 1.5 m, and for a vehicle travelling at 15 m/s (34 mph) represents a fading rate of about 10 Hz, which thus produces 'worst-case' conditions of sub-audio fading. At greater speeds, intelligibility improves somewhat until 'speech-shift' becomes troublesome, and at lower speeds, although it may improve to the extent that some of the message is decipherable, more of the message will be lost and hence overall intelligibility is not likely to improve. Envelope fading can therefore have very serious effects on mobile radio systems unless corrective measures are taken. The distortion can be very severe, even in high-signal areas. In static operation it is possible for the receiving antenna to be positioned in a fade, and hence contact is completely lost with the base station, unless the receiver is moved a short distance.

It is apparent from the discussion above that one method of correcting for the effects of envelope fading in narrowband systems is to transmit a carrier, or pilot tone, that can be used for gain and/or frequency control in the receiver. The assumption is that, in such systems, where the bandwidth of the transmission is less than the coherence bandwidth of the channel, the carrier or pilot-tone will be affected in the same way as the modulation. Automatic gain control circuits are present in all AM receivers, and if the fading is slow can provide a large measure of compensation and hence reduce the fading depth. The choice of an AGC time-constant which will reduce the fading without distorting the message information becomes more difficult, however, as the transmission frequency is increased and the spectra of fading and modulation overlap.

SSB is a form of amplitude modulation, and standard SSB receivers cannot distinguish between intended signal amplitude variations and those introduced by the fading characteristics of the channel. Solutions to this difficulty have also sought to address the problem, referred to earlier, of oscillator stability, and the additional need to provide the receiver with means for sustaining AGC functions during gaps in speech (when no signal is output from the transmitter). These all rely on introducing a low-level 'pilot' tone into the transmitted SSB signal. Various schemes have been proposed which place the pilot tone at the edge of the transmitted RF band, but these are not entirely satisfactory—tones at the SSB band edges are vulnerable to ill-defined attenuation and phase effects from the skirts of the IF crystal filters. An alternative has been proposed and researched extensively by McGeehan and Bateman [12, 13] in which the pilot tone is introduced in the middle of the speech band. To remove the continuous inband tone from the receiver output requires that a segment of the recovered baseband be removed by filtering—such a procedure has little effect on speech quality even if the filtered segment approaches 1 kHz in bandwidth, but it does create problems for data transmission. This procedure can be dispensed with, however, if the baseband signal is separated into two halves and a gap of a few hundred hertz introduced between, where the pilot tone is added. This technique, termed 'transparent tone in band' (TTIB), is shown in Figure 4.19 and has proved highly successful, particularly when used in conjunction with a procedure known as 'feed forward signal regeneration' (FFSR). The essence of this process is that the receiver uses the pilot tone not only to obtain a frequency reference for

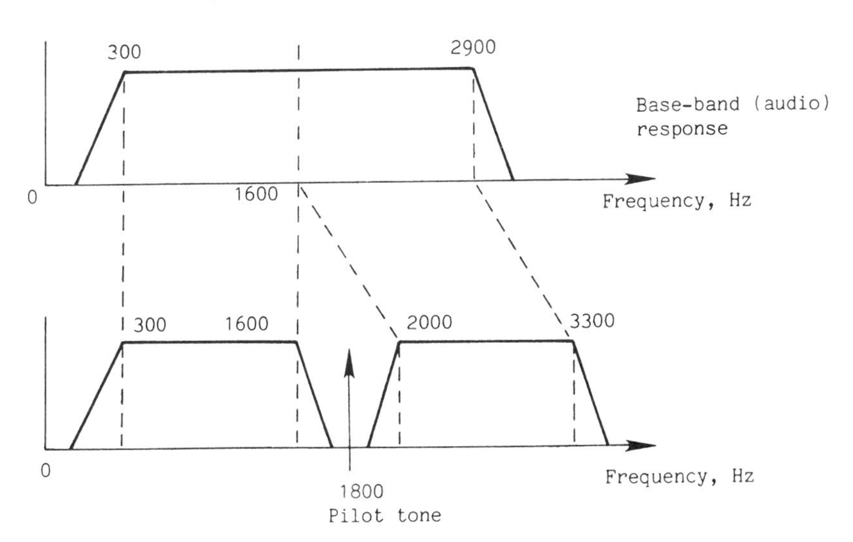

Figure 4.19 Transparent tone in-band SSB base-band response.

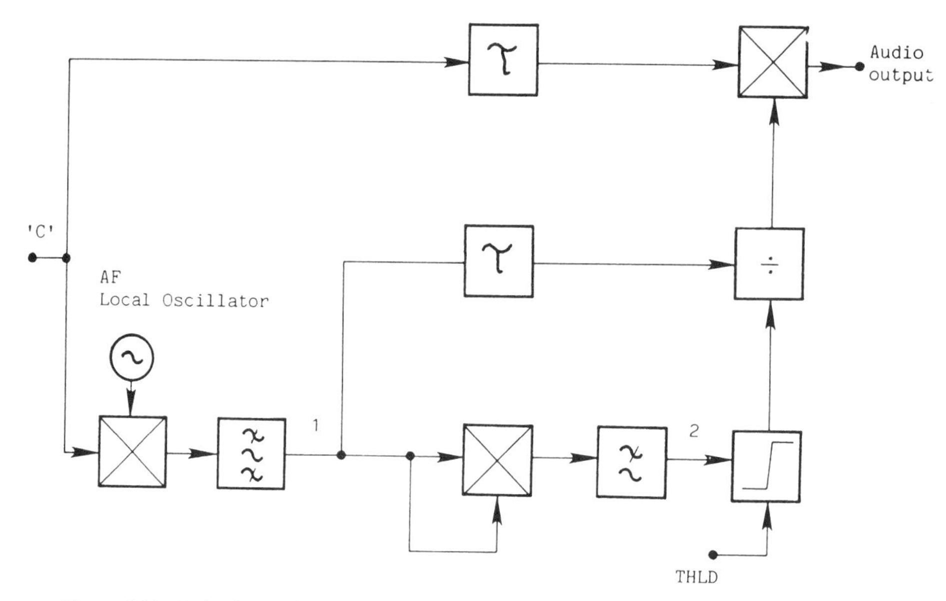

Figure 4.20 Polar form of fading correction (FFSR).

demodulation and as a known signal for AGC reference, but also to act as a basis for re-establishing the amplitude and phase features of the original transmitted SSB signal by compensating for the effects of Rayleigh fading. The receiver interprets short-term fluctuations in the pilot-tone amplitude and phase as due to Rayleigh fading, and corrects the received sidebands in a manner which maintains constant pilot tone conditions. The basic process is illustrated in Figure 4.20. One point worthy of note is that high-quality reception can be achieved only if the correlation between the pilot tone and the message frequencies is very high, typically 0.9999. For a channel with a time delay spread of 1 μs this requires a frequency separation no greater than 2.25 kHz, so if the carrier of a 3 kHz voice channel is used as a pilot, the higher modulation frequencies would be separated from the carrier by more than this critical amount and distortion would increase. The TTIB system avoids this difficulty, since a tone near the centre of a 3 kHz channel is never more than 1.5 kHz from the band edges. Furthermore, it is less susceptible to interference from adjacent channels. Systems in which coherent demodulation is necessary have to be given very careful consideration for mobile radio applications, since the extraction of a suitable carrier from the fading signal is not an easy matter.

In FM systems, the information is carried in the zero-crossings of the signal, so steps taken to reduce envelope fading do not affect the modulation. Indeed, all FM receivers contain limiters which have a large dynamic range, and these help to reduce fading effects. Distortion due to random FM remains, but techniques such as diversity, which will be treated in Chapter 6, can reduce or even eliminate this effect. Because the bandwidth of an FM signal is under the designer's control and because other effects, such as capture effect, characteristic of non-linear systems, are present, the mobile radio scene has been largely dominated by FM systems. In any case, the random FM is essentially concentrated at frequencies less than 2 to 3 times the maximum Doppler frequency, and outside this range, as shown in Chapter 2, the spectrum falls off at a rate of $1/f$. It has been shown [14] that random FM caused by fading can be considered a secondary effect in the reception of mobile radio signals.

4.11.1 Practical SSB systems

Clearly the potential of current SSB systems as spectrally-efficient means for radio service provision is great, even though the benefits obtainable in the particular case of cellular radio systems are expected to be marginal. In the context of individual systems operating in the private mobile radio (PMR) environment, the strategies of TTIB, FFSR and amplitude companding are now sufficiently mature in signal processing technology to be implemented in commercial systems at modest cost, and trials are now in progress in the UK to establish the ability of ACSSB systems to sustain high performance in a radio environment dominated by carrier-based radio schemes.

Two further features of new SSB technology are important. First, it will be seen later that transmission of digital signals over the Rayleigh channel presents particular difficulties, however using TTIB and FFSR techniques permits use of multilevel carrier modulation schemes for digital signals and this offers potentially a greatly improved throughput of digital signals in a given bandwidth.

Secondly, whilst it has been noted that transmitter linearity poses particular problems to the implementation of SSB in congested radio environments, one situation in which the problem is greatly eased is in satellite mobile systems. The poor linearity of satellite high power amplifications (HPAs), usually travelling wave tubes, is a long-standing problem, but dispersal of signal energy outside the alloted SSB channel is not so severe a problem when all the channels originate from one source, the satellite. Link budgets in satellite systems are always carefully adjusted to economize on spacecraft RF output power (which is extremely expensive to provide), and the dispersed intermodulation energy from SSB channels at 30 dB below inband signals would not be detectable on the ground by mobiles with modest antenna gains. At the time of writing, ACSSB is a strong contender for both the Canadian M-SAT and Australian AUSSAT satellite mobile projects intended to inaugurate services in the early 1990s.

4.12 Demodulation of data signals

Data signals can be transmitted using amplitude, frequency or phase modulation, so the methods of demodulation that have already been described apply equally well to these transmissions. For example, it is clear that a binary FSK signal, which is merely an FM signal using only two instantaneous

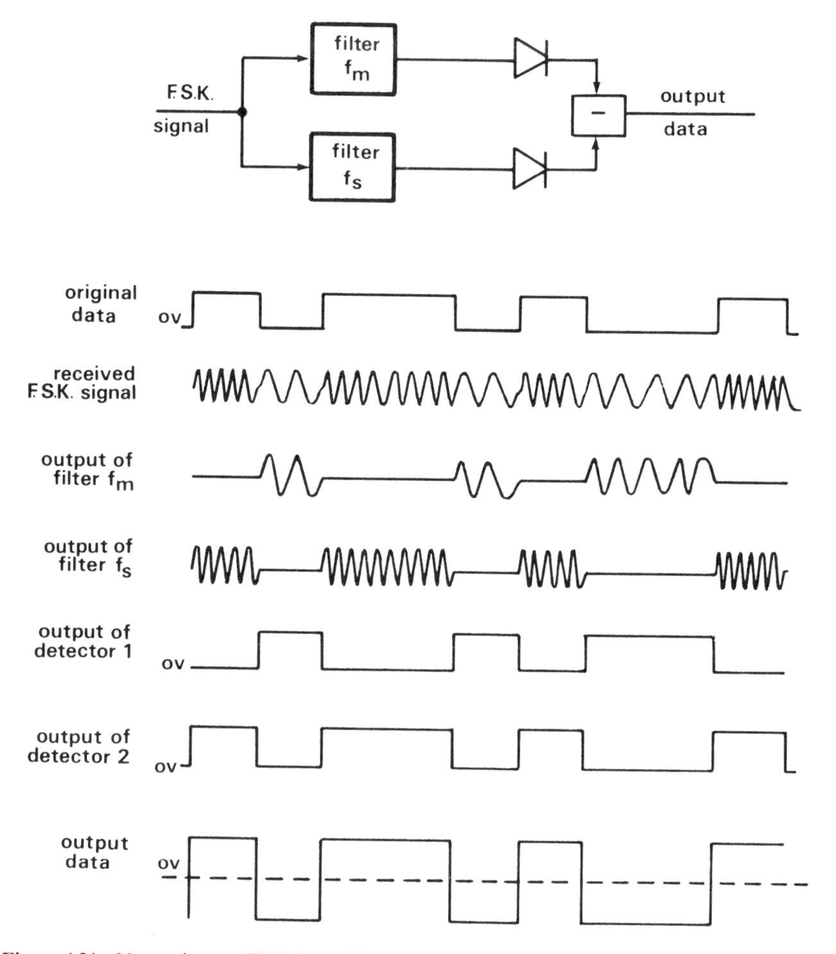

Figure 4.21 Non-coherent FSK demodulator.

frequencies, can be demodulated using any of the methods discussed in section 4.8. These are non-coherent demodulation methods which do not require the use of a re-injected carrier at the receiver. A further non-coherent method, illustrated in Figure 4.21, is often used, particularly if the frequency deviation is larger than can be conveniently handled by a conventional discriminator. The two bandpass filters have centre frequencies equal to the mark and space frequencies, and the outputs of the two envelope detectors are subtracted to give a bipolar data signal as shown.

Alternatively, coherent demodulation of an FSK signal is possible using the system shown in Figure 4.22. The injected frequencies are those corresponding to the transmitted mark and space frequencies, and the low-pass filters remove unwanted demodulation products. We will see later that coherent demodulation produces a superior performance in the presence of noise, but it is

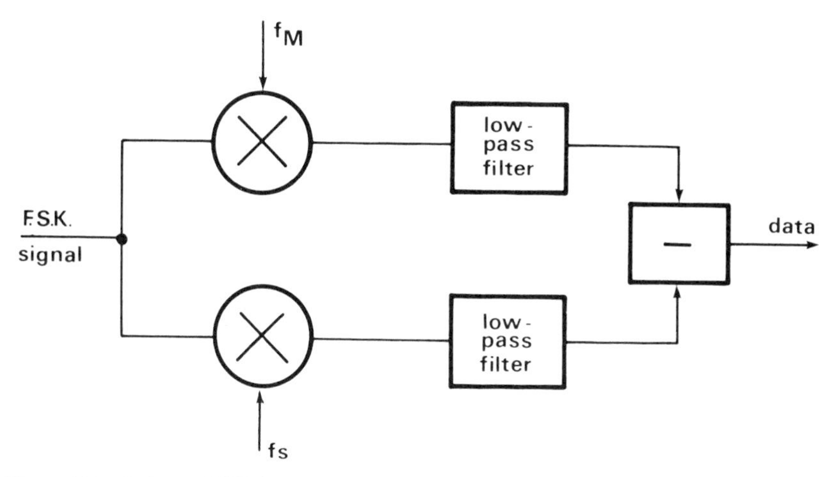

Figure 4.22 Coherent FSK demodulator.

expensive to implement because of the need for two oscillators and does not find application in mobile radio systems.

On the other hand, with PSK signals there is no alternative to coherent demodulation. Binary PSK with 180° phase reversals (often called phase reversal keying or PRK) is identical to DSBSC modulation with polar binary data, so the demodulator shown in Figure 4.8 must be used. In section 4.6.1, we concluded that successful demodulation of DSBSC signals is possible only if both the phase and frequency of the locally-generated carrier correspond to those used at the transmitter. The question of how this is to be achieved was not discussed earlier but is, of course, very important. If the spectrum of the transmitted signal contains a component at the carrier frequency, then a PLL can be used to lock on to that component and hence recover the carrier. In SSB transmissions, a 'pilot' tone is introduced into the transmission, and this can be used to aid demodulation and for automatic gain control purposes. However, in PSK and DSBSC transmissions this is not necessary, since the sidebands are located symmetrically about the suppressed carrier. Consequently, the squaring or frequency-doubling loop shown in Figure 4.23 can be used to derive a carrier reference, since for each spectral component at $(f_c + f_m)$ there exists a similar component at $(f_c - f_m)$ and multiplication produces a component at $2f_c$. This can be used in a PLL to produce a signal at f_c, the carrier frequency.

Alternatively, a more sophisticated carrier recovery technique is possible using a Costas loop [15]. The system is not dependent on the modulation and uses the in-phase and quadrature signal components. The arrangement shown in Figure 4.24 uses an in-phase (processing) channel to strip the modulation effects from the quadrature (error) channel. The Costas loop is less sensitive to drifts in carrier frequency, and is often preferred to the frequency-doubling

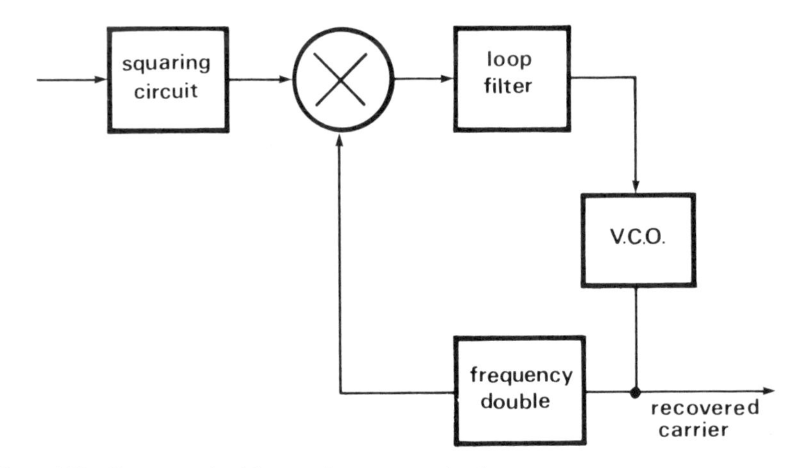

Figure 4.23 Frequency-doubling carrier-recovery circuit.

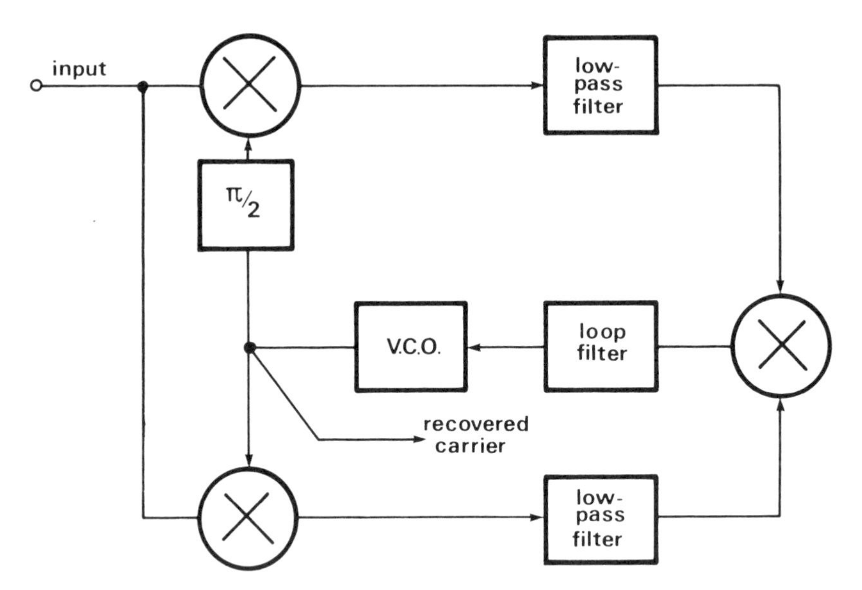

Figure 4.24 Costas loop for carrier recovery.

loop. Fairly large frequency drifts can be tolerated without increasing the bandwidth of the low-pass filters, because the VCO can track and remove slow frequency drifts before the signal reaches the low-pass filters. It is important, however, to match the group delays in the two processing channels to obtain good performance. The tracking property does not exist in the frequency-doubling loop, and the band-pass filter therefore needs to be wide enough to accommodate any frequency drifts.

In practice, there is always some phase jitter, due to noise, which degrades

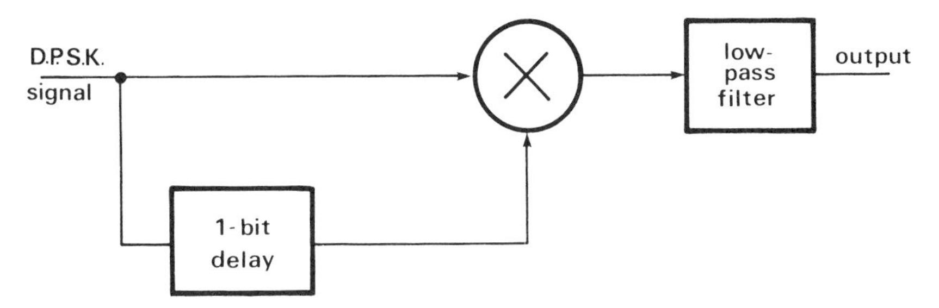

Figure 4.25 Simple DPSK demodulator.

the performance of these carrier-recovery loops. In addition, it is possible for them to lock up with a 180° phase error which must be resolved to obtain proper data recovery. In some circumstances, this can be achieved fairly simply [16].

4.13 Differentially encoded PSK (DPSK)

The difficulty of recovering a carrier signal, and the need to resolve the 180° phase ambiguity inherent in the methods described, can be overcome by differential encoding. By comparing the current bit with the previous bit it is possible to generate a new data stream in which each bit represents change or no-change from the previous bit.

In the receiver, the original data can be recovered by comparing each received bit with the previous one, and in the DPSK demodulator shown in Figure 4.25, the received data stream is multiplied by a one-bit delayed version of itself. The decoder must reverse the process of encoding so that a change in phase ($0 \rightarrow \pi$ or $\pi \rightarrow 0$) must be interpreted as a binary 1, no change as a binary 0.

The performance of DPSK is somewhat inferior to that of ideal coherently-demodulated PSK. Explicit carrier recovery is not employed, so that in the demodulator, instead of the noisy data stream being multiplied by a 'clean' recovered carrier, it is multiplied by a delayed version of itself which is also noisy. Because two successive bits are necessary to make a decision, errors tend to occur in pairs. An attractive technique to improve the performance uses coherent detection of differentially encoded data. This is known as differentially coherent PSK (DCPSK) [17].

4.14 The effect of noise on data communication systems

The discussion in the previous section has shown that the methods for demodulating analogue AM, FM and PM signals are also useful when data is being transmitted over the channel. We have also introduced other methods specifically intended for data communication receivers. There is, however, a

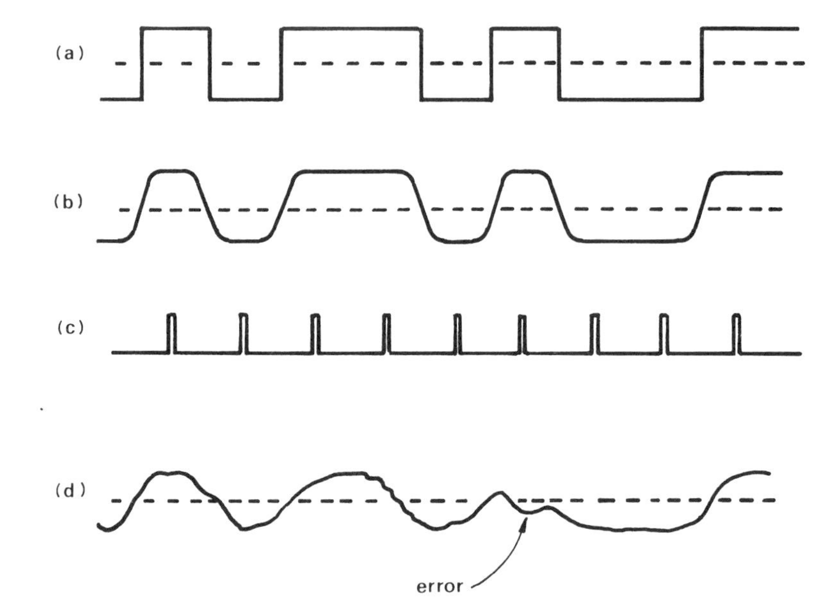

Figure 4.26 Illustrating how noise causes errors in digital communication systems.

further stage necessary following the reception of a data signal which is not needed with speech transmissions. After demodulation, it is necessary to examine the waveform and reach a decision as to whether at any particular time the waveform represents a 'one' or a 'zero'. In practice, various techniques are available for arriving at this decision. In mobile radio it is common to examine the received waveform at times nominally at the centre of the received data symbols and we will restrict attention to a brief discussion of this method.

Consider Figure 4.26 which illustrates the process, with the original data stream shown in Figure 4.26*a*. The demodulated signal which has passed through transmitter and receiver filters has lost the sharp edges of the original waveform, but if the accompanying noise is fairly small it is represented as in Figure 4.26*b*. The decisions are taken at times nominally at the centre of the received digits by using the 'sampling' waveform in Figure 4.26*c*. There is a zero threshold, and the decision is taken on the basis that if the voltage sample is positive, the symbol is a one, and if negative, a zero. If the noise accompanying the signal is large, the recovered waveform becomes more distorted, and it is then possible for errors to occur, as shown in Figure 4.26*d*, because at the sampling instant the recovered waveform is negative although a positive signal (a one) was transmitted. We now see an essential difference between the way in which noise affects analogue and data signals of the type described. In the previous sections we have considered signal and noise powers to derive a signal-to-noise ratio, which is an adequate measure of system performance for speech systems. We now conclude that this is not adequate for data transmissions of the type described—we have to consider whether, at the sampling instant, the signal voltage exceeds the noise voltage or vice versa. Since we cannot describe the noise voltage in a deterministic way, we can only calculate the probability of error, i.e. the probability that the signal voltage is greater or smaller than the noise voltage.

4.15 Carrier transmissions

In radiocommunication systems, the information is sent by modulating a carrier, and it is then useful to express the error probability in terms of the predetection SNR which, for constant envelope modulation methods, becomes the CNR, γ. Calculation of P_e is treated in several textbooks [e.g. 1] and only the results will be given here.

For PSK, which must be coherently demodulated, and coherently-demodulated FSK, the expression for P_e contains an error function, and we have

$$\text{Ideal PSK:} \qquad P_e = \tfrac{1}{2}\operatorname{erf}c(\sqrt{\gamma}) \tag{4.24}$$

$$\text{Coherent FSK:} \qquad P_e = \tfrac{1}{2}\operatorname{erf}c(\sqrt{\gamma/2}) \tag{4.25}$$

For DPSK and non-coherently demodulated FSK, P_e is evaluated in terms of exponential functions as

$$\text{DPSK:} \qquad P_e = \tfrac{1}{2}\exp(-\gamma) \tag{4.26}$$

$$\text{NCFSK:} \qquad P_e = \tfrac{1}{2}\exp(-\gamma/2) \tag{4.27}$$

It can be shown that

$$\gamma = \frac{\text{signal energy per bit}}{\text{noise spectral density}}$$

$$= \frac{\text{signal power}}{\text{noise power in the 'bit-rate bandwidth'}}$$

The 'bit-rate bandwidth' is a bandwidth numerically equal to the bit rate, i.e. x Hz in a system operating at x bits/s. The performance of the various systems is illustrated in Figure 4.27, from which it can be seen that PSK is the most robust system in white Gaussian noise.

4.16 The influence of multipath fading on data transmissions

Theoretical analyses of the error performance of data communication systems in fading conditions are usually based on the assumptions of slow, non-selective Rayleigh fading and noise which also varies slowly with respect to the symbol period. These assumptions imply that the channel does not introduce any pulse lengthening, and that during any given symbol period the noise can

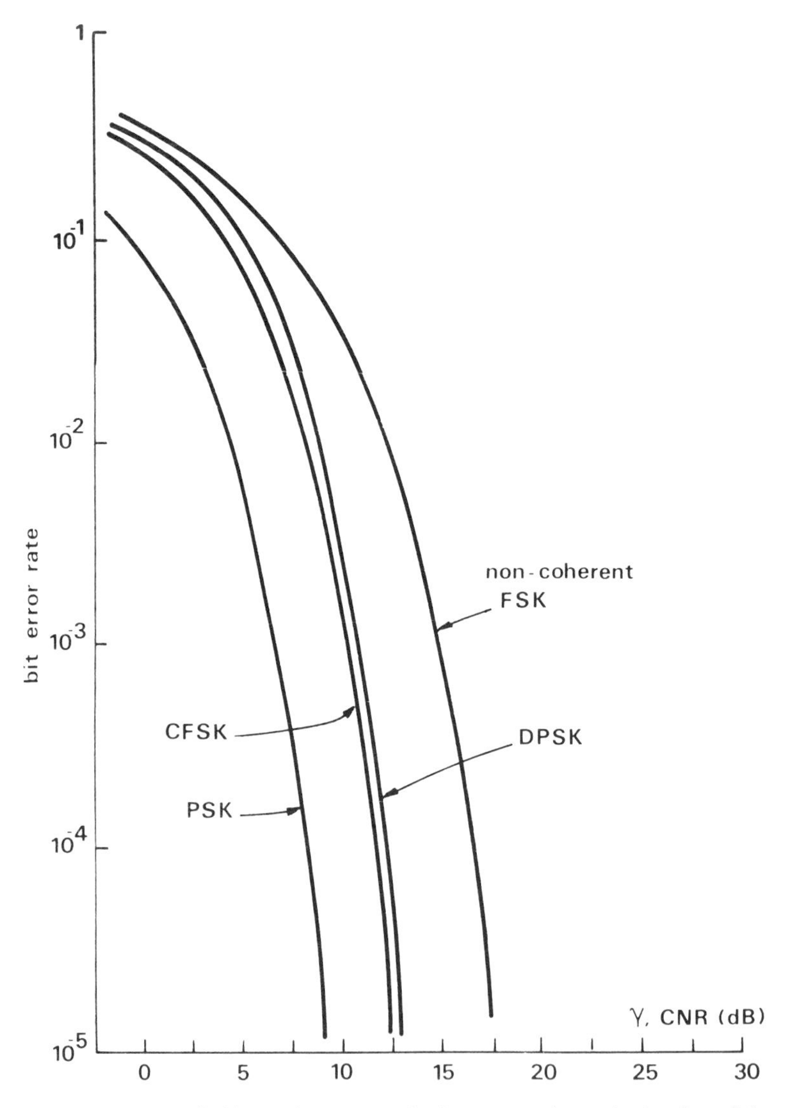

Figure 4.27 Error rates in binary data communication systems in steady signal conditions.

be regarded as effectively constant. Of course, intersymbol interference can arise due to the receiver filter and other causes, but this is normally considered negligible. In the type of narrowband channels used in conventional mobile radio systems, the time delay spread is small compared with the symbol rate, so these assumptions, while not strictly true, are nevertheless justified for the purpose of analysis. On this basis we can calculate the probability of error in a fading channel by adapting the results for the probability of error in Gaussian noise. In the fading case, the formulas given by eqns (4.24)–(4.27) can be taken to describe the probability of error for an individual symbol, conditional on the value of the multiplicative noise over that symbol, hence they describe the conditional probability of error caused by Gaussian noise. To determine the average performance in fading conditions then requires averaging this conditional probability over the ensemble of possible values of the multiplicative factor, i.e. over the Rayleigh fading.

It is intuitively clear that when the signal is subject to fading the CNR varies with time due to changes in the signal level. Since we are interested in CNR, it is convenient to find the probability density function of γ. If a Rayleigh distributed signal with an envelope r is received in the presence of Gaussian noise of mean power N, then the mean carrier power averaged over one RF cycle is

$$\gamma = \frac{r^2}{2N}$$

and we define

$$\gamma_0 = \bar{\gamma} = \frac{\text{mean carrier power}}{\text{mean noise power}} = \frac{b_0}{N}$$

To find $p_\gamma(\gamma)$ we change the variable by writing

$$p_\gamma(\gamma)\,\mathrm{d}\gamma = p_r(r)\,\mathrm{d}r$$

However, $\mathrm{d}\gamma = (r/N)\mathrm{d}r$ and also

$$\frac{r^2}{2b_0} = \frac{2N\gamma}{2N\gamma_0} = \gamma/\gamma_0$$

thus

$$p_\gamma(\gamma) = \frac{1}{\gamma_0}\exp(-\gamma/\gamma_0) \tag{4.28}$$

The probability that γ is less than some specified value γ_s is then

$$\begin{aligned}\text{prob}[\gamma \leqslant \gamma_s] = P(\gamma_s) &= \frac{1}{\gamma_0}\int_0^{\gamma_s} \exp(-\gamma/\gamma_0)\,\mathrm{d}\gamma \\ &= 1 - \exp(-\gamma_s/\gamma_0)\end{aligned} \tag{4.29}$$

4.16.1 Non-coherent FSK and DPSK

To find the error rate with non-coherent FSK we return to eqn (4.27), which gives the probability of error for one value of γ, and integrate it over all possible values of γ weighting the integral by the PDF of γ, thus

$$\begin{aligned}P_e &= \int_0^\infty \tfrac{1}{2}\exp(-\gamma/2)\frac{1}{\gamma_0}\exp(-\gamma/\gamma_0)\,\mathrm{d}\gamma \\ &= \frac{1}{2\gamma_0}\int_0^\infty \exp\left[-\gamma\left(\tfrac{1}{2}+\frac{1}{\gamma_0}\right)\right]\mathrm{d}\gamma \\ &= \frac{1}{2+\gamma_0}\end{aligned} \tag{4.30}$$

Similarly, for DPSK

$$P_e = \tfrac{1}{2}\int_0^{\infty} \exp\left[-\gamma\left(1+\frac{1}{\gamma_0}\right)\right]d\gamma$$

$$= \frac{1}{2(1+\gamma_0)} \tag{4.31}$$

4.16.2 Coherent systems

For ideal PSK and coherently demodulated FSK, the expressions for P_e involve the integral of an error function, which is somewhat more difficult to evaluate. For example, the value of P_e for ideal PSK is

$$P_e = \tfrac{1}{2}\int_0^{\infty} \operatorname{erf}c(\sqrt{\gamma})\frac{1}{\gamma_0}\exp(-\gamma/\gamma_0)\,d\gamma$$

This integral is not straightforward, but its value is listed together with many others in [18], the result being

$$P_e = \tfrac{1}{2}\left[1 - \frac{1}{\sqrt{1+1/\gamma_0}}\right] \tag{4.32}$$

and similarly for coherent FSK

$$P_e = \tfrac{1}{2}\left[1 - \frac{1}{\sqrt{1+2/\gamma_0}}\right] \tag{4.33}$$

4.17 System performance

Figure 4.28 shows how the error probability varies as a function of γ_0, the mean CNR, for various communication systems. The severe degradation due to fading is obvious; to maintain an error probability of 10^{-3} in a channel subjected to Rayleigh fading, the mean CNR must be 20 dB above that required in the absence of fading. An additional 10 dB is required to reduce the error probability by a further order of magnitude. The asymptotic performance for large CNR in Rayleigh fading is obtainable from eqns (4.24)–(4.27), and is given by

$$\begin{aligned}
&\text{Coherent PSK:} && P_e \to \frac{1}{4\gamma_0}\\
&\text{Coherent FSK:} && P_e \to \frac{1}{2\gamma_0}\\
&\text{DPSK:} && P_e \to \frac{1}{2\gamma_0}\\
&\text{Non-coherent FSK:} && P_e \to \frac{1}{\gamma_0}
\end{aligned} \tag{4.34}$$

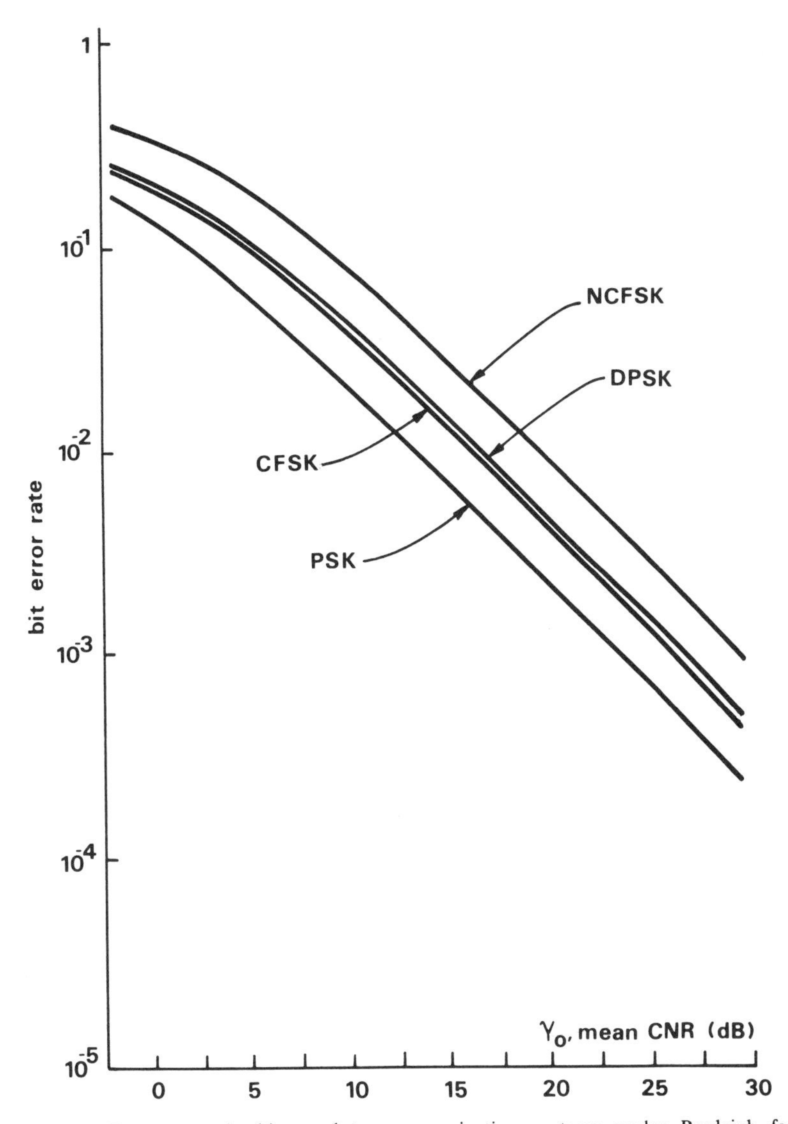

Figure 4.28 Error rates in binary data communication systems under Rayleigh fading conditions.

Thus at large CNR the error probability for all these systems is inversely proportional to the CNR, as opposed to the exponential change in the steady-signal case. This indicates the nature of the additional design burden imposed on systems that are required to operate in fading conditions and the resulting importance of techniques for overcoming the problem. One such technique, diversity reception, will be discussed in Chapter 6.

4.18 The fully digital approach

So far, in treating data transmission over mobile radio channels, the emphasis has been on relatively low bit-rate requirements to permit transmission of small amounts of numerical data for short, coded, telex-type messages. This permits the use of a straightforward approach to achieving data transmission, i.e. the use of a subcarrier in the audio band which is digitally modulated using one of the methods described earlier and then applied to the RF carrier in the same way as a speech signal. This has the added advantage that it gives the user a choice of speech or data over the channel. For both FM and AM systems, the modulation process produces two sets of sidebands, and since each sideband set is derived from the complete digitally modulated subcarrier, this technique effectively halves the available channel bandwidth for the data. As a result, subcarrier data systems are usually restricted to a transmission rate of 1200 bits per second. AM (envelope-modulated) subcarrier systems are particularly problematic because of the difficulties experienced by the non-coherent envelope detector in distinguishing modulation amplitude fluctuations from those introduced by the Rayleigh channel. As a consequence, data users who are constrained to use envelope modulation (the emergency services operated by the UK Home Office are a case in point) are restricted to substantially lower bit-rates, typically 100 or 300 bits per second.

Whilst this type of system is adequate for primitive data transmissions such as vehicle status messages and so on, the growing demands of users for data capability approaching that of the line telephone modem has stimulated much effort towards achieving direct digital modulation of the RF carrier to produce dedicated radio-connected data networks. This development has coincided with steady progress towards achieving low bit-rate digitally-encoded speech and a new all-digital approach to speech and data service provision has emerged. Successful implementation of the all-digital approach requires the use of digital modulation methods which are spectrally efficient but are nevertheless robust in the Rayleigh fading environment. The quest for an optimum modulation has by no means reached a conclusion and is the subject of vigorous research effort, but much has been accomplished, and systems are being implemented based on methods which have emerged over the last five years or so.

4.18.1 The principal compromises

Recalling earlier sections, there are clearly a limited number of carrier parameters which can be modulated. The carrier amplitude can be varied while maintaining constant carrier frequency and phase (ASK or OOK), the carrier frequency can be varied keeping the amplitude constant (FSK), the phase can be varied, (PSK), or some combination of amplitude and phase can be used, (e.g. m-ary QAM). The spectral characteristics of these forms and their performance in additive white Gaussian noise (AWGN) are dealt with at length in a number of standard texts to which the reader seeking more detail is referred [19–21]. However, it is easy to appreciate a number of salient features.

Spectral efficiency in terms of the number of hertz of bandwidth required to transmit a digitally-modulated carrier with a given bit error rate is not a feature of the modulation process only; it is also influenced by the performance of the hardware used to transmit the modulated signal and of the channel over which it is transmitted. These aspects are readily appreciated by reference to the technique much used in fixed-link systems, quaternary phase shift keying (QPSK). In a channel without any bandwidth restriction the QPSK-modulated carrier has a uniform envelope but a wide spread of sideband energy. If the signal is now filtered to constrain it to a bandwidth of, say, twice the data bit-rate, then the transmission is still acceptable to receivers, but the effect of filtering has been to introduce carrier amplitude variations. If the signal is transmitted through a nonlinear power amplifier in the transmitter, however, then the amplitude variations result in the generation of intermodulation products which effectively expand the bandwidth again and cannot be removed by filtering at the transmitter output. Minimum spectral dispersion and a uniform envelope amplitude for the modulated carrier are, therefore, important targets to aim for. QPSK appears to be non-ideal on both these counts, and there is a further consideration; the multipath propagation process produces not only large and rapid amplitude changes at the receiver, but also phase changes, since the multipath components travel over different path lengths from transmitter to mobile. Are modulation schemes based on phase shift keying, therefore, likely to be vulnerable to high error rates in the Rayleigh environment? The implication of this question is that frequency shift keying might possibly be a more satisfactory basis but, in fact, because of the relationship between frequency and phase (instantaneous frequency being the time-derivative of phase), the distinction between frequency and phase modulation methods is by no means clear in many cases. To understand the reasons for this, it is necessary to investigate those features of digital modulation schemes which result in dispersed spectra.

4.18.2 Quadrature amplitude modulation

Taking amplitude modulation schemes first, the arguments in favour of double sideband suppressed carrier modulation which were advanced for speech transmission, apply equally to data, and the ability of amplitude modulation schemes to transmit more bits per hertz of bandwidth can be increased by summing two DSBSC signals in quadrature. If binary signals are used, this results in the modulated carrier occupying one of four states, and the result is the same as modulating the carrier phase, i.e. 4-state QAM is identical to QPSK. However, if multilevel digital signals are used, then the number of possible carrier phases and amplitude states can be increased to very large numbers. Increasing the number of signal states in the constellation increases the capacity in bits per hertz of bandwidth but, since the distances between signal states become smaller as their number increases, the signal-to-noise ratio required to sustain low bit error rates increases, since the receiver must distinguish between states with increasing accuracy.

Returning to the issue of spectral containment, it has already been noted that QPSK has a wide spectrum. The reasons are apparent from Figure 4.29*a*, which shows an unfiltered QPSK waveform. Changes in signal state are seen to be abrupt and unequal—sometimes 180°, sometimes 90°. These transients

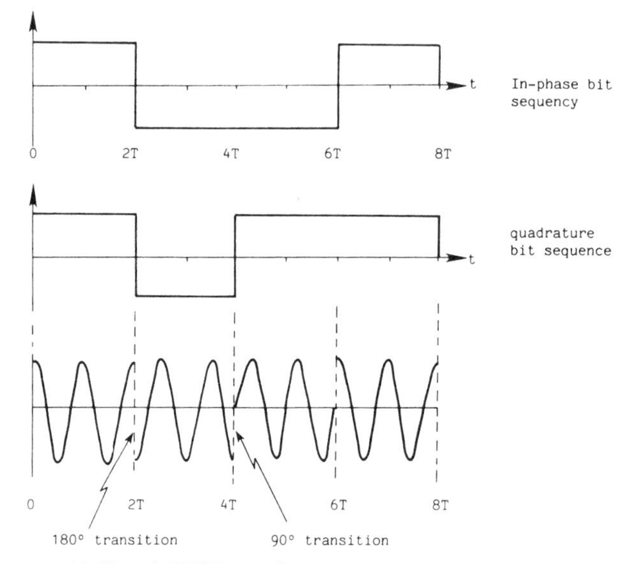

(*a*) Unfiltered QPSK waveform

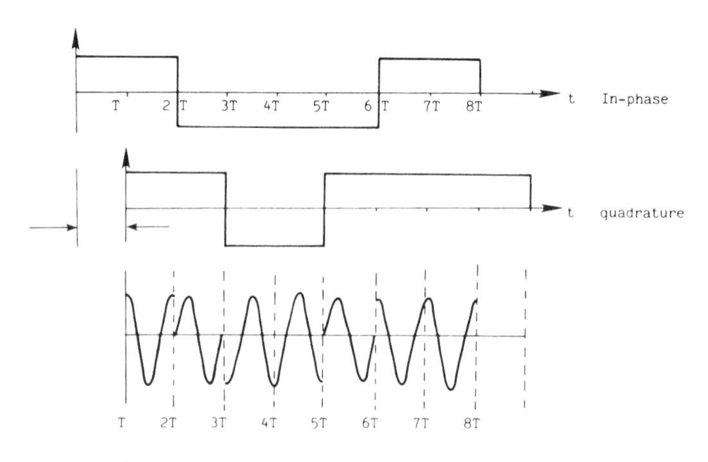

(*b*) Unfiltered OK-QPSK waveform

Figure 4.29 Comparison of QPSK and OK-QPSK.

are responsible for the large spectral dispersion. Matters can be improved by offsetting one of the quadrature bit streams as shown in Figure 4.29*b* to constrain transistions to be 90° maximum (offset keyed or OK-QPSK). Abrupt phase transitions remain, however, and to remove these requires that the baseband signal be filtered to produce a more graceful transition from one binary state to another. If this is done so as to pre-shape the baseband bits sinusoidally, the result is termed Minimum Shift Keying (MSK).

4.18.3 Frequency modulation schemes

In its simplest form, binary frequency shift keying functions by assigning one carrier frequency to one binary state and a second frequency, f Hz away, to the

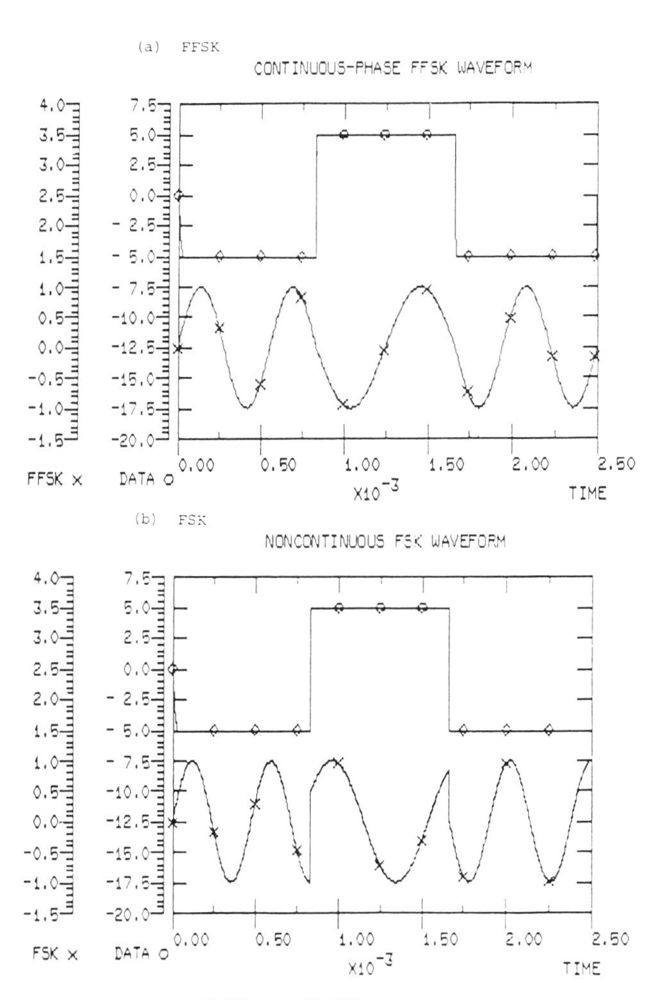

Figure 4.30 Comparison of (*a*) FFSK and (*b*) FSK.

other binary state. Defining a modulation index, d, as

$$d = f \times T$$

where T is the symbol duration (the inverse of the bit rate), spectrum containment as a function of d can be optimized. Again, unless special precautions are taken, abrupt changes in carrier phase can occur, as shown in Figure 4.30*a*, as transitions take place between the two frequency states. This can be overcome by constraining frequency changes to take place at carrier zero crossings so that phase continuity is preserved (continuous phase or CP-FSK), but this implies a relationship between frequency separation and symbol duration. With $d = 0.5$, this condition is satisfied, and the resulting modulation is termed Fast Frequency Shift Keying (FFSK), Figure 4.30*b*.

Although phase continuity has been achieved, there remains a discontinuity in frequency at each baseband symbol transition, and to remove this again requires pre-shaping of the baseband bit stream with an appropriate filter.

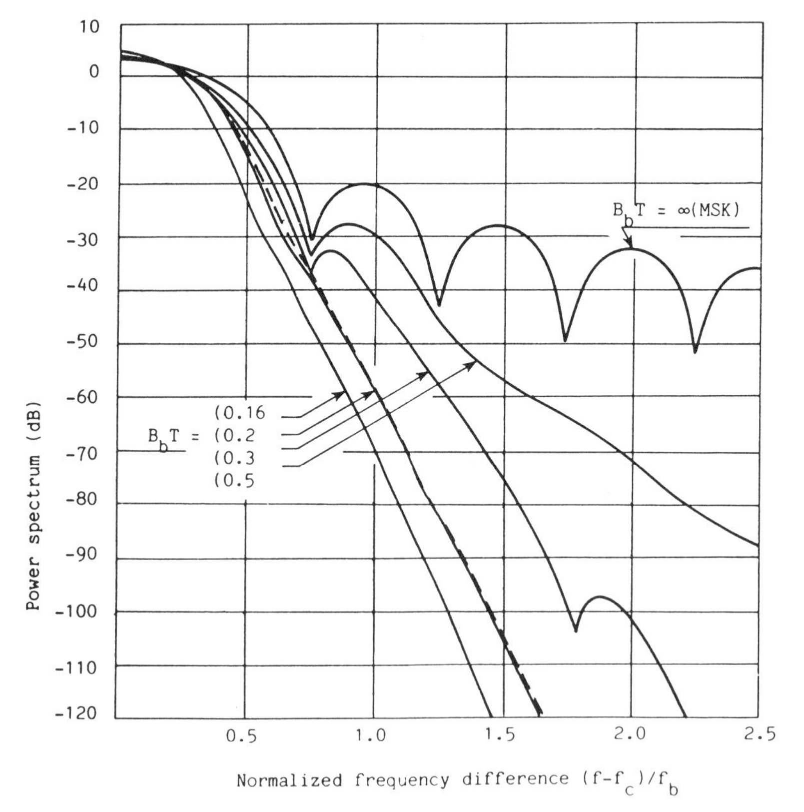

Figure 4.31 Comparison of spectral dispersions in GMSK with various premodulation filter characteristics. ----: TFM; B_bT: normalized bandwidth of premodulation Gaussian bandpass filter (GMSK); f_c: carrier frequency (Hz); f_b: bandwidth (Hz).

However, pre-filtering to smooth out the frequency transitions results in an output indentical to minimum shift keying (MSK), which emphasizes the point made earlier that frequency and phase modulation schemes are closely related.

Recalling the further requirement, in addition to spectral containment, of uniform envelope, use of a pre-modulation low-pass filter with a Gaussian characteristic in conjunction with the MSK approach achieves this, and GMSK (Gaussian MSK) has been adopted as the modulation standard for the new pan-European digital cellular network to be introduced in 1991 (see Chapter 9).

A similar scheme has been developed involving pre-filtering combined with an algorithm for selecting the carrier phase shift according to the original data, values of $\pm\pi/2$, $\pm\pi/4$ and 0 radians being allowed. This scheme, known as Tamed Frequency Modulation (TFM), has spectral containment characteristics similar to GMSK. Figure 4.31 shows comparisons.

Of course, in removing abrupt carrier state transitions, something is lost from the transmitted information, and this manifests itself in a reduction in inter-symbol interference margins. Nevertheless, acceptable compromises among all the parameters can be made, and all-digital systems operating in 25 kHz channel spacing have been convincingly demonstrated.

References

1. Stremler, F.G. *Introduction to Communication Systems.* Addison-Wesley, New York (1982).
2. Shanmugam, K.S. *Digital and Analog Communication Systems.* Wiley, New York (1979).
3. Betts, J.A. *Signal Processing, Modulation and Noise.* Edinburgh University Press (1970).
4. Clark, A.P. *Principles of Digital Data Transmission.* Pentech, London (1983).
5. Oetting, J.D. A comparison of modulation techniques for digital radio. *IEEE Trans.* **COM-27** (12) (1979) 1752–1762.
6. Gosling, W. and Petrovic, V. Area coverage in mobile radio by quasi-synchronous transmissions using DSB-DC modulation. *Proc. IEE,* **120** (1973) 1469.
7. Macario, R.C.V. A VHF surveillance receiver adapted for the narrowband reception of suppressed carrier double sideband transmissions. *Radio & Electronic Engr* **43** (July 1973) 407.
8. Carson, J.R. and Fry, T.C. Variable frequency electric circuit theory with application to the theory of frequency modulation. *Bell Syst. Tech. J.* **16** (1937) 513–540.
9. Weaver, D.K. Jr. A third method of generation and detection of single-sideband signals. *Proc. IRE* **44** (1956) 1703–1705.
10. Petrovic, V. and Gosling, W. A radically new approach to single sideband transmitter design. *IEE Conf. on Radio Transmitters and Modulation Techniques,* March 1980, 110–119.
11. Armstrong, E.H., see ref. 1, Chapter 6.
12. McGeehan, J.P. and Bateman, A.J. Phase-locked transparent tone-in-band (TTIB): a new spectrum configuration particularly suited to the transmission of data over SSB mobile radio networks. *IEEE Trans.* **COM-32** (1) (1984) 81–87.
13. McGeehan, J.P. and Bateman, A.J. Theoretical and experimental investigation of feedforward signal regeneration (FFSR) as a means of combatting multipath effects in pilot-based SSB mobile radio systems. *IEEE Trans.* **VT-32** (1983) 106–120.
14. Davis, B.R. FM noise with fading channels and diversity. *IEEE Trans.* **COM-19** (6) (1971) 1189–1200.
15. Stiffler, J. *Theory of Synchronous Communications.* Prentice-Hall, New York (1971) 251–256.
16. Pongsupaht, A. and Parsons, J.D. Novel technique for eliminating the ambiguity in PSK signal detection. *Electronics Lett.* **15** (13) (1979) 385–386.

17. Schwartz, M., Bennett, W.R. and Stein, S. *Communication Systems and Techniques*. McGraw-Hill, New York (1966), Chapter 7.
18. Ng, E.W. and Geller, M. A table of integrals of the error functions. *J. Res. NBS-B* **73B** (1) January (1969) 1–20.
19. Peebles, P.Z. Jr. *Communication System Principles*. Addison-Wesley, New York (1976).
20. Proakis, J.G. *Digital Communications*. McGraw-Hill, New York (1983).
21. Murota, K. and Hirade, K. GMSK modulation for digital mobile radio telephony. *IEEE Trans.* **COM-29** (7) (1981) 1044–1050.

5 Man-made noise

5.1 Introduction

The characteristics of the transmission medium play an important part in determining the performance of a communications system. We have already discussed the influence of the channel over the mobile radio signal in Chapters 2 and 3, but we also have to consider noise. Although research engineers are interested in understanding the nature of the noise so that it can be satisfactorily characterized, the more important aspect from a practical point of view is the ability to predict the performance of communication systems that have to operate in the presence of noise.

The effect that Gaussian noise has on communication systems was described in Chapter 4. As far as mobile radio systems are concerned, the external ambient noise is often more important than internal noise that arises from the receiving system itself. Atmospheric and galactic noise exist, but at VHF and above are negligible compared with the 'man-made' noise that arises from electrical equipment of various kinds, including the electrical systems of motor vehicles. Indeed, as far as noise is concerned, in urban areas ignition interference is often the dominant source of degradation to mobile communication systems. Man-made noise is distinguished from thermal (Gaussian) noise by its impulsive nature, which gives it quite different characteristics. As we will see later, time is an important factor in characterizing impulsive noise; for white Gaussian noise, it is usually sufficient to know the mean power (or the rms value). This is often expressed in dB relative to kT_0B (see p. 152).

As far as degradation of communications systems is concerned, thermal noise produces a 'hiss' on voice channels, but intelligibility is maintained provided that the level is not too high. Impulsive noise produces 'clicks' which can also degrade intelligibility if the level or rate is high. As far as digital transmissions are concerned, we have already discussed (Chapter 4) some of the effects of thermal noise on data communications systems. In this chapter we will discuss the methods of characterizing impulsive noise and go some way towards describing how this kind of noise affects digital transmissions.

5.2 Characterization of pulses

A true impulse is a transient that exists for an infinitesimally short time. It has an instantaneous uniform spectrum over the frequency band for which it is defined, and this implies that all frequencies are present, of equal amplitude and in phase over the frequency band of interest. Practical impulse generators do not achieve this ideal; their output consists of a train of uniformly-spaced constant-amplitude pulses. These are often of very short duration (approximately 0.5 ns) and contain spectral components up to very high frequencies; they may be termed pseudo-impulses. However, if these pseudo-impulses are applied to a circuit having a bandwidth much smaller than the reciprocal of the duration of the pseudo-impulse, the circuit response is, for all intents and purposes, indistinguishable from that of a true impulse. In practice there is no point in pursuing the distinction.

Broadband signals such as pulses are often characterized in terms of a quantity known as spectrum amplitude. This has the units of V/Hz (or more often μV/MHz). Spectrum amplitude is defined in terms of the Fourier transform $V(f)$ of a time-domain signal $v(t)$ as

$$S(f) = 2|V(f)| \tag{6.1}$$

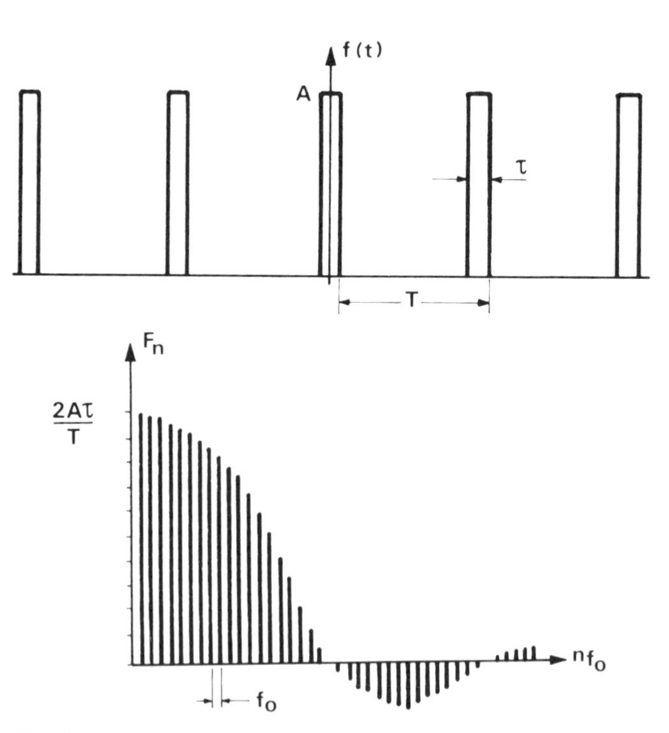

Figure 5.1 A uniform pulse train and its spectrum amplitude.

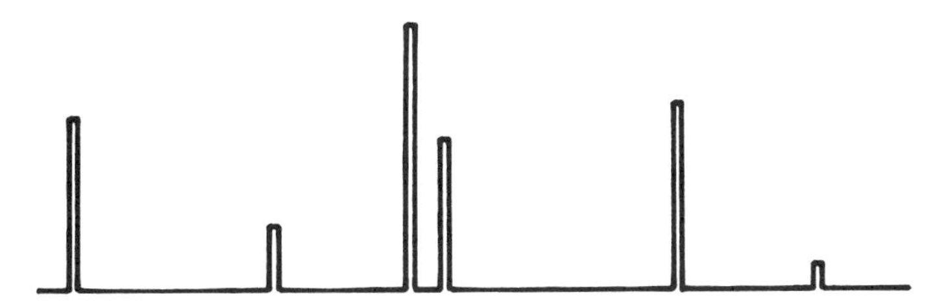

Figure 5.2 Elementary model of impulsive noise.

and is often expressed in logarithmic units, i.e. dB above $1\mu V/MHz$. As an example, the spectrum amplitude of a uniform train of pulses is shown in Figure 5.1. Commercial impulse generators are usually calibrated in terms of the spectrum amplitude at low frequencies, it being assumed that this value is usable up to a frequency where the output has fallen by 3 dB.

Impulse noise consists of a train of impulses, as shown in Figure 5.2. The amplitude and occurrence time are both random, and the overall response of a narrow-band receiver to the noise may be such that separation of the responses to individual impulses is not possible.

5.3 Characterization of impulsive noise

Examination of the extensive literature on impulsive noise does not reveal any real agreement about the techniques of measurement or the relative importance of the various measured parameters. In many cases the noise is non-stationary in the statistical sense, and a measurement made at a certain location may produce parameters quite different from those measured at the same location at a different time. One possible technique is to correlate the measured parameters with the conditions under which the measurement was carried out; in the context of mobile radio, these conditions include traffic density (number of vehicles per minute), the position of the monitoring receiver with the respect to the traffic flow, and the proximity of road features such as traffic lights and roundabouts.

If we examine the simple model portrayed in Figure 5.2, we can list factors that can be measured and give useful information about the amplitude of the noise.

(i) Mean or average voltage
(ii) Peak voltage
(iii) Quasi-peak voltage
(iv) Rms voltage
(v) Impulsiveness ratio
(vi) Amplitude probability distribution (APD)
(vii) Pulse height distribution (PHD)

(viii) Noise amplitude distribution (NAD)
(ix) Level-crossing rate (LCR)
(x) Pulse duration distribution (PDD)
(xi) Pulse interval distribution (PID).

Some of these quantities, for example (i)–(v), are expressed as a single value, whilst others, such as (vi)–(xi), can be presented only in graphical form. The list illustrates that there are a number of possibilities, the parameter that is most useful in any given situation depending on that situation. For example, peak voltage can be used to establish whether or not a particular area or object is a source of impulsive noise. It is of no use for characterizing ignition noise, which is postulated to consist of numerous peaks of random amplitude. In any case, the peak voltage measured is dependent on the measurement period; the longer the period the more likely it is that an even larger value will be observed. There is no method of using peak voltage to predict BER in data communication systems, although it has been observed that steps taken to reduce the peak value often result in reduced subjective annoyance on voice channels.

The quasi-peak value is the parameter most widely used to measure and specify regulatory levels of radiated or conducted noise. It is used by most national radio administrations for this purpose. Originally the quasi-peak meter was developed as an instrument that would be sensitive to pulse rate (which the peak reading voltmeter is not) and the specifications for quasi-peak meters to be used in various frequency bands have been internationally agreed and published in CISPR literature [1, 2, 3]. Basically, they have a charge time-constant much longer than that appropriate for a peak detector, and a discharge time constant much shorter, so that their output is a function of both impulse strength and impulse rate. The quasi-peak method of measurement recognizes time as an important factor in the characterization of impulsive noise.

There are two other methods of characterization which recognize time as a important factor. The amplitude probability distribution (APD) mentioned above is usually plotted on Rayleigh graph paper and shows the percentage of time for which the detected noise at the output of a receiver exceeds any particular level. The ordinate is usually expressed in dB above kT_0B. The reason for choosing kT_0B and this type of graph paper is because detected receiver noise has a probability distribution which plots as a straight line on this type of paper. In order to interpret APD it is necessary to know the characteristics of the receiver used to make the measurement, since bandwidth and filter response can affect the APD shape. A typical APD is shown in Figure 5.3, together with the receiver noise line from a receiver with a noise figure of 10 dB. It can be seen that there are two distinct regions to this curve, one of low slope at low amplitude levels (the background noise), and the other of high slope at high amplitude levels (the impulsive noise). The APD gives the

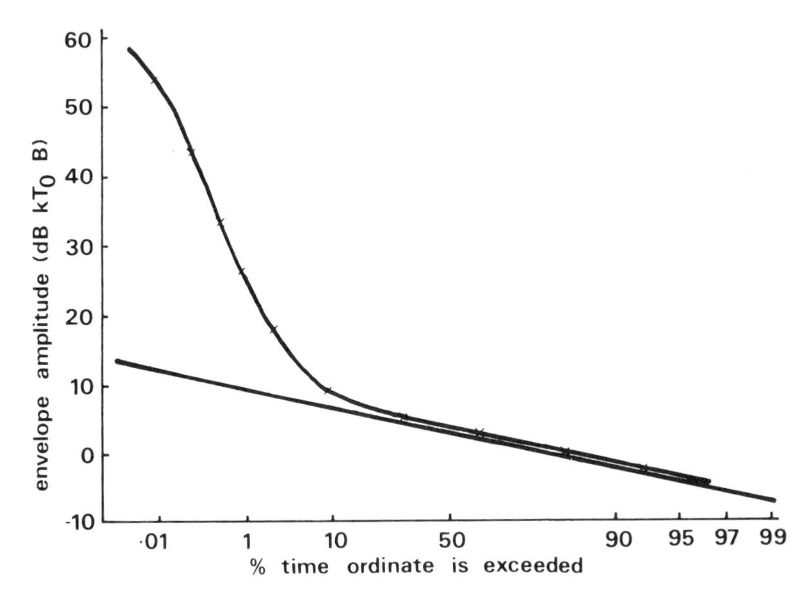

Figure 5.3 Typical amplitude probability distribution.

'first-order' statistics of impulsive noise, that is, it allows a determination of the overall fraction of time for which the noise exceeds any particular value. It gives no information about how this time was made up, i.e. whether the value was exceeded by one pulse or by many. This information is given by the noise amplitude distribution (NAD).

The NAD is a method of presenting impulsive noise in a form which gives a substantial increase in information over that provided by the quasi-peak detector. It provides a method of estimating the noise at the input of the receiver, rather than at the output, this estimate being independent of the bandwidth and, to a large extent, the characteristics of the equipment used to make the measurement. The NAD concept also provides an empirical method for determining the susceptibility of a communications receiver to impulsive noise [4]. The information given by the NAD is the number of pulses per second which exceed a given strength. The ordinate is spectrum amplitude, expressed in μV/MHz (or dB above 1μV/MHz), and the abscissa is pulse rate, as shown in Figure 5.4. There are advantages in expressing the amplitude in μV/MHz; firstly, as has already been mentioned, this is the unit normally found on impulse generators, and secondly, since it is normalized with respect to bandwidth, it allows a direct comparison with results for other bandwidths. The NAD is not a probability distribution: it was originated [4, 5] as a means to extract information from radio noise in a form which allowed the evaluation of the effect of that noise on land mobile communications systems.

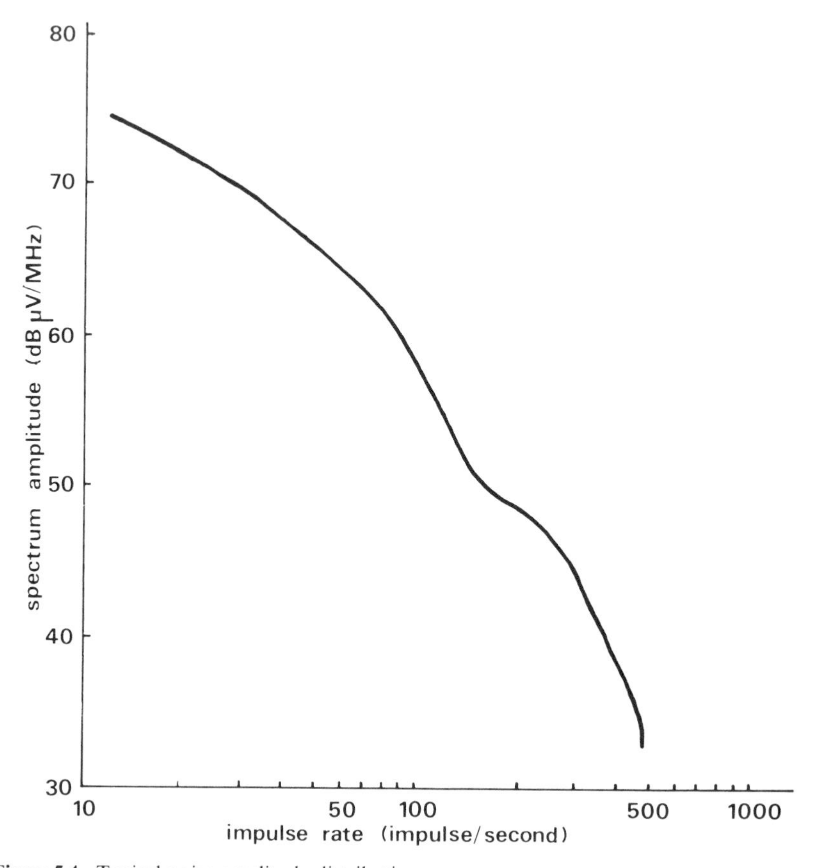

Figure 5.4 Typical noise amplitude distribution.

5.4 Measuring equipment

Having outlined some of the parameters that can be measured, it remains to specify the measuring equipment. It is almost inevitable in practice that an instrument designed to measure a quantity as complex as impulsive noise will influence the quantity that is being measured, but even if this cannot be avoided it is important to know the extent to which the measured parameters are affected. Most important are the bandwidth, sensitivity, impulse response and type of detector.

5.4.1 Bandwidth

We have seen previously that impulsive noise occupies a very wide frequency range, but an instrument designed to measure in a very wide bandwidth would be quite unrealistic. Any attempt to use a very wideband receiver in a field-trial situation, for example to measure the noise at a mobile or base-station site, is doomed to failure, since there would inevitably be coherent transmissions occuring within the bandwidth of the receiver. Secondly, unless the information from the receiver is processed on-line, it will need to be recorded for subsequent off-line processing, and this would inevitably lead to a need for costly, wideband recording equipment. It seems that the aim should be to use a reasonably wide bandwidth, certainly wider than the likely bandwidth of the communications system under consideration, and for an instrument designed to measure in the current mobile radio bands a bandwidth of approximately 100 kHz is reasonable. An instrument to be described later uses a bandwidth of 120 kHz, which is the value specified by CISPR for quasi-peak meters operating in the frequency range 30–300 MHz.

5.4.2 Dynamic range

The dynamic range requirement of a suitable measuring system is a difficult parameter to access, since it depends on the bandwidth and the frequency of measurement. From several sources [6, 7], it has been reported that the mean magnitude of the VHF man-made noise decreases as the measurement frequency increases, and in the UK this mean level is

$$N(\text{dB}) = 67 - 28 \log f \quad (f \text{ in MHz}) \tag{6.2}$$

It has also been reported that in the HF band, using a receiver with a bandwidth of 10 kHz, the dynamic range between the 0.001 and 99 percentiles exceeds 80 dB. It is clear that the larger the bandwidth the larger the dynamic range required, so quite clearly, unless very expensive, highly linear, equipment is available, some compromise is necessary. In practice, a dynamic range of 50–60 dB is obtainable with relatively inexpensive equipment [8], and this is usually regarded as satisfactory. This compromise enables most of the high peaks to be accommodated whilst ensuring that the lower values are not obscured by receiver noise.

5.4.3 Sensitivity and noise figure

Although it is apparent from the discussion above that the dynamic range must be wide enough to cope with the high noise peaks and at the same time ensure that the lower values are not lost in receiver noise, it is not usually necessary to use a very sensitive receiver. In general terms, we are not principally interested in those low levels of noise that occur with high probability, and a commercial receiver with a noise figure of 10 dB is usually adequate. Moreover, there is no difficulty, with modern low-noise devices, in designing a receiver able to respond to signals below −110 dBm, and this is adequate for most purposes.

5.4.4 Impulse response considerations

Impulse response is probably the most important parameter of the impulsive noise measurement system. Limitations in the other parameters, such as

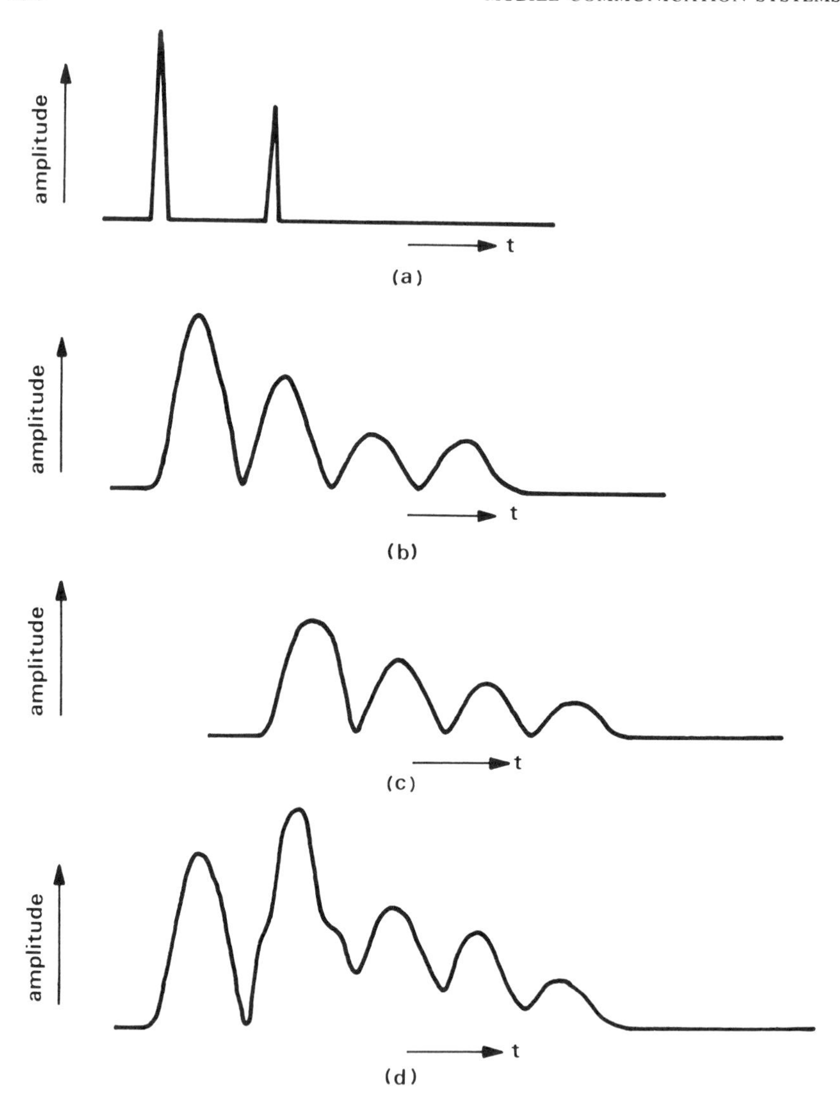

Figure 5.5 The effect of time sidelobes on the ability of the system to resolve closely-spaced impulses. (*a*) Input impulses; (*b*) filter response to first impulse; (*c*) filter response to second impulse; (*d*) possible reconstructed output due to both impulses.

dynamic range and bandwidth, still allow results to be obtained, but limitations in system impulse response result in fundamental modifications to the noise parameters themselves. The ideal impulse response for a measurement system is one in which there is no ringing, such as may be obtained from a Gaussian filter. The presence of ringing, or time sidelobes, may seriously affect the capability of the system to resolve low-strength impulses which follow very high-strength impulses. This is illustrated in Figure 5.5. The various noise parameters are affected differently, and the manner in which they are modified depends on the statistical properties of the noise. The extent to which errors occur in the measurements depends on the probability of occurrence of high-amplitude impulses, and the way in which the individual parameters are affected is discussed below.

If a time sidelobe exists in the impulse response, then the measured probability of exceeding lower levels will be increased. This arises because there will be not only a contribution from pulses at that level, but an additional contribution from the sidelobes due to previous high-strength impulses. The extent of the additional contribution may be estimated from a knowledge of the actual impulse response and the relationship between the numbers of high and low-strength impulses (the noise slope). If the noise slope is high, for example if the probability of pulses exceeding 60 dB kT_0B is 1% of the probability of exceeding 30 dB kT_0B, then the measured probabilities at lower levels are enhanced by only a small amount, and errors are negligible. Noise slope therefore plays an important part in determining the required impulse response, and if the noise slope is high, an inferior impulse response can be tolerated. In practice, several researchers [7, 9] have made measurements which indicate that, provided time sidelobes are about 30 dB below the main response, errors in measurement are unlikely to be serious.

Because the effect of a poor impulse response is to cause secondary responses generated by ringing, and because these are indistinguishable from responses to genuine lower-strength impulses, there is an increase in the crossing rate at low threshold levels. This increase will depend on the crossing rate at very high levels and the number of significant time sidelobes. Nevertheless, provided the increase at low levels is masked by the number of genuine impulses at these levels, the error in measurement is not significant. However, the average crossing rate (ACR) for high threshold levels is not affected by the poor impulse response and in most cases it is these data which are of greatest interest. Another effect of the time sidelobes present in the impulse response is to increase the measured pulse duration and decrease the time interval between pulses. Again, the lower threshold levels are most affected but, assuming a well-behaved noise slope, the errors are insignificant.

5.5 Practical measuring systems

A practical measuring system for man-made noise can be built around commercially available equipment. If APD is to be the principal measure of noise amplitude, then it is possible to record the detected noise on an instrumentation tape recorder and to analyse the recording to obtain other parameters of interest such as LCR, PDD and PID. However, the receiver selected must be used in a fixed-gain mode, that is, with no AGC, and account must be taken of any limitations in dynamic range and impulse response.

There are several ways to achieve an improvement in dynamic range, the

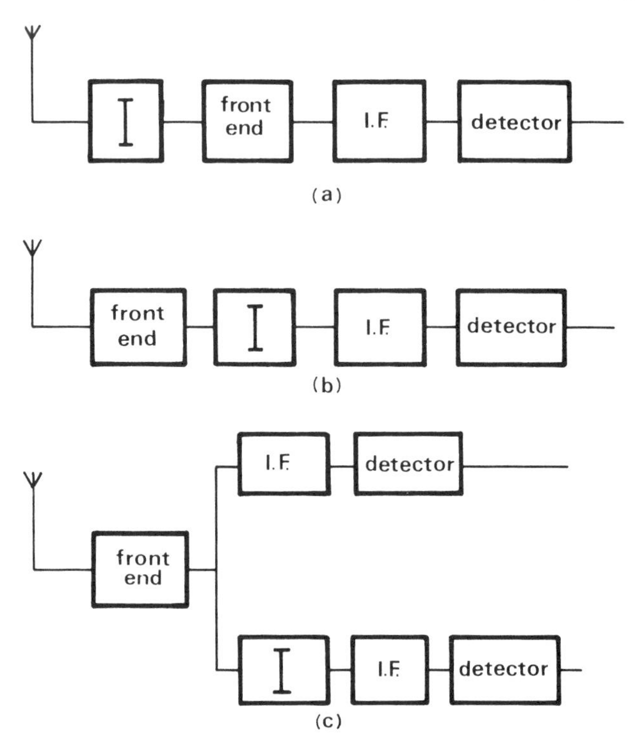

Figure 5.6 Methods to improve the system dynamic range. (*a*) Attenuator between antenna and front-end; (*b*) attenuator between front-end and IF; (*c*) dual IF system.

most attractive being to insert an attenuator either between the antenna and the receiver or in the IF stages when measuring high-level noise. These alternatives are illustrated in Figure 5.6. Effectively, they amount to looking at the noise data through two 'windows', but from an economic and technical point of view, the best method seems to be to place the attenuator in the IF section of the receiver. Placing it in this position is attractive because the receiver front-end often has an adequate dynamic range, and the use of IF attenuation causes only a marginal degradation in the receiver noise figure. By using a variable attenuator, we can effectively slide a 'window' with a width equal to the unmodified dynamic range of the system up and down, and hence select any convenient portion of the input signal for reception. The limitation is that the inserted attenuation must not approach the value of the front-end gain, otherwise the effective front-end gain is reduced, with a consequent degradation in the receiver noise figure. Provided this limit is not approached, the absolute range over which signals can be received is equal to the inherent dynamic range of the receiver plus the inserted attenuation.

When a receiver is subjected to impulsive noise with a dynamic range far in excess of the inherent receiver dynamic range, the high-strength impulses tend to saturate the receiver. Although this is not serious in itself, time sidelobes in the receiver impulse response will appear in the range being measured and can cause problems. However, if, as previously suggested, the number of pulses due to time sidelobes is swamped by the number of genuine pulses at these lower levels, measurement errors will be insignificant. For example, if a time sidelobe exists 30 dB below the mean peak of the receiver response, and at this level the impulse rate is 100 times the rate at a level 30 dB higher, the measurement error is 1%. When there is inserted attenuation, the limiting factor is noise generated by those parts of the system which follow the attenuator, and very often, when recordings are being made for subsequent analysis, the most important is tape-recorder noise, which affects the lower level signals. This is overcome by arranging for the two 'windows' to overlap slightly and then ignoring information from the lower end of the upper window.

Basically, the impulse response problem can only be solved by using better filters, and Gaussian impulse response filters are ideal for this purpose. The point in the system where these filters should be inserted is a matter for further consideration. Single-conversion receivers often have a high IF (21.4 or 10.7 MHz is typical), and it is difficult to design narrow-band Gaussian filters at these frequencies. The two possible approaches to the problem are either to convert the signal to an IF not greater than about ten times the required bandwidth and then use Gaussian filters, or to use wider bandwidth filters at the existing IF, and to improve the impulse response at baseband by a suitable low-pass filter. In normal circumstances the second approach is more attractive. A measuring system as described would normally be calibrated in dB kT_0B using a suitable Gaussian noise source, and a block diagram is shown in Figure 5.7.

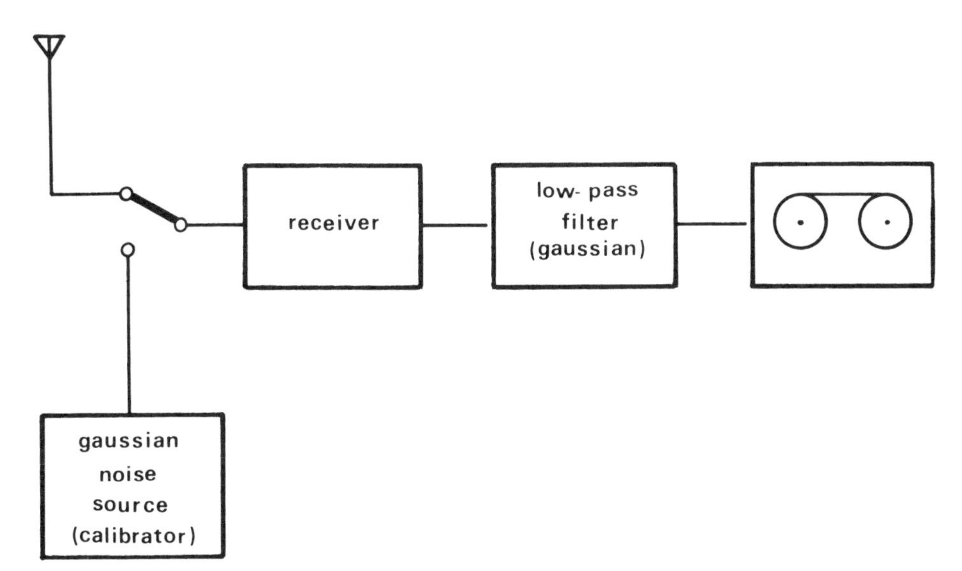

Figure 5.7 Equipment for measuring APD.

5.6 Measurement of noise amplitude distribution

Equipment designed to measure NAD should be capable of detecting all the noise impulses likely to affect a communications receiver, and be able to distinguish between successive impulses which occur at the highest likely rate.

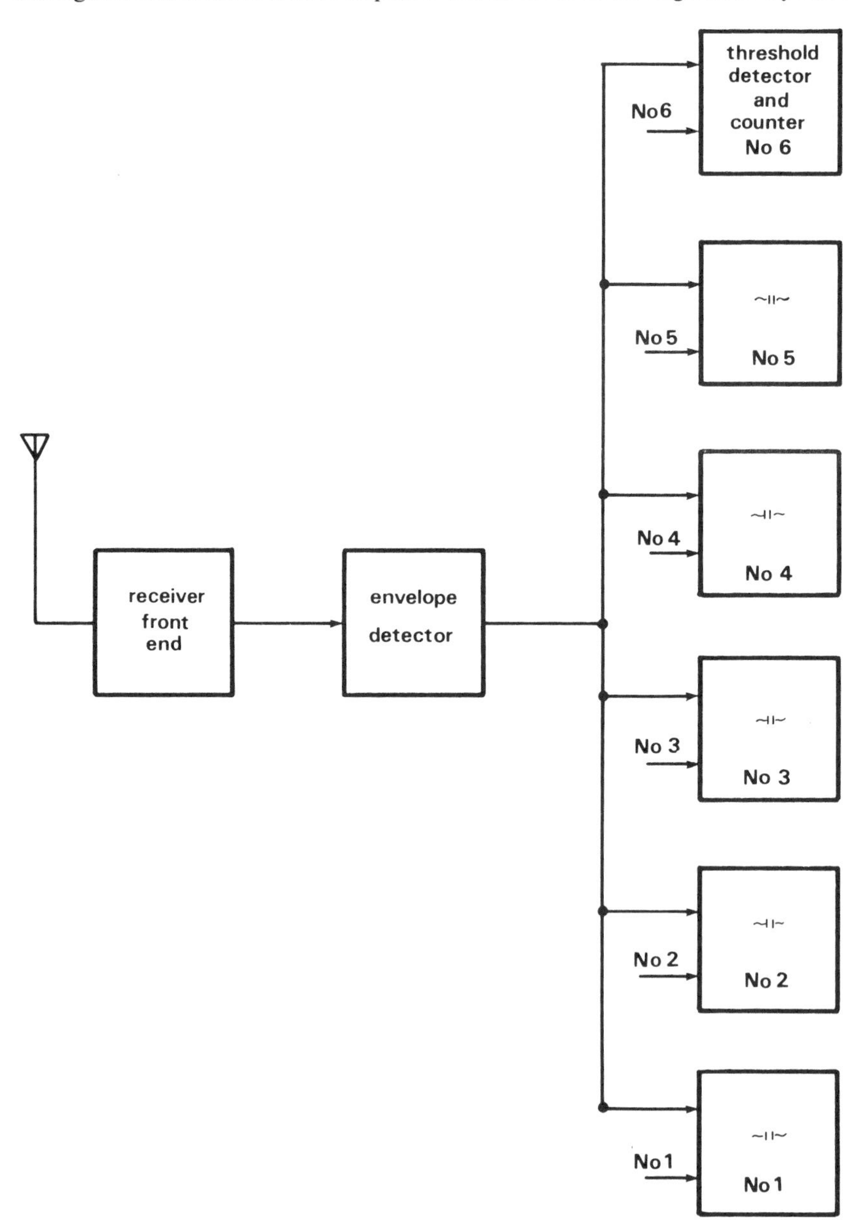

Figure 5.8 System for measuring NAD.

At the very minimum, for conventional land mobile radio systems this implies a dynamic range of 60 dB and a RF bandwidth of 30 kHz. It is highly desirable that the equipment has a well-behaved impulse response, so that its effect on the measured results is minimal.

Basically the equipment consists of two parts, a high-gain receiver having as its output a band-limited detected signal, and a pulse height analyser which is such that each noise pulse is simultaneously compared with several preset threshold levels. Each threshold detector is coupled to a digital counter and store, which accumulates the total number of pulses which exceed that particular threshold level. This process is illustrated in Figure 5.8, for a system with six thresholds. Practical realizations of the techniques have been described [10, 11]. The latest of these uses sophisticated digital technique so that the instrument can measure other parameters such as APD, pulse intervals and pulse durations.

Calibration of the equipment in terms of impulse strength at the input is straightforward, but in this case an impulse generator is used rather than a Gaussian noise source. The reference level to each threshold detector is set to maximum, and the impulse generator is set to the lower limit of the required measurement range and connected to the receiver input. The reference level of branch 1 is then adjusted until counter 1 just starts to count. The output of the impulse generator is then increased to the next threshold level (say 10 dB higher), and the reference level for detector 2 is adjusted until counter 2 starts to count. This procedure is repeated for all levels until the instrument is completely calibrated.

In use, the instrument is connected to an appropriate antenna (in urban areas this is often a vehicle antenna), all counters are set to zero, and a measurement is made, typically for 15 minutes. The readings of the various counters at the end of this time allow the NAD to be plotted. It is, of course, essential to measure at a frequency which is free from coherent interference, and if a suitable receiver is available it is wise to monitor the channel during the measurement period.

5.7 Statistical characterization of noise

If a measuring system as illustrated in Figure 5.7 is used to measure and record noise, then the detected noise envelope is an undirectional time-varying voltage. The exact shape of the detected waveform depends on a number of factors including the noise input, the receiver impulse response and the receiver bandwidth. The amplitude of the noise can be represented by the amplitude probability distribution (APD) which is obtained by fixing a particular level and then summing all the individual durations above that level to find the total time for which the level is exceeded. The APD gives information only about the overall percentage of time for which any particular level is exceeded, but there is no information about how the time was made up, that is, how many pulses exceeded the value in question.

In order to provide more detail, the APD can be supplemented by a graph showing the average number of times per second any particular level is crossed. Average crossing rate (ACR) is usually expressed in the form of a graph of amplitude against crossing rate, on a logarithmic scale.

ACR curves give only first-order information about the number of times each level is crossed, and when taken in conjunction with APD curves can only produce information about the average pulse duration and the average pulse interval. Further information about the pulse duration can be provided by a graph of the pulse duration distribution (PDD), i.e. the probability that any given duration is exceeded. To do this, it is necessary to count the individual pulse durations at a given amplitude level. In a similar way, further information about the intervals between pulses is given by a graph of the pulse interval distribution (PID), which is obtained by counting all the individual time intervals between successive positive-going crossings at a given level.

The most straightforward method of obtaining these parameters from a recorded tape is to replay the tape into a computer via an ADC, able to convert at a rate appropriate to the bandwidth of the recorded signal, to store the digitized data within the computer and to find the parameters of interest by means of software. However, this is often not practicable and data reduction techniques are commonly used [12].

5.8 Impulsive noise measurements

Measurements of impulsive noise derived from experiments conducted at roadside locations must be interpreted with some care. The levels of noise depend on the volume of traffic and the influence of road features such as traffic lights and junctions. However, the results are useful in two different ways. Firstly, they can be used to test the validity of existing mathematical models representing impulsive noise, or as a basis for proposing new models. Secondly, the data can be used to predict the performance degradation of communication systems subjected to impulsive noise.

Noise data measured at various frequencies between 40 MHz and 900 MHz [13], using a receiver with an IF bandwidth of 20 kHz, is shown in Figures 5.9–5.11. At suburban locations, traffic generally flows quite freely without significant influence from road features such as roundabouts or traffic lights. The traffic density varies between 15 and 20 vehicles per minute, 15 being a typical value. In urban areas traffic flows in a constant stream at a much steadier rate than at suburban locations, 30 vehicles per minute being a typical rate. The APD curves for the two types of location reflect this increase in traffic density, the difference being between 2 and 4 dB at all frequencies. It is also clear that the overall level of noise decreases with an increase in frequency. For example, in suburban areas, the value exceeded with 10^{-4} probability

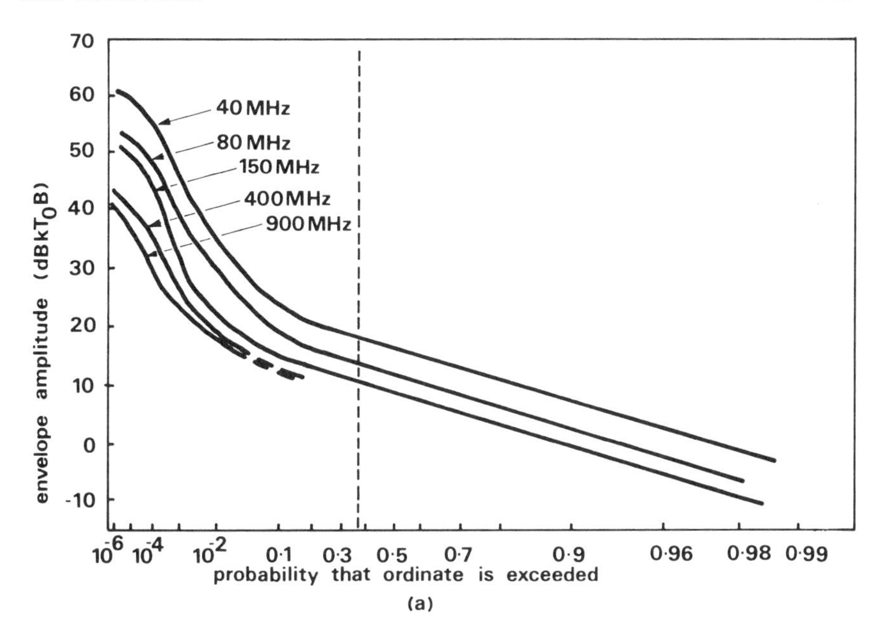

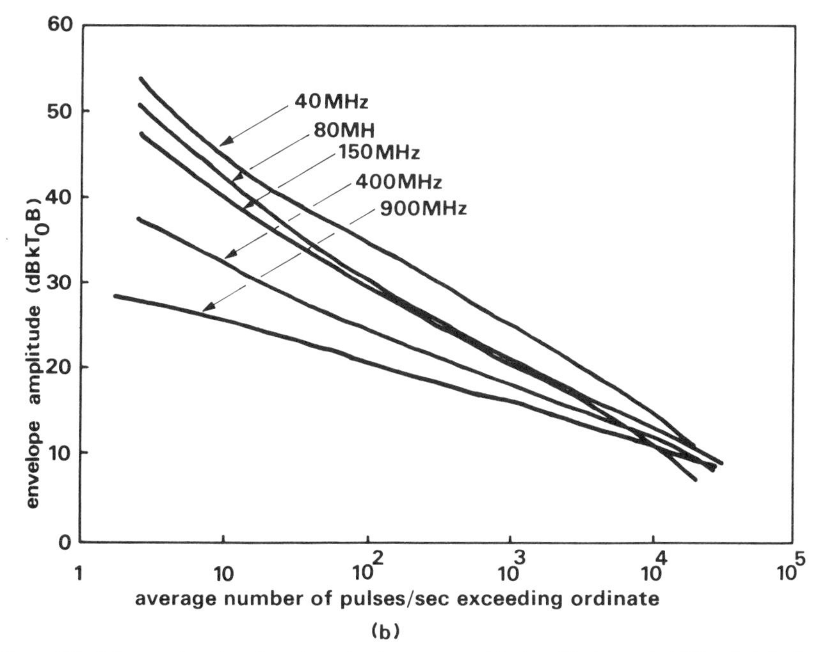

Figure 5.9 Measured noise data in suburban areas. (*a*) Amplitude probability distribution; (*b*) average crossing rate.

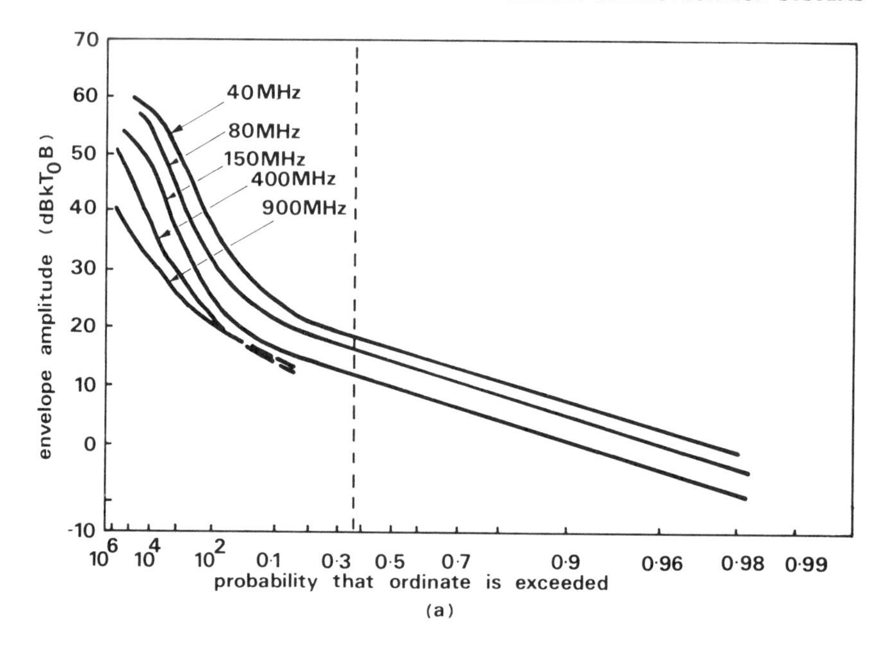

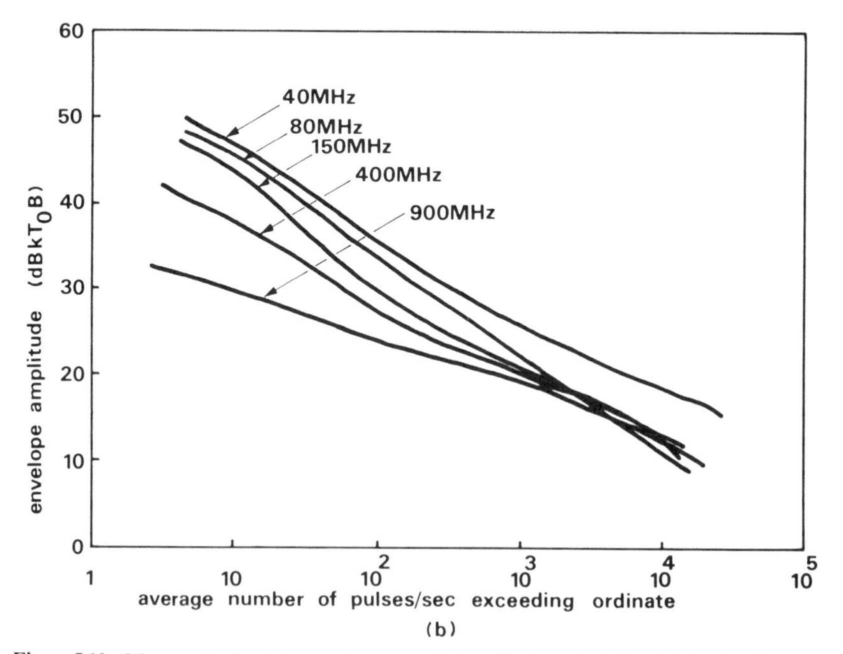

Figure 5.10 Measured noise data in urban areas. (*a*) Amplitude probability distribution; (*b*) average crossing rate.

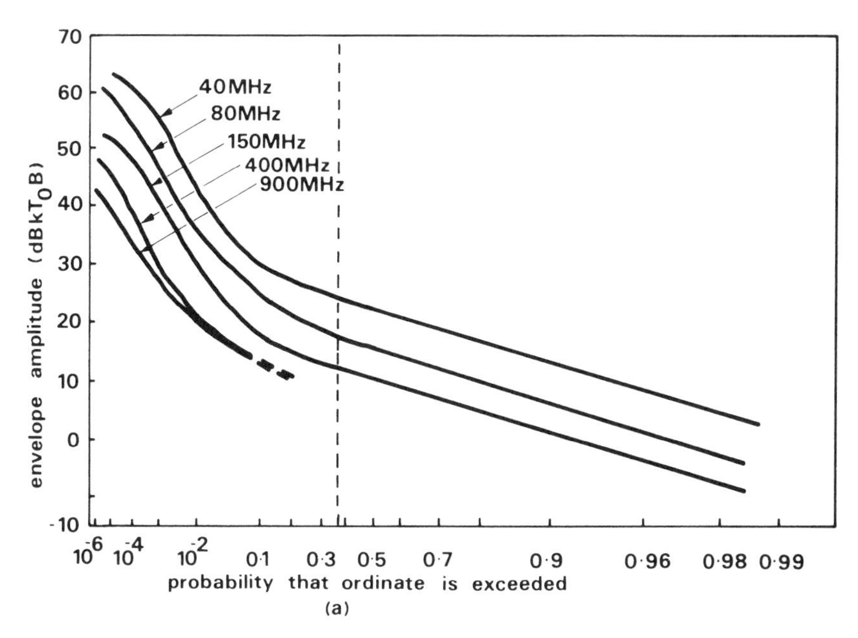

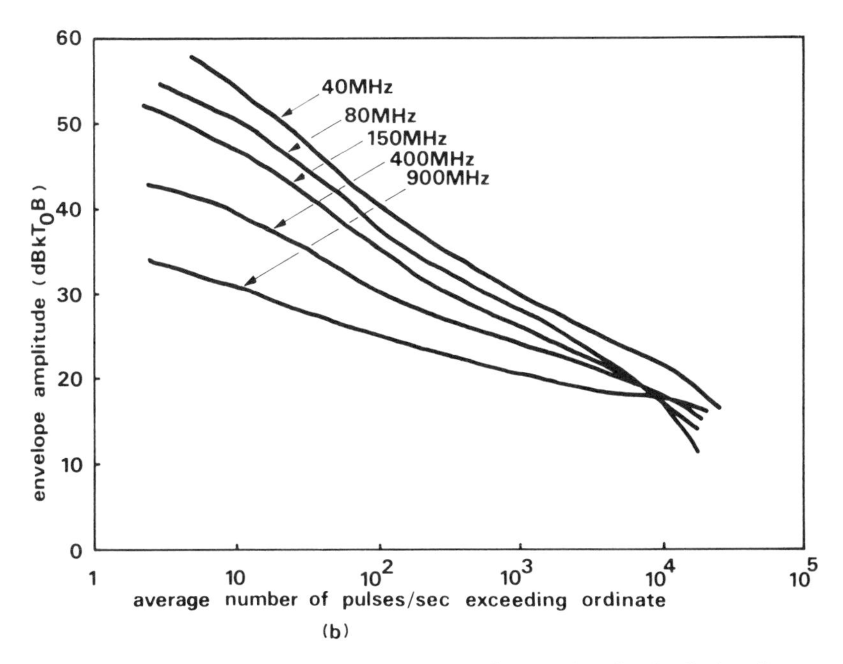

Figure 5.11 Measured noise data in a city centre. (*a*) Amplitude probability distribution; (*b*) average crossing rate.

decreases from 46 dB kT_0B at 40 MHz to 30 dB kT_0B at 900 MHz. At 40 MHz, the median value of the curve representing background noise lies approximately 12 dB above the corresponding point on the line representing receiver noise. Looking at the difference another way, the level 11 dB kT_0B is exceeded by the receiver noise for only 10^{-2}(1%) of the time, whereas it is exceeded by the external noise for above 80% of the time, so it is fairly safe to conclude that at this frequency the external noise is very dominant. On the other hand, at 400 and 900 MHz the external noise is much lower, and for most of the time receiver noise is dominant.

A further comparison between receiver and man-made noise may be made by noting that at 40 MHz man-made noise in urban areas exceeds 11 dB kT_0B for 84% of the time. For 1% of the time at 80 MHz receiver noise exceeds 11.5 dB kT_0B (receiver noise figure, 4 dB), whereas the external noise exceeds this level for about 65% of the time. At the higher frequencies, receiver noise is still very important.

The ACR curves in Figure 5.10*b* also indicate the higher level of noise measured in urban areas. There does not appear to be any significant change in the shape of the graphs with respect to the corresponding curves in Figure 5.9*b* for suburban areas. However, the tendency for the curves to become lower and flatter as frequency increases is evident in both cases. The number of pulses with high amplitude is greater in urban areas, although not excessively so. Taking the 25 dB kT_0B level, there are about 70 pulses/s at 900 MHz that exceed this level, compared with 12 in suburban areas. However there is no corresponding increase at the lower frequencies. There are still about 300 pulses/s exceeding this level at 150 MHz, with just over 1000 at 40 MHz.

A misleading idea of the influence of traffic in the central business area can be obtained if the only factor taken into account is the number of vehicles passing the observation point. In the city centres, vehicles move slowly under the control of road features such as traffic lights, and although the apparent traffic density is low, 14–24 vehicles/min in Figure 5.11, there is always a large number of vehicles, moving and stationary, sufficiently close to the receiver to influence the measured noise parameters.

The APD curves in Figure 5.11*a* illustrate this in a convincing way. The noise level here is higher than at the other two types of location. The level exceeded at 10^{-4} probability is now 61 dB at 40 MHz, 39 dB at 400 MHz and 34 dB at 900 MHz, all these values being 2–3 dB above the value measured at urban locations even though the apparent traffic density is significantly lower. In city centre areas at 40 MHz, man-made noise is completely dominant, since the 11 dB kT_0B level, which the receiver noise exceeds for only 1% of the time, is exceeded by the man-made noise for about 93% of the time. At 80 MHz, the value exceeded by receiver noise for 1% of the time (11.5 dB kT_0B) is exceeded by external noise for 65% of the time, similar to the figure for urban areas, and at 150 MHz (receiver noise figure 3 dB) the 1% level (12.5 dB kT_0B) is exceeded by external noise for 38% of the time. The ACR curves in Figure 5.11*b* confirm the above conclusions. Once again, they tend to be lower and flatter at the higher frequencies, and there are more pulses at very high levels. At the 25 dB kT_0B level there is an increase in the number of pulses at 900 MHz, from 70 to 100, but at the same level the number of pulses at 40 MHz has increased to 3500. The tendency for the curves to cross over at a few thousand pulses/s, apparent at suburban and urban locations, is also present here. The slope of the ACR curves at 400 and 900 MHz is similar to that at other locations, but there is a tendency for the graphs at lower frequencies to be slightly steeper, about 12 dB/decade at 80 and 40 MHz.

Curves illustrating PDD and PID have been published in the literature [13]. In this case it is not meaningful to derive an 'average' set of curves from measurements, since the detailed shape is very sensitive to changes in traffic density and pattern. Instead, Figure 5.12 shows sets of graphs corresponding to test measurements which produced APD and ACR curves very close to the typical suburban curves in Figure 5.9. The PDD graphs show that pulses with a duration exceeding 100 μs (the reciprocal of the post-detection bandwidth) are not uncommon, even at the higher threshold levels, and this is indicative of overlapping receiver impulse responses due to the arrival of groups of pulses (noise bursts) too closely spaced in time to be adequately separated by a narrowband receiver. There is a general tendency for the various PDD curves to bunch together close to the 100 μs level, with the curves for lower thresholds lying to the left of those for higher thresholds below the 100 μs level, and to the right above that level.

If there was no overlapping of impulse responses, then the PID graphs would all be distinct and separate, with the curves for lower thresholds lying to the left of those for higher thresholds. However, the presence of overlapping responses leads to a set of PID curves that cross each other. Generally, overlapping of impulse responses will occur with a much higher probability at low threshold levels than at higher threshold levels, and the set of typical PID graphs for a suburban location show this tendency. The 50% exceedance probability is 0.04 ms at 11 dB kT_0B, increasing to 30 ms at the 50 dB kT_0B level.

Notwithstanding the different locations at which measurements were made and the different traffic densities involved, it is evident that the general level of man-made noise decreases as the measurement frequency is increased. The amplitude characteristics in suburban, urban and city centre areas are summarized in Table 5.1, and this effect is apparent at all probability levels. The ACR curves tend to be well separated at low pulse rates, but cross over each other in the 10^3–10^4 pulses/s region, and the curves at higher frequencies have a lower slope. A major contribution to the noise at lower frequencies is therefore in the form of fairly infrequent high amplitude pulses; this is not so at higher frequencies where the variation in pulse heights is much smaller.

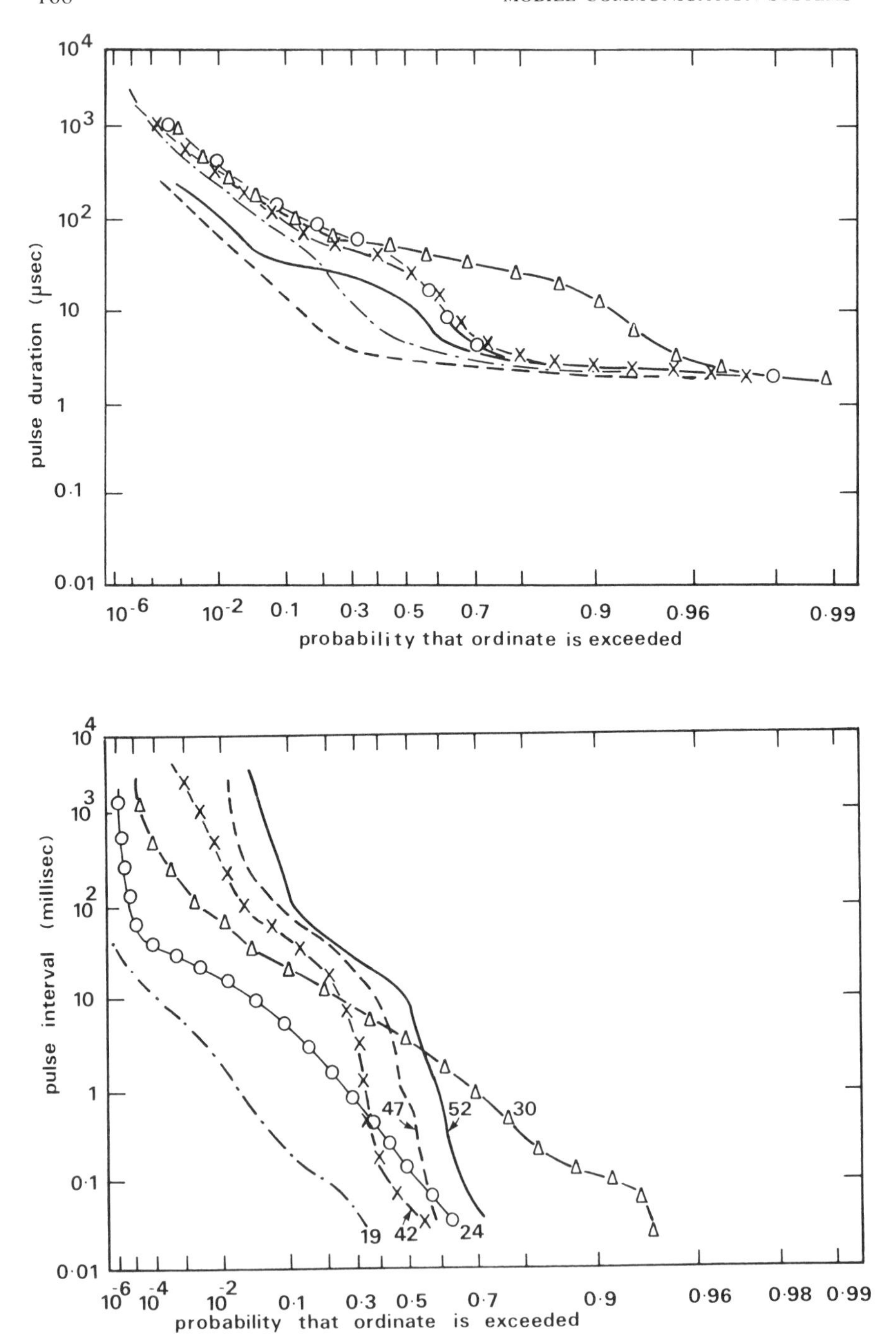

Figure 5.12 Pulse duration and pulse interval distributions in a suburban area. Labels on graphs show level in dB relative to kT_0B.

Table 5.1 Statistics of noise amplitude

Frequency (MHz)	Envelope amplitude dB kT_0B					
	City Centre		Urban Area		Suburban Area	
	$P=10^{-4}$	$P=0.5$	$P=10^{-4}$	$P=0.5$	$P=10^{-4}$	$P=0.5$
40	61	22	58	17	56	16
80	56	14	55	14	50	12
150	48	10	48	10	45	8
400	39	< receiver noise	38	< receiver noise	36	< receiver noise
900	34		32		30	

5.9 Summary

It is apparent from Figures 5.9–5.11 that, at the lower frequencies, man-made noise is high and is the dominant source of interference. However, at the higher frequencies, the APD curves fall below the receiver noise curve (noise figure, 9.5 dB) at probabilities less than 10%, even in city-centre areas, and at these frequencies receiver noise plays a much more important role. There is therefore a strong incentive to design a UHF receiver with a low noise figure because there is much to be gained by doing so. The same is not true for receivers at 40 MHz intended for use in urban areas. The amplitude levels measured at the 10^{-6} probability level (i.e. the values that would be indicated by a peak-reading meter) decrease steadily with frequency. However, they are always significantly in excess of the maximum contribution that can be expected from receiver noise, even at 900 MHz, the difference being about 20 dB in urban areas. A very misleading impression can therefore be obtained from peak measurements, and their usefulness is severely limited.

To summarize briefly, we can say that in the frequency range 40 MHz to 900 MHz there is a monotonic decrease in the level of detected noise envelope. However, there is no evidence of any significant change in the statistical character of the noise with frequency. The APD curves all have the same general shape, although the detail varies with factors such as frequency, traffic density and traffic pattern. The ACR curves show that the noise is composed of a relatively large number of low-level pulses, together with a smaller number of high-level pulses, the slope of these curves being steeper at low frequencies than at high frequencies. At low frequencies, external noise is very dominant, but this is not the case in the UHF range. The PDD and PID curves show that overlapping receiver responses are quite common, but they also occur with decreasing probability as the measurement frequency is increased.

5.10 Performance prediction techniques

Not all the parameters described in section 5.3 find use in directly evaluating communication system performance. Little attempt has been made, for example, to use the peak detector as a measure of communication system degradation, mainly because of its obvious limitations. As examples of the difficulties, it has been reported that different types of noise from power lines, giving the same peak reading, often result in widely different assessments of the degradation of TV reception. Also, a peak detector's measurement of the noise, even from a single vehicle, corresponds very poorly with the resulting degradation in the sensitivity of a radio receiver.

Quasi-peak measurements have, however, been used, indeed one of the reasons for developing the quasi-peak meter was to provide a measurement parameter, with appropriate weighting for measuring the degradation in performance of radio broadcasting receivers. However, with the trend to use higher radio frequencies, it has become obvious that the information provided by a quasi-peak meter is very limited, and there is a need to use other parameters for the prediction of system performance.

As far as performance prediction is concerned, the most important methods of characterizing noise are those which recognize time as an important factor, viz. APD and NAD. Each method has its advantages and disadvantages. The APD represents the properties of the detected noise envelope, and is therefore a function of the receiver characteristics such as bandwidth and impulse response. There is a lack of generality in such procedures, but measurements made at the IF output do at least have the advantage that they represent the total noise (ambient + impulsive + receiver) presented to the data demodulator, and may be readily used. The lack of generality is to some extent minimized in systems such as VHF and UHF land mobile radio, where the necessity to meet stringent specifications has led to the development of a range of IF filters having impulse responses that are essentially very similar.

The NAD, on the other hand, represents the noise process at the receiver input and models it in terms of a series of pseudo-impulses. The disadvantage therefore is that ambient and receiver Gaussian noise is not included, but this may not be too restrictive if impulsive noise is the dominant feature. The major advantage of using NAD is that methods based on it utilize separate formulation of the receiver and impulse noise characteristics and are therefore applicable, in principle, to any receiver whose characteristics are known.

5.10.1 The assessment of receiver performance

Assessing the performance of radio receivers in the presence of Gaussian noise is fairly straightforward, because good quality wideband Gaussian noise sources are readily available and testing methods are well established. When impulsive noise is considered, the situation is quite different. It would be possible to test mobile radio receivers by installing them at a suitable roadside location where interference is generated by passing vehicles, but in general such *ad hoc* procedures are clearly unsatisfactory. There is an obvious need to establish an experimental performance assessment technique which yields a meaningful measure of receiver system performance and which can be used either in the laboratory or at the production stage.

As a first step in assessing the susceptibility of a radio receiver to impulsive noise, it is necessary to measure the degradation in receiver performance caused by impulses having a given strength and a given repetition rate. It has been suggested that this may be expressed in the form of impulse noise tolerance curves (sometimes termed isodegradation or isosensitivity curves). For FM land mobile radio receivers, Shepherd [4] developed a set of isodegradation curves using the system shown in Figure 5.13. The first step is to determine the carrier level required to obtain 20-dB quieting of the audio, with no external noise. An X dB isodegradation curve is then obtained by increasing the carrier level by X dB and using the variable-rate impulse generator to find what combinations of impulse strength and repetition rate will yield the same 20-dB quieting of the audio. These combinations will form a smooth curve, as in Figure 5.14, which shows a family of isodegradation curves for a VHF/FM receiver. It can be seen that, for example, if the receiver is subjected to an impulse strength of 45 dB above 1 μV/MHz at its input at a rate of 16 impulses/s, then an additional 5 dB of carrier power is required to provide 20-dB quieting of the audio, i.e. the receiver sensitivity is reduced by 5 dB. If the impulse noise source operates at 4 impulses/s, then the degradation is only 2 dB.

As far as data communication receivers are concerned, it is doubtful whether the 20-dB audio quieting criterion is the best basis for comparison. It has the advantage of being easily measured, but the signal level at which it occurs varies from one receiver to the next, so that performance measured relative to the 20-dB quieting level is not an absolute measure. Even in speech communication systems, no systematic correspondence has been established between it and a subjective assessment of receiver performance. It therefore seems sensible to establish, for each type of communication system, a performance criterion that has direct relevance to the type of system being considered [14]. In data communication systems, BER is the obvious measure. In Figure 5.15 we show a measuring system for a data communication system. A pseudo-random data stream is used to modulate a carrier, and

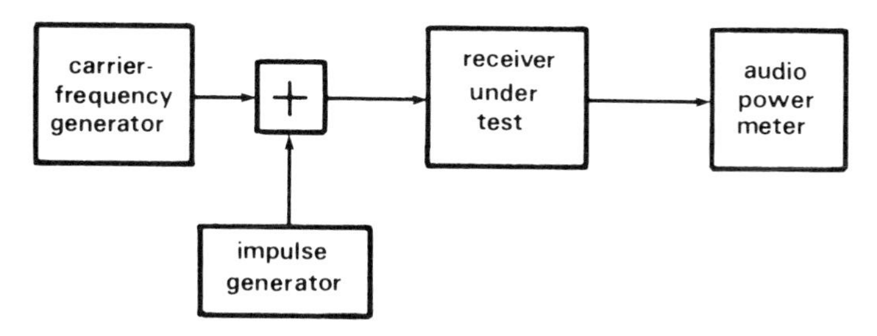

Figure 5.13 Method of obtaining isodegradation curves for FM radio receivers.

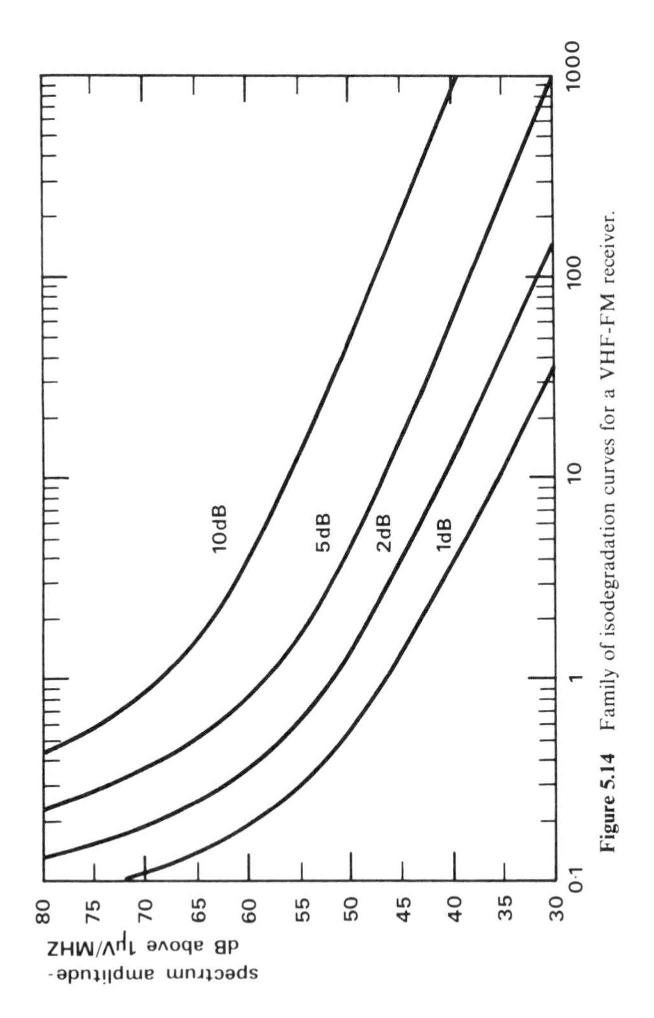

Figure 5.14 Family of isodegradation curves for a VHF-FM receiver.

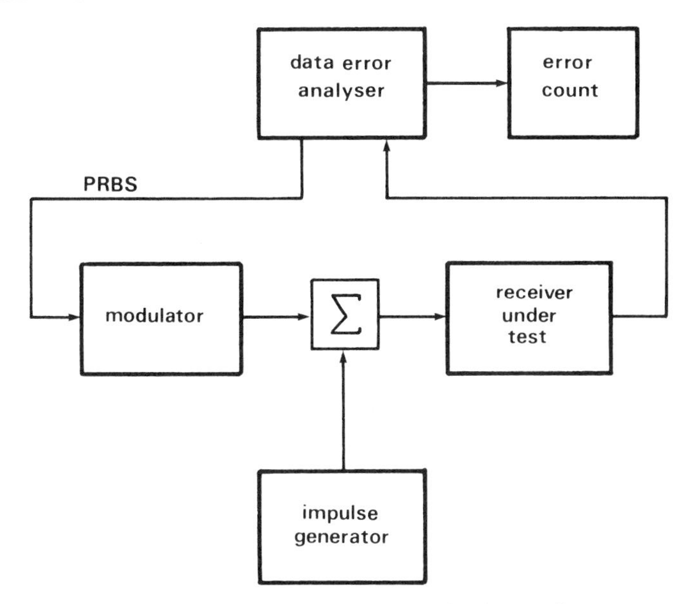

Figure 5.15 Method of obtaining isodegradation curves for data receivers.

after corruption by noise from the impulse generator, the signal passes into the receiver. Following demodulation, the output is passed into an error detector where it is compared on a bit-by-bit basis with the original data stream, and errors are counted.

The experimental procedure is to obtain a set of curves at a certain input carrier level by setting the spectrum amplitude of the impulse generator to a given value and plotting BER as a function of impulse rate. Figure 5.16 shows a family of such curves for a commercial VHF/FM receiver subjected to a periodic pulse train. They verify the work of Bello and Esposito [15] and Engel [16], who suggested that when the incoming pulses are uniformly spaced in time, then for a constant spectrum amplitude, the BER is linearly proportional to the average number of impulses/s. By repeating the experiment at several different input carrier levels it is possible to produce several sets of curves similar to those in Figure 5.16.

There are two ways in which the results can be presented. Firstly, on a single set of curves as in Figure 5.16 we can choose our performance criterion (say $\text{BER} = 10^{-4}$), and by drawing a horizontal line we can determine what combinations of spectrum amplitude and impulse rate give rise to that value of BER. Figure 5.17 shows this form of presentation, in which carrier level is the parameter. Alternatively, if we have several sets of curves similar to Figure 5.16 plotted at various carrier levels, then from each set we can obtain a curve relating spectrum amplitude to average impulse rate, at a predetermined value of BER. Figure 5.18 shows a family of curves produced in this way; note that the BER is now a parameter. This form of presentation is almost directly

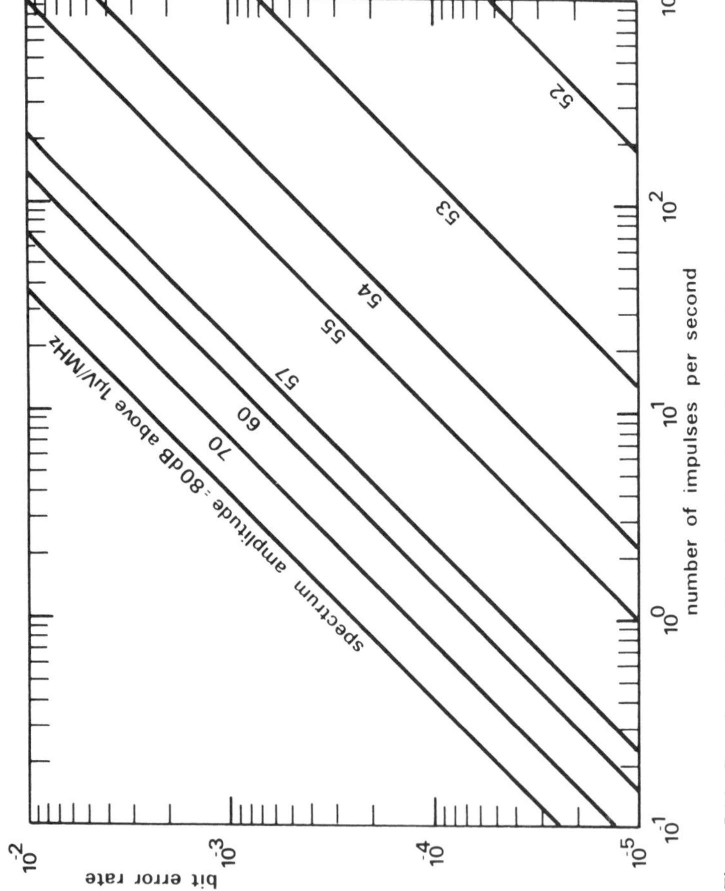

Figure 5.16 Experimental results for a VHF receiver obtained using the experimental system shown in Figure 5.15.

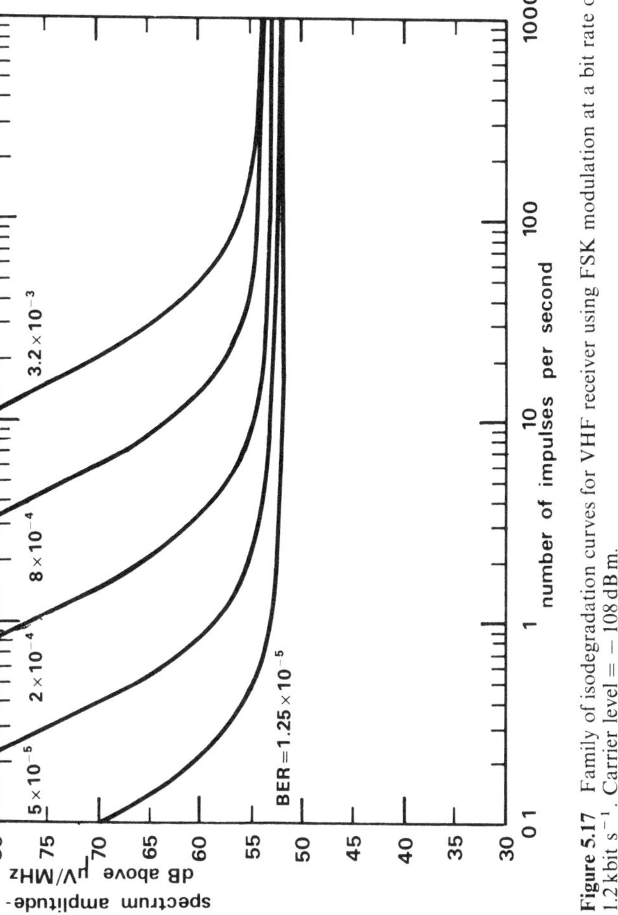

Figure 5.17 Family of isodegradation curves for VHF receiver using FSK modulation at a bit rate of 1.2 kbit s^{-1}. Carrier level = − 108 dB m.

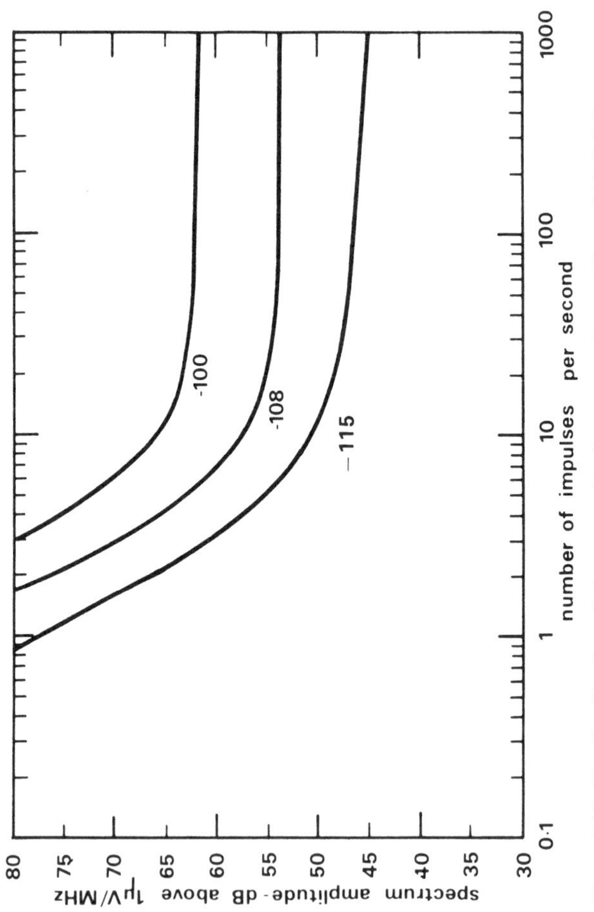

Figure 5.18 Family of isodegradation curves for the same receiver as Figure 5.17. Labels on curves represent signal level in dB m. Bit error rate $= 4 \times 10^{-3}$.

analogous to the method used by Shepherd [4], the difference being that the primary performance measure has been set at BER $= 4 \times 10^{-3}$ rather than 20-dB quieting. The first form of presentation is much preferred because it has several practical advantages. Firstly, it clearly shows a threshold level below which the noise impulses have negligible effect. Secondly, if we were to conduct an experiment to measure the BER due to a certain noise source, either in the field or in the laboratory, then the simplest way to proceed is to fix the input carrier level, add the noise at the receiver input, and measure the BER. In practice, therefore, carrier level is the obvious parameter.

5.10.2 Prediction methods using APD

The usual intermediate step between noise measurement and BER prediction is noise modelling—the models can be empirical, or related to the physical noise mechanisms involved. Noise models based on the APD usually depend on fitting some form of mathematical expression to the noise curve [17–22], the common approach thereafter being to predict BER by integrating the expression over the noise range, using numerical techniques. The expressions derived for BER cannot usually be reduced to compact forms, and are left in the form of integral representations. The integrations are then evaluated numerically for certain combinations of parameters, and various approaches have been adopted.

However, in all the BER prediction models that use APD, the measured characteristic can in principle be used to obtain the BER directly without the need to go through the intermediate stage of mathematical modelling. This brings the advantage that there is no loss of accuracy though fitting a mathematical expression, and there is a clear saving of time. This direct usage of experimental noise data is made possible by the data storage and processing capabilities of modern digital computers. The measured APD can be stored with a high degree of accuracy, and the BER can be predicted using digital processing techniques [23]. The output of the APD measuring equipment can be either digitized on-line or recorded for later off-line processing. A histogram of the samples can be built up in the computer memory with very high precision, the numbers in successive bins can be added to produce the cumulative probability, and the APD can be obtained.

BER prediction techniques based solely on APD are not always sufficient to determine accurately the error performance of a detector. However, for communication systems such as land mobile radio, where channel bandwidths are narrow, filter characteristics are well-defined, and data rates are such as to occupy the available bandwidth efficiently, accurate predictions can be made using APD alone. In systems such as that illustrated in Figure 5.19, the radiofrequency carrier is directly modulated by the data stream, and in the receiver, decisions are made by sampling the demodulated signal at times nominally at the centre of the data symbols, there being a minimum amount of postdetection filtering. In the majority of cases, the prediction procedure

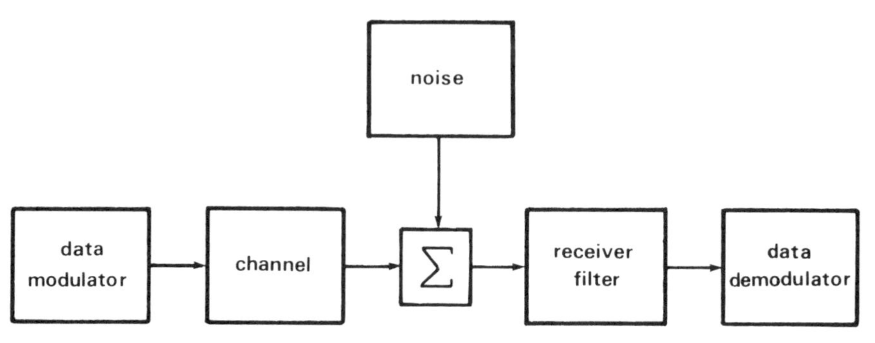

Figure 5.19 Simple data communication system.

amounts to evaluating a one-dimensional integral of the form

$$P_e = \int_a^b f_r(r) P[e|(r/s)] \, \mathrm{d}r \tag{5.3}$$

where $f_r(r)$ is the probability density function of the noise envelope and $P[e|(r/s)]$ is the conditional probability of error in the receiver given r, the noise envelope, and s, the signal envelope.

In practice, for different modulation techniques $P[e|(r/s)]$ can either be obtained theoretically, or approximated. The integral can be evaluated by exploiting the inherent precision in measuring the APD using a large number of histogram bins in the computer. If eqn (5.3) is turned into a finite summation over the histogram bins, then

$$P_e = \sum_{i=n_a}^{i=n_b} f_r(r_i) P[e|(r_i/s)] \, \Delta r_i \tag{5.4}$$

and, since $f_r(r_i)\Delta \simeq H_i$, this becomes

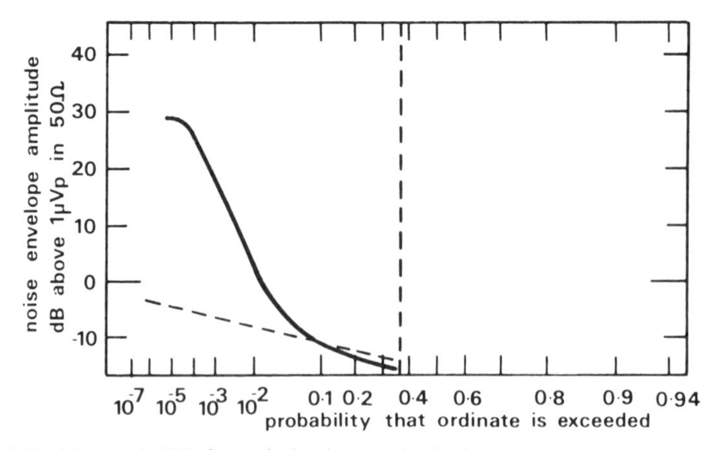

Figure 5.20 Measured APD for typical noise sample obtained using simulator.

$$P_e = \sum_{i=n_a}^{i=n_b} H_i P[e|(r_i/s)] \tag{5.5}$$

where n_a and n_b are the indices of histogram bins corresponding to a and b respectively, and H_i is the probability corresponding to the measured content of the ith histogram bin.

Using appropriate variants of this equation, bit error probabilities have been calculated for PSK and DPSK. Figure 5.20 shows a measured APD from a hardware noise simulator, and Figure 5.21 shows the predicted and

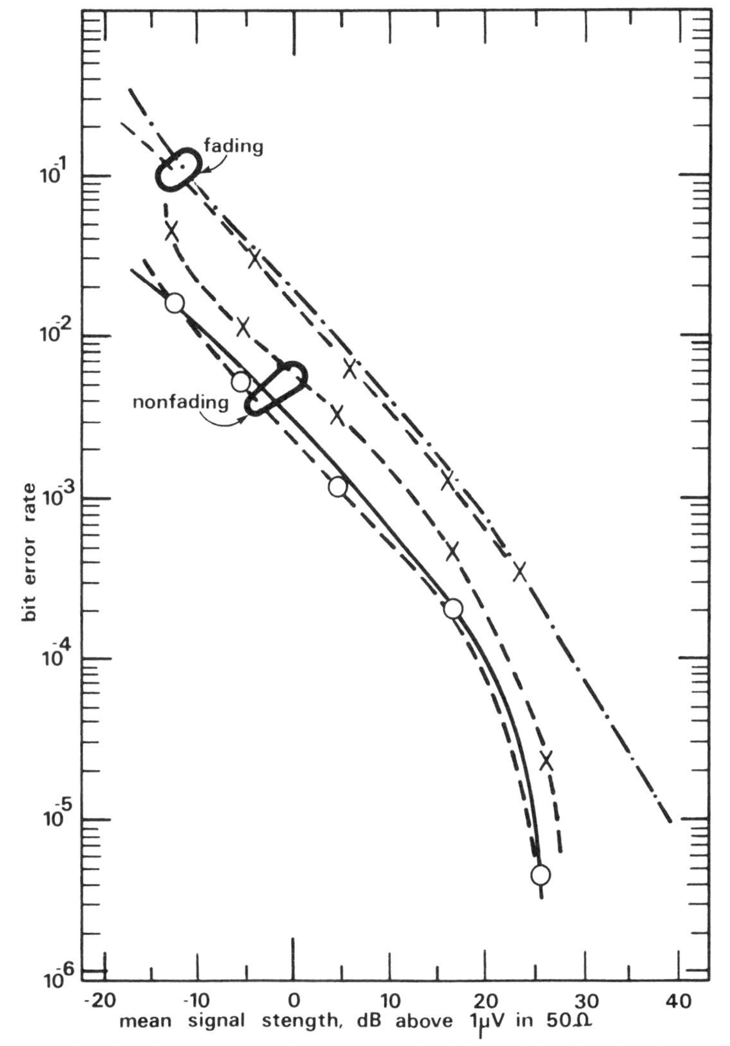

Figure 5.21 Predicted and measured bit error rates for PSK and DPSK using the noise sample represented by Figure 5.20.—PSK prediction; —·—·DPSK prediction. Experimental points: O PSK X DPSK.

measured error rates for this noise sample for both steady and Rayleigh fading signals. The measured results were obtained at the same time as the APD was measured, by allowing the generated noise to corrupt a data communication system of the appropriate kind.

5.10.3 Prediction methods using NAD

An early document relating to the NAD concept [24] also proposed an empirical method for determining the susceptibility of the radio receiver to impulsive noise. For FM radio receivers, the NAD 'overlay' technique involves overlaying the measured NAD curve for a given noise source on the isodegradation curves for the receiver under consideration, in the manner shown in Figure 5.22. It was originally suggested that the point of tangency between the NAD and one of the isodegradation curves indicates the degradation to be expected from that source. In the case of Figure 5.22, which shows two NAD curves overlaid on a set of isodegradation curves, the expected reduction in sensitivity is approximately 2 dB. It is interesting to note in this connection that it is not the relatively small number of high-amplitude noise pulses that cause the maximum degradation, nor indeed is it the large number of low-amplitude pulses. The pulses about 10 dB below the maximum which occur at a rate of a few per second contribute most. Nevertheless, it is prudent to be a little cautious in using this criterion, because although the tangency point is a good indicator of performance and may be useful in, say, comparing the degradation caused in different receivers by the same noise source, the precise relationship between the degradation indicated by the tangency point and that actually measured under experimental conditions has not been theoretically established. It would appear intuitively that the relative slopes for the two sets of curves must have an influence. For example, the NAD curve marked (*b*) in Figure 5.22 is likely to cause greater degradation than curve (*a*), even though the tangency point is the same, because there are a larger number of high-amplitude impulses present.

One possible limitation associated with the overlay technique is that the isodegradation curves are plotted for a train of impulses uniformly spaced in time, whereas the NAD for the measured noise source is likely to consist of pulses with random arrival times. Clearly, an accurate prediction is possible only if overlapping of impulse responses at the receiver IF output is negligible. The expected effect is small, but the question has been investigated [14] by arranging for a pulse train having a Poisson distribution of arrival times to be applied to the 'external trigger' socket of an impulse generator used in the experimental configuration of Figure 5.15. The only observed effects were at high average impulse rates, where there is a significant probability, due to overlapping responses, that one data symbol is affected by more than one noise pulse. The method of measurement using pulses uniformly spaced in time is therefore adequate for almost all practical purposes.

As far as data communication systems are concerned, the overlay technique

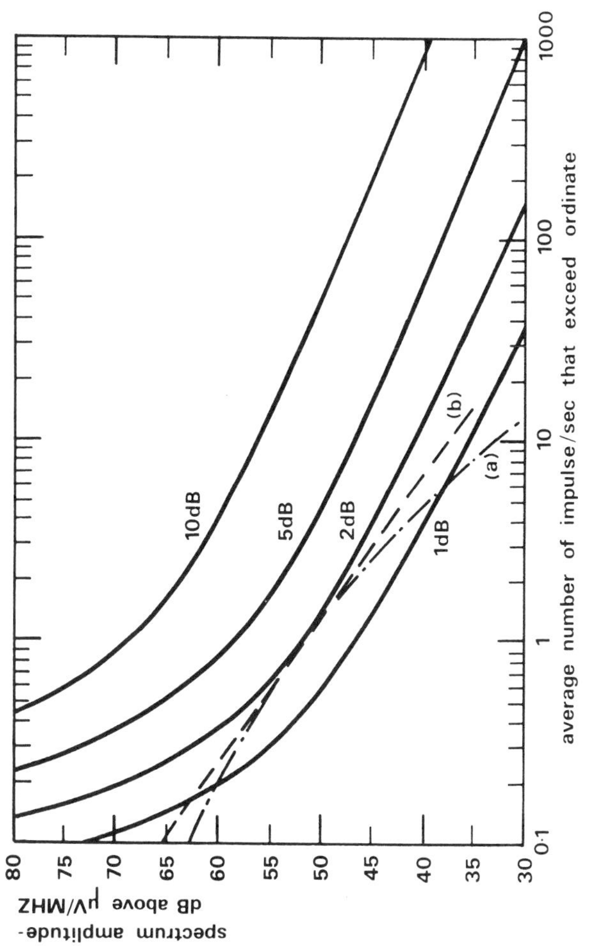

Figure 5.22 Illustrating how measured NAD curves and isodegradation curves can be overlaid to obtain an estimate of system degradation.

can be extended and used in association with isodegradation curves plotted as in Figure 5.17. The method is illustrated in Figure 5.23, from which it can be seen that for the given receiver at a signal level of -108 dBm the point of tangency between the NAD and the isodegradation curves corresponds to an error of approximately 2×10^{-4}.

Experiments [25] have established an empirical relationship between the degradation indicated by the tangency point (termed the tangential degradation) and experimentally-measured error rates at different radio locations. The results indicate that a multiplication factors in the range $2-2.5$ is appropriate, depending upon whether the vehicle is moving in traffic, or is stationary.

5.10.4 The Bello and Esposito technique

The separate formulation of receiver and input noise characteristics provides what is potentially a very powerful and flexible technique for BER prediction. One such method has been described by Bello and Esposito [26], in which it is assumed that each received noise pulse causes an appropriate receiver response, and the output noise is the sum of these responses. Accordingly, the noise output of the receiver 'front-end' was modelled as a summation of impulse responses, each multiplied by a complex amplitude factor related through a constant delay to the arrival time of an input noise pulse. It was further assumed that, in order to determine the probability of error in a particular received digit, it was necessary to take account only of receiver responses to those noise pulses falling within a finite time interval (termed the effective noise occurrence interval) associated with that digit.

A general analysis was given for the appearance of k pulses ($k = 1, 2, 3, \ldots$) within the relevant interval and the probability of error was established for each event. These error probabilities amount to the product of the event probability and the conditional error probability, given the event. This latter probability is derived from a receiver impulse characteristic which is the conditional probability of error given specific values of noise pulse strength and occurrence time, and the joint probability density function of pulse strengths and occurrence times. The phase characteristic of the noise is assumed to be uniformly distributed in the interval $0-2\pi$. The total probability of error is then the sum of the individual probabilities of order k, and can be expressed mathematically by an equation of the form

$$P_e = \sum_{k=1}^{\infty} P_k \int_{\Gamma} \int_{\partial} R_k(e|\mathbf{\Gamma}, \partial) W_k(\mathbf{\Gamma}, \partial) \, \mathrm{d}\mathbf{\Gamma} \, \mathrm{d}\partial \tag{5.6}$$

where

P_k = probability that k noise pulses occur in the effective noise occurrence interval

$\mathbf{\Gamma}$ = vector representing the strengths of noise pulses within the effective noise occurrence interval

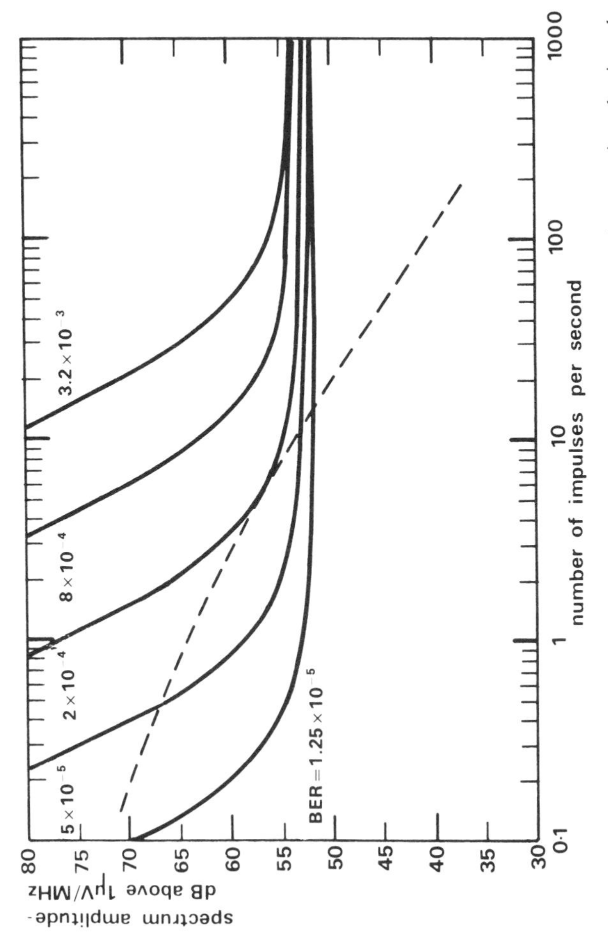

Figure 5.23 Illustrating the application of the overlay technique to a data receiver having the isodegradation curves shown in Figure 5.17.

∂ = vector representing pulse occurrence times
$R_k(e|\mathbf{\Gamma}, \partial)$ = kth order receiver characteristics
$W_k(\mathbf{\Gamma}, \partial)$ = joint probability density function of $\mathbf{\Gamma}$ and ∂ in the kth order case.

Bello and Esposito restricted detailed analysis to the first-order case only, by assuming a maximum of one noise pulse per information digit, that is, they limited analysis to a class of noise processes, symbol durations and receiver configurations for which only the first term in eqn (5.6) is significant. Illustrative computations for binary PSK with and without hard limiting, DPSK and FSK were given, always under the assumption that the receivers used contained post-detection integrate-and-dump circuits [26, 27]. However, no experimental verification of the prediction technique was attempted.

For BER prediction purposes, appropriate practical distributions of impulsive noise strength can be obtained from measurements of the noise amplitude distribution (NAD) with a suitable wideband receiver [28], and the measured NAD can be approximated by a standard statistical distribution [5, 29]. The time distribution was assumed to be adequately described by Poisson statistics.

5.10.5 Application to other receivers

The method has been adapted to a 'sampling' type of data receiver by Edwards *et al.* [30]. A block diagram of the type of receiver considered is shown in Figure 5.19. Unlike the receivers considered by Bello and Esposito, it has no post-detection integrate-and-dump filter, the data extraction decision being taken after sampling the discriminator output at an instant corresponding to the midpoint of each received digit. The occurrence of errors therefore depends on the relative strengths of signal and noise at that instant.

A limiter is included, preceding the discriminator, and as a basis for obtaining an expression for the probability of error in this system it was assumed that two modes of operation are possible, the choice of mode being dependent on the relative magnitudes of signal and noise envelopes at the limiter input. If, at this point, the signal is greater than the noise at the instant of sampling, then it is assumed that no error occurs. On the other hand, if the noise is greater than or equal to the signal, then it is assumed that marks and spaces occur at random with equal probability. If $Q(e|A)$ is the conditional probability of error given the noise magnitude at the limiter input, then the error performance of the non-linear part of the receiver may be expressed as

$$\begin{aligned} Q(e|A) &= 0 \quad & A < S \\ &= 0.5 \quad & A \geqslant S \end{aligned} \tag{5.7}$$

The probability of error depends also, of course, on the amplitude distribution of noise at the limiter input $p_A(A)$. If a single noise pulse occurs within the relevant time interval with probability P_1, then the first-order probability of error is

$$P_e = P_1 \int_A Q(e|A)p_A(A)\,dA = \tfrac{1}{2}P_1 \int p_A(A)\,dA \tag{5.8}$$

Now the noise waveform at the limiter input can be related to the amplitude and phase parameters of a given receiver input noise by the impulse response $h(t)$ of the linear part of the receiver. If, for convenience, the instant of sampling is taken as the time origin and if constant time delays are ignored, then the limiter input noise at the instant of sampling is

$$\Gamma \exp(j\psi)h(\delta)$$

The magnitude of the noise at sampling is therefore $\Gamma|h(\delta)|$.

The required function, $p_A(A)$, can be found by integration of the joint density function of A and δ, so that, assuming δ to be uniformly distributed within the effective noise occurrence interval d_1,

$$p_A(A) = 1/d_1 \int_{d_1} p(A|\delta)\,d\delta \tag{5.9}$$

In practice, A cannot approach infinity, and if it extends only to some maximum value A_m, then,

$$\begin{aligned} P_e &= \frac{P_1}{2d_1} \int_S^{A_m} \int_{d_1} p(A|\delta)\,d\delta\,dA \\ &= \frac{P_1}{2d_1} \int_S^{A_m} \int_{d_1} \frac{1}{|h(\delta)|} p\Gamma\left[\frac{A}{|h(\delta)|}\right] d\delta\,dA \end{aligned} \tag{5.10}$$

For the purpose of computation, it is convenient to normalize the magnitude variable in terms of S and to ensure unity gain for a steady-state signal by the introduction of the gain g. Then

$$P_e = \frac{P_1}{2d_1} \int_1^{a_m} \int_{d_1} \frac{S}{g|h(\delta)|} P_r\left[\frac{aS}{g|h(\delta)|}\right] d\delta\,da \tag{5.11}$$

Further, it is also convenient to normalize the time variable in terms of T, the duration of one information digit, so that

$$P_e = \frac{P_1 T}{2d_1} \int_1^{a_m} \int_{d_1/T} \frac{S}{g|h(T\tau)|} P_r\left[\frac{aS}{g|h(T\tau)|}\right) dr\,da \tag{5.12}$$

Assuming a Poisson occurrence distribution, of average rate v, the event probability in this expression is given by

$$P_1 = vd_1 \exp(-vd_1) = vd_1 \quad \text{(for small } vd_1\text{)} \tag{5.13}$$

In order to provide a specific solution, it is necessary to model the receiver impulse response during the important interval d_1 by a suitable analytical expression. For convenience, Bello and Esposito chose a double-sided exponential which, although it does not accurately represent the envelope shape, is easy to handle mathematically. Edwards *et al.* chose the same function, and hence obtained an expression for P_e which could be computed if the distribution of impulse strengths at the receiver input was known, the solution of course applying only for the particular impulse response assumed. An experimental verification was attempted using specific noise distributions (Rayleigh and log normal). Agreement between prediction and measurement was good at low average impulse rates, but deteriorated as the average impulse rate was increased.

The most probable explanation for this discrepancy between theory and experiment is the inadequacy of the assumed receiver impulse response, and this point has also been investigated [31] by comparing measurements with BER predictions made using different approximations to real filter responses. The basis on which the comparison was made is that the filters should all treat the signal in the same way, i.e. have the same gain at the carrier frequency, and that they should all have the same ratio between the signal response and the peak impulse response.

The impulse response of a narrow band receiver containing a crystal filter is often characterized by 'ringing', and the impulse envelope has a shape similar to part of a sinx/x function. In practical filters, of course there is no response for negative values of time, neither is the relationship between the amplitudes of the various time side-lobes exactly that of the sin x/x function. Nevertheless, this provided a suitable starting point. Because the sin x/x function presents certain difficulties with regard to numerical integration, a simpler approximation in which each lobe is represented by a cosine function was also considered. The aim was not necessarily to find a model which accurately represents the shape of the impulse response envelope; it was considered more important to find a simple, mathematically-tractable model which led to BER predictions of acceptable accuracy. Gaussian and double-sided exponential responses were also considered. Comparison of the predictions with the measured results shows that although all the theoretical models produce predictions having the correct trend, the Gaussian and double-exponential models badly underestimate the true BER value and cannot be considered satisfactory.

References

1. Specifications for CISPR radio interference measuring apparatus for the frequency range 0.15 Mc/s to 30 Mc/s. *CISPR Publication* **1**, IEC, Geneva (1972).
2. Specifications for CISPR radio interference measuring apparatus for the frequency range 25 Mc/s and 300 Mc/s. *CISPR Publication* **2**, IEC, Geneva (1975).
3. Specifications for CISPR radio interference measuring apparatus for the frequency range 10 kHz to 150 kHz. *CISPR Publication* **3**, IEC, Geneva (1975).

4. Shepherd, R.A. *et al.* Measurement parameters for automobile ignition noise. Stanford Research Institute, NTIS PB 247766 (June 1975).
5. Deitz, J. *et al.* Man-made noise. *Rept of the FCC Advisory Committee for Land Mobile Radio Services (ACLMRS)*, Appendix A, Vol. 2, Part 2, (Working Group B-3), November 1967. (Available from NTIS, Springfield, Va., USA, PB 174–278.)
6. Deitz, J. Man-made noise. *Rept to Technical Committee of the Advisory Committee for Land Mobile Services for Working Group 3*, Federal Communication Commission, Washington DC (1986).
7. Disney, R.T. and Spaulding, A.D. Amplitude and time-statistics of atmospheric and man-made radio noise. *Environmental Science Services Administration Technical Rept* **ERL ISO-ITS** 98 (February 1970).
8. Esposito, R. and Buck, R.E. A mobile wide-band measurement system for urban man-made noise. *IEEE Trans.* **COM-21** (11) (1973) 1224–1232.
9. Shepherd, R.A. Measurement of amplitude probability distribution and power of automobile ignition noise at h.f. *IEEE Trans.* **VT-23** (3) (August 1974) 72–83.
10. Parsons, J.D. and Sheikh, A.U.H. A receiving system for the measurement of noise amplitude distribution. *Proc IERE Conf. on Radio Receivers and Associated Systems*, Southampton 1978, (*IERE Conf. Publ.* **40**), 281–287.
11. Turkmani, A.M.D. and Parsons, J.D. A wide band, high dynamic range receiving system for measuring parameters of impulsive noise. *4th Internat. Conf. on Radio Receivers and Associated Systems*, Bangor, July 1986 (*IERE Conf. Publ.* **68**) 25–33.
12. Parsons, J.D. and Sheikh, A.U.H. Statistical characterization of VHF man-made radio noise. *Radio and Electronic Engineer* **53** (3) (March 1983) 99–106.
13. Sheikh, A.U.H. and Parsons, J.D. Frequency dependence of urban man-made radio noise. *Radio and Electronic Engineer* (March 1983), **53**, (3) 92–98.
14. Parsons, J.D. and Turkmani, A.M.D. A new method of assessing receiver performance in the presence of impulsive noise *IEEE Trans.* **COM-34** (11) (1986) 1156–1161.
15. Bello, P.A. and Esposito, R.A. New method for calculating error probabilities due to impulsive noise, *IEEE Trans.* **COM-17** (1969) 367–379.
16. Engel, J.S. Digital transmission in the presence of impulsive noise. *Bell System Tech. J.* **44** (October 1965) 1699–1743.
17. Spaulding, A.D. Determination of error rates for narrowband communication of binary-coded messages in atmospheric radio noise. *Proc. IEEE* **52** (1964) 220–221.
18. Conda, A.M. The effect of atmospheric noise on the probability of error for an NCFSK system. *IEEE Trans.* **COM-13** (1965) 281–284.
19. Shepelavey, B. Non-Gaussian atmospheric noise in binary data, phase coherent communication systems. *IEEE Trans.* **CS-11** (1963) 280–284.
20. Ziemer, R.E. Character error probabilities for M-ary signalling in impulsive noise environment. *IEEE Trans.* **COM-15** 32–44.
21. Spaulding, A.D. and Middleton, D. Optimum reception in an impulsive interference environment—Part I: Coherent direction. *IEEE Trans.* (1977) 910–923.
22. Spaulding, A.D. and Middleton, D. Optimum reception in an impulsive environment—Part II: Incoherent reception. *IEEE Trans.* **COM-25** (1977) 924–934.
23. Parsons, J.D. and Reyhan, T. Prediction of bit error rate in the presence of impulsive noise: a numerical approach using measured noise data. *IEE Proc. Part F*, **132** (5) (1985) 334–342.
24. Shepherd, N.H. Impulsive radio noise measurements (man-made and atmospheric). International Electrotechnical Commission. Technical Committee No. 12, 12F/WGI Part 9, Section 1, June 1978.
25. Turkmani, A.M.D. and Parsons, J.D. Empirical prediction of BER in FSK data communication systems subjected to impulsive noise. (submitted to IEEE).
26. Bello, P.A. and Esposito, R. Error probabilities due to impulsive noise in linear and hard-limited dpsk systems. *IEEE Trans.* **COM-19** (February 1971) 14–20.
27. Bello, P.A. and Esposito, R. The effect of impulsive noise on FSK digital communication. *Arch. Electron. and Ubertragunstech.* **27** (1973) 25–29 (in English).
28. Shepherd, N.H. Noise measurement and degradation of land mobile receiver performance—test program. Automobile Manufacturing Association, Detroit, Michigan (1971).
29. Bruckert, E.J. and Sangster, J.H. The effects of fading and impulsive noise on digital transmission over a land mobile radio channel. Presented at IEEE vehicular technology conference, Dec. 1969.

30. Edwards, J.A., Parsons, J.D. and Albert-Osaghae, V.K.I. Error rate prediction in binary FSK communication systems subjected to impulsive noise. *IEE Proc. Part F.* **128** (6) (1981) 337–341.
31. Turkmani, A.M.D. and Parsons, J.D. Prediction of error rate in FSK data communication systems subjected to impulsive noise. *IEE Proc. Part F* **7** (Dec. 1987) 633–642.

6 Diversity reception

6.1 Introduction

We have seen in Chapter 2 that the various buildings and other obstacles in built-up areas act as scatterers of the signal, and because of the interaction between the several incoming waves, the resultant signal at the antenna is subject to rapid and deep fading. The fading is most severe in heavily built-up areas such as city centres, and the signal envelope often follows a Rayleigh distribution over short distances in these heavily cluttered regions. As the degree of urbanization decreases, so the fading becomes less severe, and in rural areas it is significant only when there are obstacles such as trees close to the vehicle.

A receiver moving through this spatially varying field experiences a fading rate which is proportional to its speed and the frequency of transmission, and because the various components arrive from different directions there is a Doppler spread in the received spectrum. It has been pointed out in Chapter 2 that the fading and the Doppler spread are not separable, since they are both manifestations (one in the time domain and the other in the frequency domain) of the same phenomenon.

One well-known method of reducing the effects of fading is by the use of diversity reception techniques. In principle, such techniques can be applied either at the base station or at the mobile, although different problems have to be solved. The basic idea underlying diversity reception is that if two or more independent samples of a random process are taken, then these samples will fade in an uncorrelated manner. It follows that the probability of all the samples being simultaneously below a given level is very much less than the probability of any individual sample being below that level. In fact, the probability of m samples all being simultaneously below a certain level is p^m, where p is the probability that a single sample is below the level, so it can be seen that a signal composed of a suitable combination of the various samples will have much less severe fading properties than any individual sample alone.

Two questions must be answered. The first of these is how we can obtain these independent samples. Potentially there are several ways in which this can be done; for example, we could use the fact that the scattered path lengths are different at different frequencies to obtain independent samples from transmissions at different frequencies. However, frequency diversity, as it is called, is not a viable proposition for conventional mobile radio, because the coherence bandwidth is quite large (several MHz in some circumstances), and in any case the pressure on spectrum utilization is such that multiple-frequency allocations cannot be made. Other possibilities are polarization diversity, relying on the scatterers to depolarize the transmitted signal, and field diversity, which uses the fact that the electric and magnetic components of the field at any receiving point are uncorrelated, as shown in Chapter 2. Both these methods have their difficulties, however, since it does not appear that there is sufficient depolarization along the transmission path for the former method to be successful, and there are difficulties with the design of antennas suitable for field diversity.

Space diversity (obtaining signals from two or more antennas physically separated from each other) appears to be by far the most attractive and convenient method of diversity reception for mobile radio. There are no antenna problems, since conventional whips can be used. It has been shown in Chapter 2 that, with isotropic scattering, the autocorrelation coefficient of the signal envelope falls to a low value at distances greater than about a quarter wavelength, and so almost independent samples can be obtained at a mobile from antennas sited this far apart. At VHF and above, the distance involved is less than a metre and this is easily obtained within the dimensions of a normal vehicle. At 900 MHz and above, it may be feasible even on hand portable equipment.

6.2 Basic diversity methods

Having obtained the necessary samples, we now come to the second question which is how to process these signals to obtain the best results. We will consider the various possibilities, but must bear in mind that what is 'best' really amounts to deciding what method gives us the optimum improvement, taking into account the complexity and cost involved.

For the majority of communication systems the possibilities reduce to methods which can be broadly classified as 'linear' combiners. In linear diversity combining, the various signal inputs are individually weighted and then added together. If addition takes place after detection, the system is known as a 'post-detection combiner', and if before detection, as a 'pre-detection combiner'. In the latter case, it is necessary to provide a method of co-phasing the signals before addition.

Assuming that any necessary processing of this kind has been done, we can express the output of a linear combiner consisting of N branches as

$$s(t) = a_1 s_1(t) + a_2 s_2(t) \ldots\ldots + a_N s_N(t)$$

$$s(t) = \sum_{k=1}^{N} a_k s_k(t) \tag{6.1}$$

where $s_k(t)$ is the envelope of the kth signal to which a weight a_k is applied.

The analysis of combiners is usually carried out in terms of signal-to-noise ratio (SNR), with the following assumptions [1]:

(i) the noise in each branch is independent of the signal and is additive
(ii) the signals are locally coherent, implying that, although their amplitudes change due to fading, the fading rate is much slower than the lowest modulation frequency present in the signal
(iii) the noise components are locally incoherent and have zero means, with a constant local mean square (constant noise power)
(iv) the local mean-square values of the signals are statistically independent.

Depending on the choice of a_k, different realizations and performances are obtained, and this leads to three distinct types of combiners, namely scanning and selection combiners, equal-gain combiners and maximal-ratio combiners. These are illustrated in Figure 6.1. In the scanning and selection combiners, only one a_k is equal to unity at any time, while the others are all zero. The method of choosing which a_k is set to unity distinguishes the scanning from the selection combiner. In the former, the system scans through the possible input signals until one greater than a preset threshold is found. The system then uses that signal until it drops below the threshold, when the scanning procedure restarts. In the latter, the branch with the best SNR is always selected. Equal-gain and maximal-ratio combiners accept contributions from all branches simultaneously. In equal-gain, all a_k are unity and in maximal-ratio, a_k is proportional to the root mean square signal and inversely proportional to the mean square noise in the kth branch.

While scanning and selection diversity do not use assumptions (ii) and (iii), equal-gain and maximal-ratio combining rely on the coherent addition of the signals against the incoherent addition of noise. This means that both equal-gain and maximal-ratio show a better performance than scanning or selection, provided the four assumptions hold. It can also be shown that, in this case, maximal-ratio, among all linear combiners, gives the maximum possible improvement in SNR, the output SNR being equal to the sum of the SNRs from the branches [2]. However, this is not true when either assumptions (ii) or (iii), or both, do not hold (as might be the case with ignition noise, which tends to be coherent in all branches), in which case selection or scanning can outperform maximal-ratio and equal-gain, especially when the noises in the branches are highly correlated.

In the remainder of this section we briefly review some of the fundamental

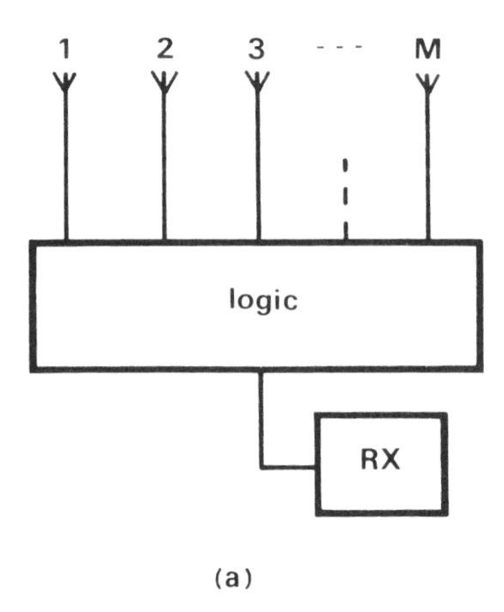

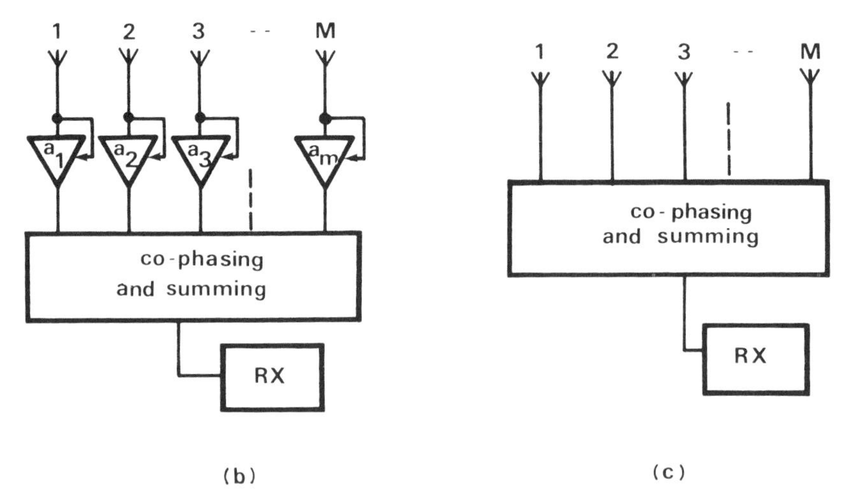

Figure 6.1 Diversity reception systems. (a) Selection diversity; (b) maximal ratio combining ($a_k = r_k/N$); (c) equal gain combining.

results for different diversity schemes. The subject is fully treated by Jakes [3], so the mathematical treatment is not reproduced here.

6.2.1 Selection diversity

Conceptually, and sometimes analytically, selection diversity is the simplest of all the diversity systems. In an ideal system of this kind, the signal with the highest instantaneous SNR is used, so the output SNR is equal to that of the best incoming signal. In practice, the system cannot function on a truly instantaneous basis, so to be successful it is essential that the internal time constants of a selection system are substantially shorter than the reciprocal of the signal fading rate. Whether this can be achieved depends on the bandwidth available in the receiving system. Practical systems of this type usually select the branch with the highest signal-plus-noise, or utilize the scanning technique mentioned in the previous section. This technique will be described in more detail later.

For the moment we examine the ideal selector as illustrated in Figure 6.1*a* and state the properties of the output signal. We assume that the signals in each diversity branch are uncorrelated narrow-band Gaussian processes of equal mean power, so that their envelopes are Rayleigh distributed, and, following the analysis in Chapter 4, the PDF of the SNR can be written as

$$p(\gamma) = \frac{1}{\gamma_0} \exp(-\gamma/\gamma_0)$$

The probability of the SNR on any one branch being less than or equal to any given value γ_g is

$$P[\gamma_k \leqslant \gamma_g] = \int_0^{\gamma_g} p(\gamma_k)\,\mathrm{d}\gamma_k = 1 - \exp(-\gamma_g/\gamma_0) \tag{6.2}$$

and hence the probability that the SNRs in all branches are simultaneously less than or equal to γ_g is given by

$$P_M(\gamma_g) = P[\gamma_1 \ldots \gamma_M \leqslant \gamma_g] = [1 - \exp(-\gamma_g/\gamma_0)]^M \tag{6.3}$$

This expression gives the cumulative probability distribution of the best signal taken from M branches.

The mean SNR at the output of the selector is also of interest and can be obtained from the probability density function of γ_s, as

$$\bar{\gamma}_s = \int_0^{\infty} \gamma_s \cdot p(\gamma_s)\,\mathrm{d}\gamma_s \tag{6.4}$$

where

$$p(\gamma_s) = \frac{\mathrm{d}}{\mathrm{d}\gamma_s} P(\gamma_s) = \frac{M}{\gamma_0}[1 - \exp(-\gamma_s/\gamma_0)]^{M-1} \exp(-\gamma_s/\gamma_0) \tag{6.5}$$

and the suffix S is used to denote selection.

Substituting this into (6.4) gives

$$\begin{aligned}\bar{\gamma}_s &= \int_0^{\infty} \frac{\gamma_s m}{\gamma_0}[1 - \exp(-\gamma_s/\gamma_0)]^{M-1} \exp(-\gamma_s/\gamma_0)\,\mathrm{d}\gamma_s \\ &= \gamma_0 \sum_{k=1}^{M} \frac{1}{k}\end{aligned} \tag{6.6}$$

The cumulative probability distribution of the output SNR is plotted in Figure 6.2 for different orders of diversity. It is immediately apparent that there is a 'law of dimishing returns' in the sense that the greatest gain is achieved by increasing the number of branches from one (no diversity) to two. Moreover, the improvement is greatest where it is most needed i.e. at low

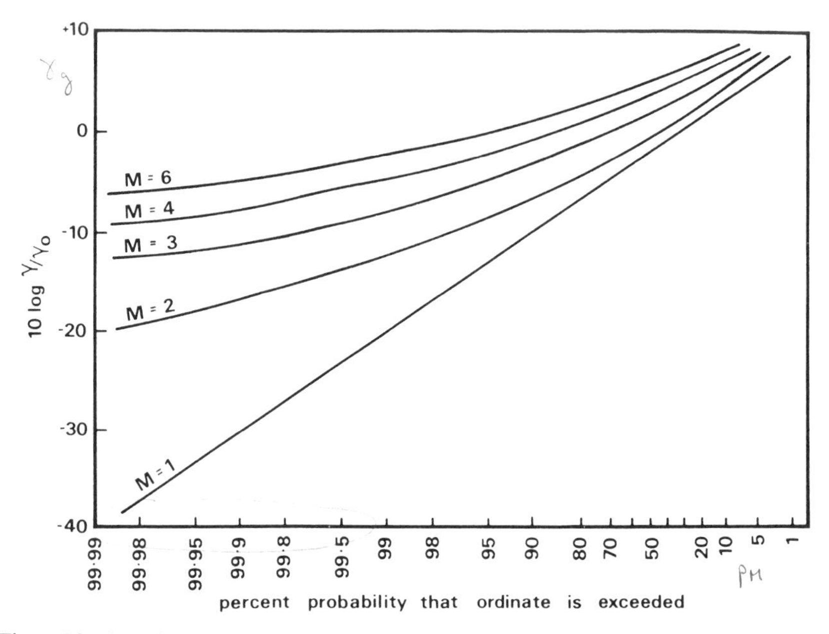

Figure 6.2 Cumulative probability distribution of output SNR for selective diversity systems.

values of SNR. Increasing the number of branches from two to three produces some further improvement, and so on, but the increased gain becomes less for larger numbers of branches. Figure 6.2 shows a gain of 10 dB at the 99% reliability level for two-branch diversity and about 14 dB for three branches.

6.2.2 Maximal-ratio combining

In this method, each branch signal is weighted in proportion to its own signal-voltage to noise-power ratio before summation, as shown in Figure 6.1*b*. When this takes place before demodulation, it is necessary to co-phase the signals prior to combining, and various techniques for achieving this will be described later. Assuming that this has been done, the envelope of the combined signal is

$$r_R = \sum_{k=1}^{M} a_k r_k \tag{6.7}$$

where a_k is the appropriate branch weighting and the suffix R is used to identify maximal-ratio. In a similar way we can write the sum of the branch noise powers as

$$N_{tot} = N \sum_{k=1}^{M} a_k^2$$

so that the resulting SNR is

$$\gamma_R = \frac{r_R^2}{2N_{tot}}$$

Maximal ratio combining was first proposed by Kahn [2], who showed that if the various branches are weighted in the ratio signal-voltage/noise-power (i.e. $a_k = r_k/N$), then γ_R will be maximized and will have a value

$$\gamma_R = \frac{(\sum r_k^2/N)^2}{2N\sum(r_k^2/N)^2} = \sum_{k=1}^{M} \frac{r_k^2}{2N} = \sum_{k=1}^{M} \gamma_k \tag{6.8}$$

This shows that the output SNR is equal to the sum of the SNRs of the various branch signals, and this is the best that can be achieved by any linear combiner.

The probability density function of γ_R is

$$p(\gamma_R) = \frac{\gamma_R^{M-1}\exp(-\gamma_R/\gamma_0)}{\gamma_0^M (M-1)!} \qquad \gamma_R \geqslant 0 \tag{6.9}$$

and the cumulative probability distribution function is given by

$$P_M(\gamma_R) = 1 - \exp(-\gamma_R/\gamma_0)\sum_{k=1}^{M} \frac{(\gamma_R/\gamma_0)^{k-1}}{(k-1)!} \tag{6.10}$$

It is a simple matter to obtain the mean output SNR from (6.8) by writing

$$\bar{\gamma}_R = \sum_{k=1}^{M} \bar{\gamma}_k = \sum_{k=1}^{M} \gamma_0 = M\gamma_0 \tag{6.11}$$

and thus it can be seen that $\bar{\gamma}_R$ varies linearly with the number of branches M. The cumulative distributions for various orders of maximal ratio diversity, plotted from eqn (6.10), are shown in Figure 6.3.

6.2.3 Equal-gain combining

Equal-gain combining is similar to maximal-ratio, but there is no attempt at weighting the signals before addition. A block diagram of the scheme is shown in Figure 6.1*c*. The envelope of the output signal is given by eqn (6.7) with all $a_k = 1$, the suffix E being used to identify equal-gain.

$$r_E = \sum_{k=1}^{M} r_k$$

and the output SNR is therefore

$$\gamma_E = \frac{r^2}{2NM}$$

Of the diversity systems so far considered, equal gain is analytically the most difficult to handle, because the output r_E is the sum of M Rayleigh-distributed variables. The probability distribution function of γ_E cannot be expressed in terms of tabulated functions for $M > 2$, but values have been obtained by numerical integration techniques using a computer. The distribution curves lie in between the corresponding ones for maximal ratio and selection systems, and in general are only marginally below the maximal ratio curves.

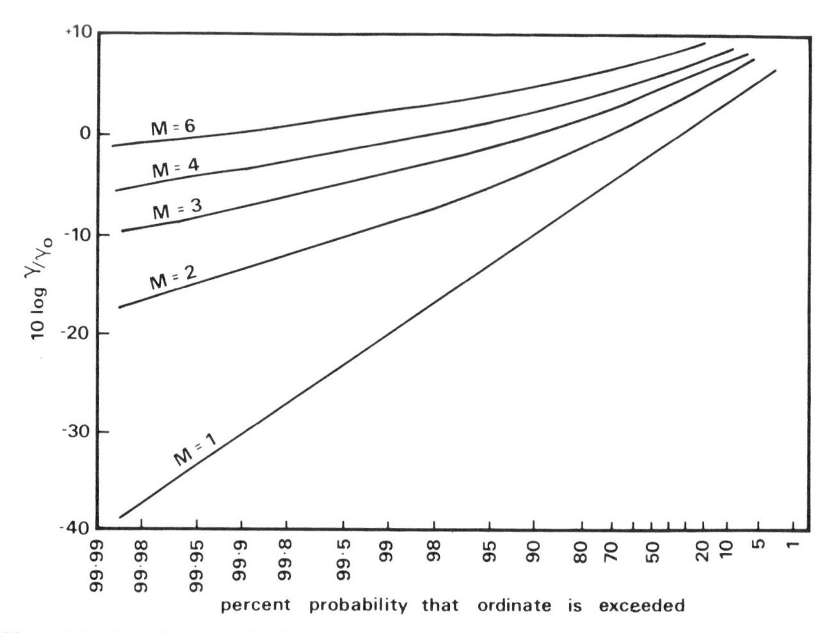

Figure 6.3 Cumulative probability distribution of output SNR for maximal-ratio diversity systems.

The mean value of the output SNR, $\bar{\gamma}_E$, can be obtained fairly easily as

$$\bar{\gamma}_E = \frac{1}{2NM}\overline{\left(\sum_{k=1}^{M} r_k\right)^2}$$

$$= \frac{1}{2NM}\sum_{j,k=1}^{M}\overline{(r_j r_k)} \tag{6.12}$$

We have seen in Chapter 2 that $\overline{r_k^2} = 2b_0 \overline{r_k} = \sqrt{\pi b_0/2}$. Also, since we have assumed the various branch signals to be uncorrelated, $\overline{r_j r_k} = \overline{r_j}\,\overline{r_k}$ if $j \neq k$. In this case (6.12) becomes

$$\bar{\gamma}_E = \frac{1}{2NM}\left[2Mb_0 + M(M-1)\frac{\pi b_0}{2}\right]$$

$$= \gamma_0\left[1 + (M-1)\frac{\pi}{4}\right] \tag{6.13}$$

6.3 Improvements obtainable from diversity

There are various ways of expressing the different improvements obtainable from diversity techniques. Most of the theoretical results have been obtained for the case when the branches have signals with independent Rayleigh-fading envelopes and equal mean SNR.

One useful way of obtaining an overall ideal of the relative merits of the various diversity methods is to evaluate the improvement in average output SNR relative to the single branch SNR. For Rayleigh fading conditions, this quantity, $\bar{D}(M)$, is easily obtained in terms of the number of branches M, from eqns (6.6), (6.11) and (6.13), the results being:

selection: $$\bar{D}(M) = \sum_{k=1}^{M} \frac{1}{k} \quad (6.14)$$

maximal ratio: $$\bar{D}(M) = M \quad (6.15)$$

equal gain: $$\bar{D}(M) = 1 + \frac{\pi}{4}(M - 1) \quad (6.16)$$

These functions have been plotted in the literature (see ref. 3, Chapter 5) and it is apparent that selection has the poorest performance and maximal-ratio the best. The performance of equal-gain combining is only marginally inferior to maximal-ratio, the difference between the two being always less than 1.05 dB (this is the difference when M tends to infinity). It is also clear that the additional improvement obtained decreases as the number of branches is increased, being a maximum when going from a single branch to dual diversity.

Equations (6.14)–(6.16) show that the average improvements in SNR obtainable from the three techniques do not differ greatly, especially in systems using low orders of diversity, and the extra cost and complexity of the combining methods cannot be justified on this basis alone. Looking back at section 6.2, we see that with selection diversity, the output SNR is always equal to the best of the incoming SNRs, whereas with the combining methods an output with an acceptable SNR can be produced even if none of the inputs on the individual branches is itself acceptable. This is a major factor in favour of the combining methods.

6.3.1 Envelope probability distributions

The few dB increase in average SNR which diversity provides is relatively unimportant as far as mobile radio is concerned. If this was all diversity did, we could achieve the same effect by increasing the transmitter power. Of far greater significance is the ability of diversity to 'smooth out' the fades, and reduce the variations in the signal, so that the receiver AGC or limiter can cope. In statistical terms we can use diversity to reduce the variance of the signal envelope and this is something which cannot be achieved just by increasing the transmitter power.

To show this effect we need to look at the first-order envelope statistics of the signal, i.e. the way the signal behaves as a function of time. Cumulative probability distributions of the composite signal have been calculated for Rayleigh-distributed individual branches with equal mean SNRs in the previous paragraphs. For two-branch selection and maximal-ratio systems, the appropriate cumulative distributions can be obtained from (6.3) and (6.10), and for $M = 2$ an expression for equal-gain combining can be written in terms of tabulated functions. The normalized results are of the following form:

selection: $$P_2(\gamma_n) = [1 - \exp(-\gamma_n)]^2 \quad (6.17)$$

maximal ratio: $$P_2(\gamma_n) = 1 - (1 + \gamma_n)\exp(-\gamma_n) \quad (6.18)$$

equal-gain: $$P_2(\gamma_n) = 1 - \exp(-2\gamma_n) - \sqrt{\pi\gamma_n}\exp(-\gamma_n)\mathrm{erf}\sqrt{\gamma_n} \quad (6.19)$$

where γ_n is the chosen output SNR relative to the single branch mean and erf( ) is the error function.

Figure 6.4 shows these functions plotted on Rayleigh graph paper, with the single-branch mean SNR taken as reference, and the single-branch (Rayleigh) distribution also shown for comparison. It is immediately obvious that the diversity curves are much flatter than the Rayleigh curve, indicating a lower variance. To gain a quantitative measure of the improvement, we note that the predicted reliability for two-branch selection is 99% in circumstances where a single-branch system would be only about 88% reliable. This means that the coverage area of the transmitter is far more 'solid', and there are fewer areas in which signal flutter causes problems. This may be a very significant improvement, especially when data transmissions are being considered. To achieve a comparable result by altering the transmitter power would involve an increase of about 12 dB. Apart from the cost involved, such a step would be

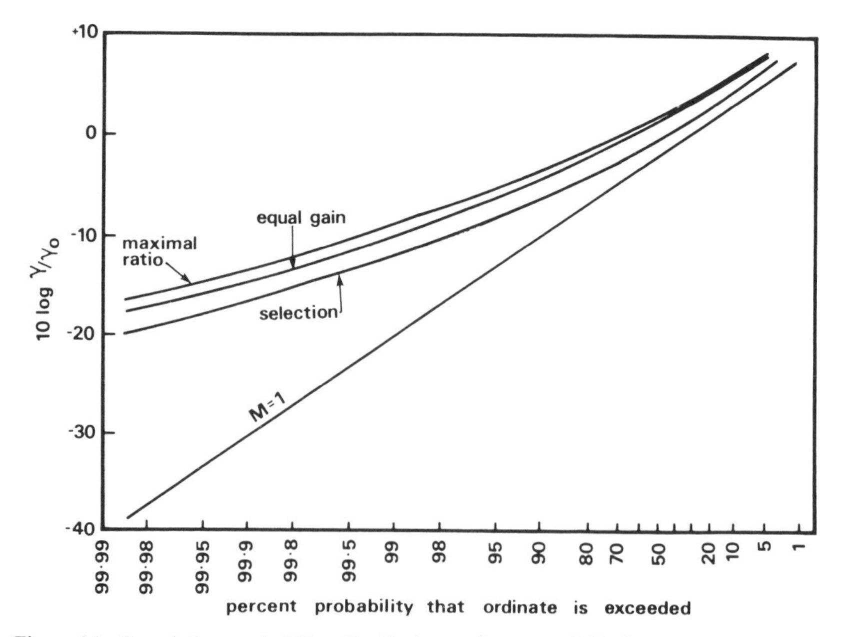

Figure 6.4 Cumulative probability distributions of output SNR for two-branch diversity systems.

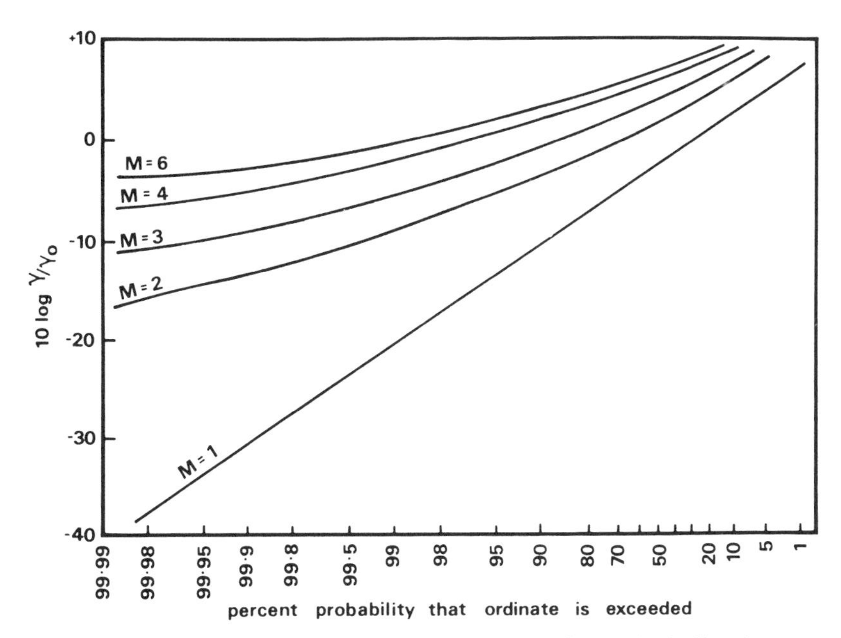

Figure 6.5 Cumulative probability distributions of output SNR for equal-gain diversity systems.

undesirable, since it would approximately double the range of the transmitter and hence make interference problems much worse. Neither would it change the statistical characteristics of the signal, which would remain Rayleigh.

We have already seen that there is a law of diminishing returns in respect of the additional advantages to be gained by increasing the number of diversity branches. This is illustrated further in Figure 6.5 which shows the cumulative probability distributions for various orders of equal-gain combining. Putting the point another way, it can be seen that whereas the use of two-branch diversity increases reliability at the -8 dB level from 88% to 99%, 3-branch increases it to 99.95%, and 4-branch to $>$ 99.99%. It is doubtful whether, at the mobile, it would be possible to economically justify the use of anything more complicated than a 2-branch system. The base station is another matter.

Although the theoretical results above have been derived for the case of uncorrelated Rayleigh signals with equal mean-square values, some attention has been given to non-Rayleigh fading, correlated signals and unequal mean branch powers. Most of the theoretical results available have been obtained for selection and/or maximal ratio systems since these are mathematically tractable, but they are believed to hold, in general terms, for equal-gain combiners.

Maximal-ratio still gives the best performance with non-Rayleigh fading. The performance of selection and equal-gain systems depends on the signal distribution, and, the less disperse the distribution (e.g. Rician with large signal-to-random-component ratio) the nearer equal-gain combining approaches maximal-ratio. In these conditions selection becomes relatively poorer. For more disperse distributions selection diversity can perform marginally better than equal-gain combining, although the average improvement $\bar{D}(M)$ of equal-gain systems is not substantially degraded.

The performance of all systems deteriorates in the case of correlated fading, especially if the correlation coefficient exceeds 0.3. Maximal-ratio continues to show the best performance, and equal-gain approaches maximal-ratio as the correlation coefficient increases, its performance relative to selection also improving. However, some diversity improvement is still apparent even with correlation coefficients as high as 0.8 and it is interesting to speculate on the reasons for this. Fundamentally, as we have already seen, diversity is useful in removing the very deep fades which cause the greatest system degradation. However, in probabilistic terms these deep fades are comparatively rare events—a Rayleigh signal is more than 20 dB below its median level for only 1% of the time! We can anticipate, therefore, that even with two signals which have a fairly high overall correlation, there is a low probability that both will be suffering a rare event (i.e. a deep fade) at the same time. It is likely that much of the diversity advantage will be retained even when significant correlation exists, and this can be seen from Figure 6.6 which shows the cumulative probability distribution function for a two-branch selection diversity system when the inputs have various degrees of correlation.

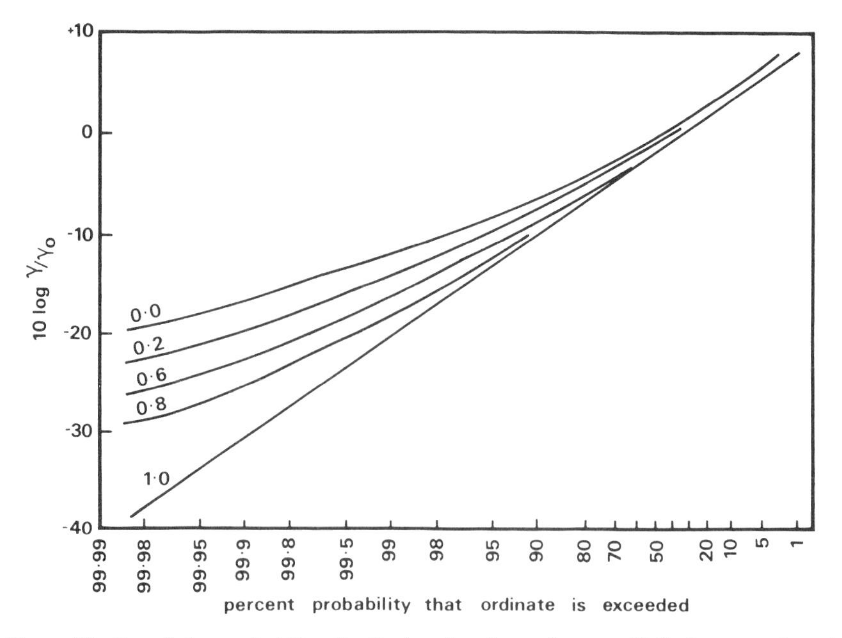

Figure 6.6 Cumulative probability distribution functions of output SNR for a two-branch selection system with various correlation coefficients.

If the signals in the various branches have different mean-square values, a diversity improvement based on the geometric mean (i.e. average of the dB values) of the signal powers is to be expected, at least in the low-probability region of the curves.

6.3.2 Level-crossing rate and average fade duration

The discussion in the previous section was concerned with illustrating the effects of diversity on the first-order statistics of the signal envelope. However some theoretical predictions can also be made about higher-order statistics such as the level crossing rate and the average fade duration.

An early analysis of this problem was due to Lee [4], in which equal-gain combining was investigated. Assuming that the envelope of the combiner output signal and its time derivative are both independent random processes, it was shown that the level crossing rate at a mobile depends on the angle between the antenna axis and the direction of vehicle motion. This can be extended to a unified analysis [5] for other two-branch prediction systems assuming Rayleigh-fading signals; correlation effects can also be included.

In the nomenclature used previously the level crossing rate $N(R)$ and the average fade duration $\tau(R)$ are given by

$$N(R) = \int_0^\infty \dot{r} p(R, \dot{r}) \mathrm{d}\dot{r}$$

$$\tau(R) = \frac{P(R)}{N(R)}$$

If we assume that the noise power N in each branch is the same, and we take this into account so that $r^2/2N$ represents the combiner output SNR, then we can exactly compare the effects of the different diversity systems on the LCR and AFD, of the combiner output r, in the following way:

$$r(t) = \begin{cases} \max\{r_1(t), r_2(t)\} & \text{SC} \\ \dfrac{r_1(t) + r_2(t)}{\sqrt{2}} & \text{EGC} \\ \sqrt{r_1^2(t) + r_2^2(t)} & \text{MRC} \end{cases} \tag{6.20}$$

and hence we obtain

$$\dot{r}(t) = \begin{cases} \begin{array}{l} \dot{r}_1(t), r_1(t) \geqslant r_2(t) \\ \dot{r}_2(t), r_1(t) < r_2(t) \end{array} & \text{SC} \\ \dfrac{\dot{r}_1(t) + \dot{r}_2(t)}{\sqrt{2}} & \text{EGC} \\ \dfrac{r_1(t)\dot{r}_2(t) + r_2(t)\dot{r}_1(t)}{\sqrt{[r_1^2(t) + r_2^2(t)]}} & \text{MRC} \end{cases} \tag{6.21}$$

It can be shown that $\dot{r}(t)$ is a Gaussian random variable, and hence the mean value $m_{\dot{r}}$ and the variance $\sigma_{\dot{r}}^2$ can be found. Hence, for independent fading signals, it can be shown that the level-crossing rate at any level R is given by

$$N_2(R) = \begin{cases} 2\sqrt{2\pi}\, f_m \rho \exp(-\rho^2)[1 - \exp(-\rho^2)] & \text{SC} \\ \sqrt{2\pi}\, f_m \rho \exp(-\rho^2)\left[\exp(-\rho^2) + \dfrac{\sqrt{\pi}}{2\rho}(2\rho^2 - 1)\operatorname{erf}\rho\right] & \text{EGC} \\ \sqrt{2\pi}\, f_m \rho^3 \exp(-\rho^2) & \text{MRC} \end{cases} \quad (6.22)$$

where $\rho = \dfrac{R}{\sqrt{2b_0}} = R/r_{\text{rms}}$.

We recall that for a signal branch

$$N(R) = \sqrt{2\pi}\, f_m \rho \exp(-\rho^2) \quad (6.23)$$

The level crossing rates given by eqn (6.22) confirm that, as expected, diversity substantially reduces the LCR at low levels, but the rate at higher levels is increased.

The effect of correlation between the signals on the two branches is to increase the LCR at low levels. For a diversity system on a mobile, using two antennas with omnidirectional radiation patterns, the correlation (and hence the LCR) depends on both the antenna spacing d and the angle α between the ling joining the two antennas and the direction of vehicle motion. When the antenna spacing is very large, the two received signal envelopes fade independently, but as d is reduced, so the correlation increases until, for very small spacings, the single-branch LCR is approached. It is worth noting that the angle α is more important for large antenna spacings than for small spacings.

As far as the average fade duration is concerned, eqn (2.12) shows that it depends on the ratio between the cumulative distribution function $P(R)$ and the level-crossing rate $N(R)$. Closed-form expressions for the CDF of selection and maximal-ratio systems are available in the literature [3]. Equal gain combining can be approximated by using the CDF for maximal-ratio, replacing the average signal power σ^2 of a single branch with $(\sqrt{3/2})b_0$ [3]. For independently fading signals, eqns (6.17)–(6.19) apply to two-branch systems, and the AFD is then given (replacing γ_n by ρ^2) by:

$$\tau_2 = \frac{[1 - \exp(-\rho^2)]^2}{2\sqrt{2\pi}\, f_m \rho \exp(-\rho^2)[1 - \exp(-\rho^2)]}$$

$$= \frac{1}{2\sqrt{2\pi}\, f_m} \cdot \frac{\exp(\rho^2) - 1}{\rho} \qquad \text{SC} \quad (6.24)$$

and similarly

$$\tau_2 = \begin{cases} \dfrac{1}{\sqrt{2\pi}\, f_m} \cdot \dfrac{\exp(\rho^2) - \exp(-\rho^2) - \sqrt{\pi}\,\rho\,\mathrm{erf}\,\rho}{\rho\exp(-\rho^2) + [2\rho^2 - 1]\sqrt{\pi/2}\,\mathrm{erf}\,\rho} & \text{EGC} \\ \dfrac{1}{\sqrt{2\pi}\, f_m} \cdot \dfrac{\exp(\rho^2) - (1 + \rho^2)}{\rho^3} & \text{MRC} \end{cases} \tag{6.25}$$

Again we recall that for a single branch

$$\tau(R) = \frac{1}{\sqrt{2\pi}\, f_m} \frac{\exp(\rho^2) - 1}{\rho} \tag{6.26}$$

It is apparent from eqns (6.24) and (6.26) that two-branch selection diversity halves the AFD for independent signals, and indeed the result can be generalized to conclude that the average duration of fades is reduced by a factor equal to the number of branches used: $\tau_M = \tau/M$. For equal-gain and maximal-ratio combiners we can infer that a similar result holds.

The effect of envelope correlation is carried through into the results for AFD [5], since these are simply related to those for LCR; however, there are considerable differences for $\alpha = 0$ and $\alpha = \pi/2$. In the latter case (when the antennas are perpendicular to the direction of vehicle motion), the antenna spacing is of far less importance than in the former case.

6.3.3 Random FM

Diversity techniques can also be effective in reducing the random FM present in the signal, but the effectiveness depends upon the manner in which the system is realized. For a single branch, the probability density function of the random FM experienced by a mobile receiver moving through an isotropically scattered field pattern was described in Chapter 2, and for the electric field is given by

$$p(\dot{\theta}) = \frac{1}{\omega_m\sqrt{2}}\left[1 + 2\left(\frac{\dot{\theta}}{\omega_m}\right)^2\right]^{-3/2} \tag{6.27}$$

The analysis leading to an expression for the random FM in a selection diversity system amounts to determining the random FM on the branch which, at any particular time, has the largest envelope, and is a rather complicated procedure. No closed-form expression for the power spectrum is obtainable, but since the baseband frequencies in a narrow band system (300–3000 Hz) are much greater than the spread of the Doppler spectrum, an asymptotic solution as $f \to \infty$ is sufficient. The asymptotic spectra for different orders of selection diversity are shown in Figure 6.7, the calculations having been made on the assumption of a uniform angle-of-arrival Doppler spectrum. To give some idea of the magnitude of the quantities involved, a two-branch

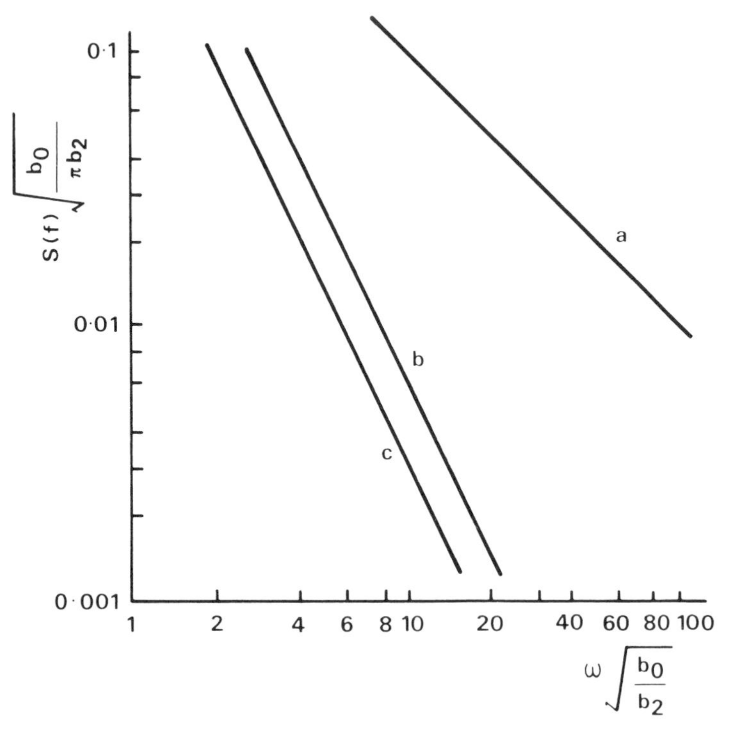

Figure 6.7 Asymptotic spectra for random FM. (*a*) No diversity; (*b*) 2-branch selection; (*c*) 3-branch selection. Note:

$$b_n = (2\pi)^n \int_{fc-fm}^{fc+fm} S(f)(f-f_c)^n \, \mathrm{d}f$$

selection diversity system has an output random FM about 13 dB lower than that of a single branch system. The use of three-branch diversity further improves this to about 16 dB. Selection diversity thus provides a significant reduction provided the highest baseband modulation frequency is much larger than the Doppler frequency.

The effectiveness of the combining methods in reducing random FM is highly dependent on the method of realization. If, during the co-phasing process necessary in predetection combiners, the signals are all co-phased to one of them, then the output random FM is the same as that of the reference branch. If the sum of all the signals is used as the reference the output random FM is reduced. In those systems where a pilot signal near the carrier frequency is transmitted along with the signal, or, if no pilot is sent, some equivalent signal processing is used in the receiver, it is possible to completely eliminate random FM. If the pilot (the carrier can sometimes be used to perform the same function) is sufficiently close to the signal so that its phase fluctuations are the same as those of the signal, then when the two are mixed together the random phase fluctuations are cancelled. Even a single branch receiver using this kind of demodulation process (see section 6.8) would have its random FM completely eliminated.

6.4 Switched diversity

A major disadvantage of implementing true selection diversity as described in section 6.2.1 is the expense of continuously monitoring the signals on all the branches. In some circumstances it is useful to employ a derivative system known as scanning diversity. Both selection and scanning diversity are switched systems in the sense that only one of a number of possible inputs is allowed into the receiver. The essential difference is that in scanning diversity there is no attempt to find the best input, just one which is acceptable. In general, the inputs on the various branches are scanned in a fixed sequence until an acceptable one, i.e. an input above a predetermined threshold, is found. This input is used until it falls below the threshold, when the scanning process continues until another acceptable input is found. Compared with true selection diversity, scanning diversity is inherently cheap to build, since, irrespective of the number of branches, it requires only one circuit to measure the short-term average power of the signal actually being used. Scanning recommences when the output of this circuit falls below a threshold. In this context 'short-term' refers to a period which is short compared with the fading period or, in the mobile radio context, the time taken by the vehicle to travel a significant fraction of a wavelength. A basic form of scanning diversity is shown in Figure 6.8, although of course it is not essential for the averaging

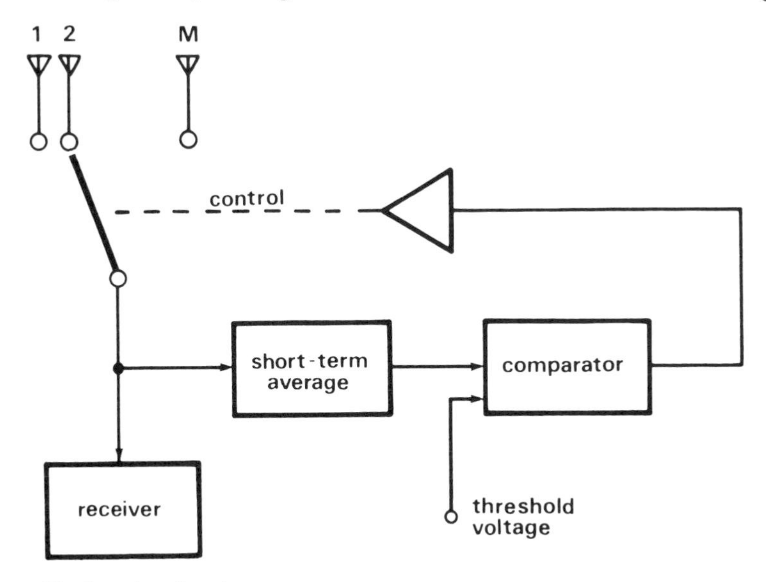

Figure 6.8 Scanning diversity.

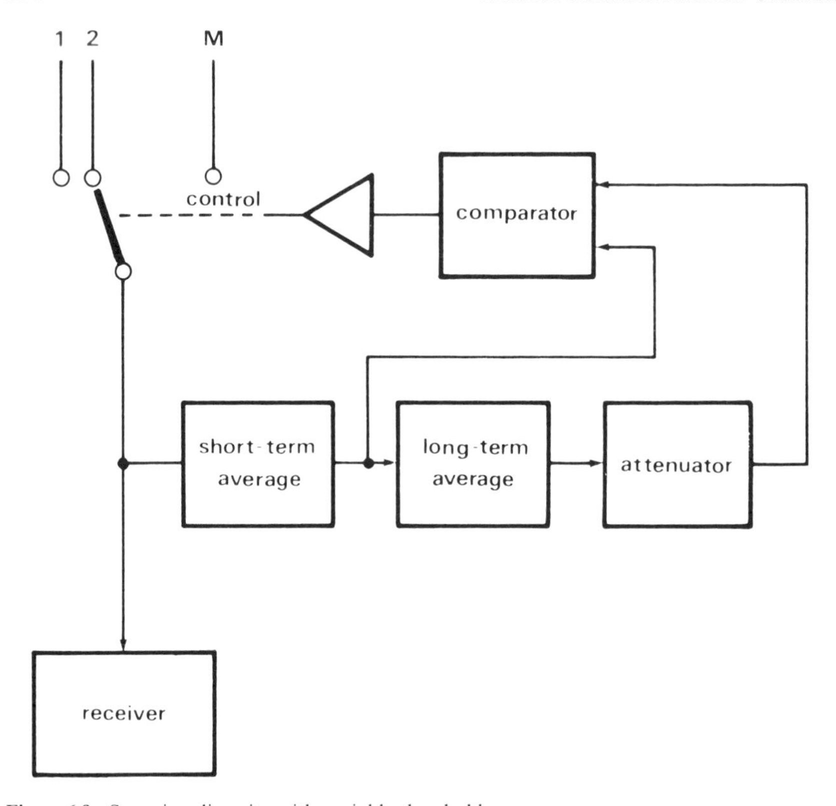

Figure 6.9 Scanning diversity with variable threshold.

circuit to be connected to the front-end of the receiver. In its simplest form, only two antennas are used, and switching from one to the other occurs whenever the signal level on the one in use falls sufficiently to activate the changeover switch. In this form it is commonly known as switched diversity. There is some practical advantage to be gained from the use of a variable threshold because a setting which is satisfactory in one area may cause unnecessary switching when the vehicle has moved to another location where the mean signal strength is different. Figure 6.9 shows a modified system in which the threshold level is derived from the mean signal-plus-noise in the vicinity of the vehicle. The long term average is computed over a period comparable with the time the vehicle takes to travel about ten wavelengths, and the attenuator setting determines the threshold in terms of the mean input level.

Basically, there are two switching strategies which can be used, and these cause different behaviours when the signals on both antennas are in simultaneous fades. Firstly, there is the 'switch-and-examine' strategy which causes the system to switch rapidly between the antennas until the input from one of them rises above the threshold, and secondly, the 'switch-and-stay' strategy, in

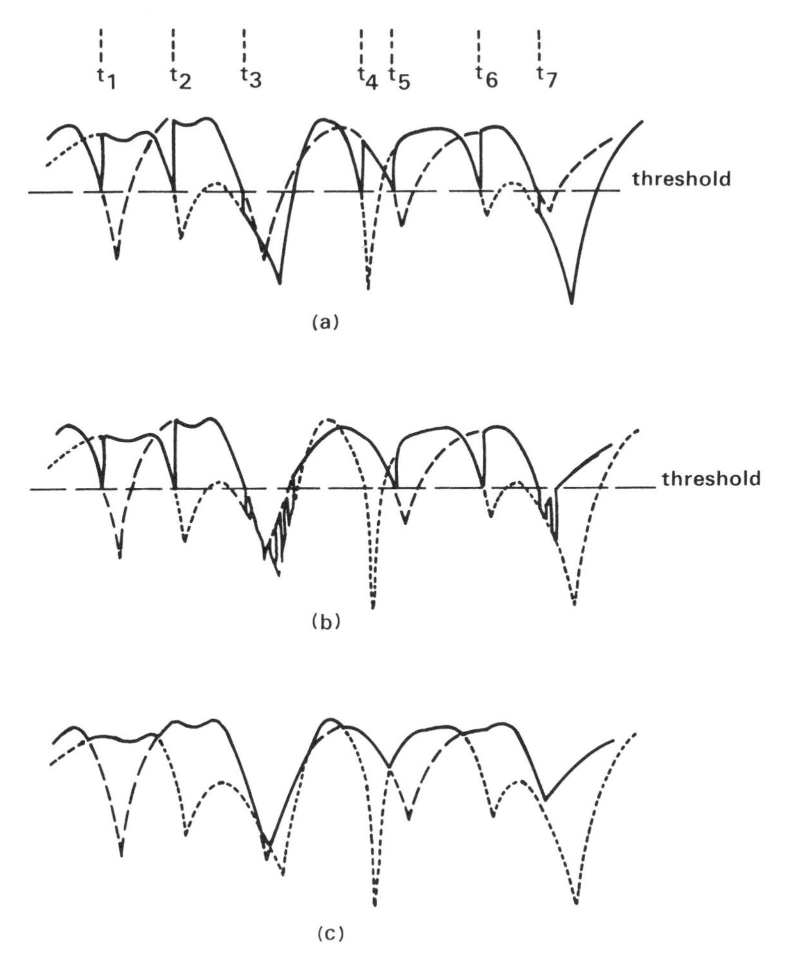

Figure 6.10 Behaviour of switched-diversity systems with different switching strategies compared with selection diversity. ----$r_1(t)$----$r_2(t)$——$r_i(t)$. (*a*) Switch-and-stay strategy; (*b*) switch-and-examine strategy; (*c*) selection diversity.

which the receiver is switched to, and stays on, one antenna as soon as the input on the other falls below the threshold, irrespective of whether the new input is acceptable or not. Figure 6.10 compares the behaviour of diversity systems using these two switching strategies with that of true selection diversity, $r_1(t)$ and $r_2(t)$ are the signal envelopes on the two branches, and $r_i(t)$ is the input signal to the receiver ($i = 1$ or 2). Selection diversity is subject to deep fading only when the signals on both branches fade simultaneously, for example between t_3 and t_4. In addition to this, deep fades can be caused in the switched systems by a changeover to an input which is already below the threshold and with the signal entering a deep fade, for example, following t_7. Although in these cases use of the switch-and-examine strategy allows a marginally quicker return to an acceptable input, it causes rapid switching with an associated noise burst, and for this reason the switch-and-stay strategy is preferable for mobile radio in normal circumstances.

Although the ability of switched systems to remove deep fades is inferior to that of selection, the difference can be made small at low signal levels (where diversity has most to offer) and its inherent simplicity therefore makes switched diversity an attractive proposition for mobile use.

6.4.1 Cumulative probability distribution

To assess the performance of a switched system and to compare it with other forms of diversity, it is necessary to analyse its performance on a statistical basis, and the cumulative probability distribution of signal level is of major interest. Consider a simple two-branch system using the switch-and-stay strategy (as Figures 6.8 and 6.10*a*) in which switchover from one antenna to the other occurs whenever the input on the antenna currently in use falls below a predetermined threshold. Assume that the signal envelopes $r_1(t)$ and $r_2(t)$ are independent Rayleigh processes with identical mean values. Equation (4.28) then describes the SNR on the individual branches of a diversity system. If the system under consideration uses the 'switch-and-stay' strategy and switching occurs at a threshold value r_T, then the plot of output SNR as a function of time is as shown in Figure 6.10*a*.

The input to the receiver, $r_i(t)$, consists of alternate segments taken from $r_1(t)$ and $r_2(t)$ as shown in Figure 6.10*a*. It can be seen that there are two kinds of transition taking place in this waveform. Firstly, there is the kind which is characterized by the fact that the value of $r_i(t)$ immediately following the transition is greater than the value immediately preceding it $(t_1 t_2 t_4 t_5 t_6)$. These we may term 'successful' transitions. Secondly, there is the kind which results in an immediate decrease in $r_i(t)$ and may be termed 'unsuccessful' (t_{3t_7}). During the whole of the segment following a successful transition, e.g. t_4 to t_5, $r_i(t)$ is above the threshold, whereas following an unsuccessful transition $r_i(t)$ is always below the threshold for part of the time, e.g. between t_3 and t_4.

To retain consistency with earlier analysis, we work in terms of the SNR $\gamma(=r^2/2N)$ and use the same suffixes as above. In the segments which follow a successful transition, the probability that $\gamma_i \leqslant \gamma_s$ amounts to the conditional probability that $\gamma \leqslant \gamma_s$ when the signal on the branch in use is above the threshold γ_T. Since $\gamma_1(t)$ and $\gamma_2(t)$ are indistinguishable, we can write

$$\text{prob}[\gamma_i \leqslant \gamma_s] = P[\gamma \leqslant \gamma_s | \gamma > \gamma_T]$$

$$= \begin{cases} \dfrac{P_\gamma(\gamma_s) - q}{1 - q} & \gamma_s \geqslant \gamma_T \\ 0 & \gamma_s < \gamma_T \end{cases} \tag{6.28}$$

where $q = P_\gamma(\gamma_T)$, i.e. q is the probability (given that a transition has taken place) that it is unsuccessful.

Segments which follow an unsuccessful transitions can be split into two parts, corresponding to the time spent below and above the threshold. For the first of these, the probability that $\gamma_i \leqslant \gamma_s$ is the conditional probability that $\gamma \leqslant \gamma_s$ when the signal on the branch in use is below the threshold, so for this period we can write

$$\text{prob}[\gamma_i \leqslant \gamma_s] = P[\gamma \leqslant \gamma_s | \gamma \leqslant \gamma_T] = \begin{cases} 1 & \gamma_s \leqslant \gamma_T \\ \dfrac{P_\gamma(\gamma_s)}{q} & \gamma_s < \gamma_T \end{cases} \tag{6.29}$$

For the remainder of the segment the SNR is above the threshold and eqn (6.28) applies.

To obtain the overall cumulative distribution it is necessary to combine eqns (6.28) and (6.29), with a weighting to account for the relative times over which they apply. Let τ_a represent the average duration of a period when $\gamma(t)$ is above the threshold. Then, since the transitions take place at random times, the average duration of a segment which follows a successful transition is $\tau_a/2$. Similarly, if τ_b represents the average duration of a period when $\gamma(t)$ is below the threshold, then the average duration of segments which follow an unsuccessful transition is $(\tau_b/2) + \tau_a$. The probabilities of successful and unsuccessful transitions are $(1 - q)$ and q, respectively, and the weighting appropriate to each distribution is proportional to probability of occurrence $\times$ average duration. Now eqn (6.28) applies following successful transitions for a time $\tau_a/2$, and following unsuccessful transitions for a time τ_a, and so the appropriate weighting is

$$(1-q)\frac{\tau_a}{2} + q\tau_a = (1-q)\frac{\tau_a}{2}$$

Similarly, eqn (6.29) applies following unsuccessful transitions for a time $\tau_b/2$, and its weighting is

$$\frac{q\tau_b}{2} = \frac{q^2\tau_a}{2(1-q)} \quad \text{since} \quad \tau_b = \frac{q\tau_a}{1-q}$$

The relative, normalized, weighting is therefore

$$\text{eqn (6.28)} : \text{eqn (6.29)} = (1-q^2) : q^2$$

We can therefore combine the distributions to give, for $\gamma_s \geqslant \gamma_T$

$$\text{prob}[\gamma_i \leqslant \gamma_s] = \frac{P_\gamma(\gamma_s) - q}{1-q}(1-q^2) + q^2 \tag{6.30}$$

and for $\gamma_s < \gamma_T$

$$\text{prob}[\gamma_i \leqslant \gamma_s] = \frac{P_\gamma(\gamma_s)}{q} \cdot q^2 \tag{6.31}$$

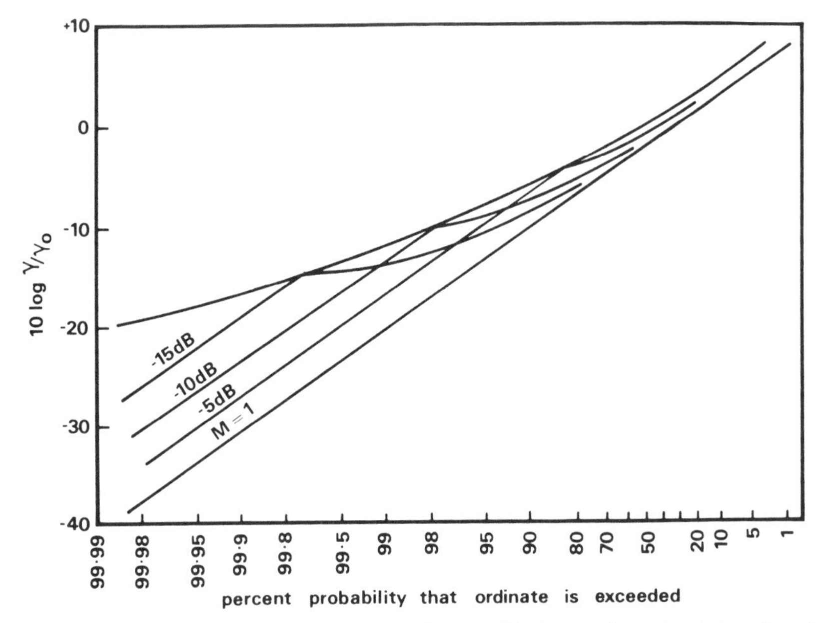

Figure 6.11 Cumulative distribution functions of output SNR for two-branch switched diversity systems using various threshold settings relative to the mean value.

so that we obtain the distribution function

$$\text{prob}[\gamma_i \leqslant \gamma_s] = \begin{cases} (1+q)P_\gamma(\gamma_s) - q & \gamma_i \geqslant \gamma_T \\ qP_\gamma(\gamma_s) & \gamma_i < \gamma_T \end{cases} \tag{6.32}$$

By differentiation, the density function (which has a discontinuity at $\gamma_i = \gamma_T$) is

$$p_i(\gamma_i) = \begin{cases} (1+q)p_\gamma(\gamma) & \gamma_i > \gamma_T \\ qp_\gamma(\gamma) & \gamma_i < \gamma_T \end{cases} \tag{6.33}$$

Theoretical cumulative distributions based on eqn (6.32) are shown in Figure 6.11, with the Rayleigh and ideal two-branch selection-diversity distributions shown for comparison. Switched diversity is, of course, a degenerate form of selection diversity which uses a fixed switching threshold (or a threshold which is simply related to the local mean value of the input). Selection diversity has an adaptive, infinitely variable threshold equal to the instantaneous value of the input not currently in use.

It can be seen that the distributions for switched diversity each touch the selection-diversity curve at the threshold level. Well above this level they merge with the Rayleigh curve, but this is of no consequence, since, at high signal levels, there is very little to be gained. Below the threshold level, the curves have an approximately Rayleigh slope. Switched diversity has most to offer in the region immediately above the threshold level, and it seems clear that the optimum threshold setting is at, or slightly above, the level corresponding to the lowest acceptable SNR. There is no point in a lower setting, because the signal would become unusable before switching took place, and a setting near this level minimizes the number of switches and maximizes the probability of successful transitions. Consider, for example, that in a certain area the signal is usable down to 10 dB below its mean value. Figure 6.11 shows that the use of switched diversity with this threshold setting increases the reliability from 92.4% (single antenna) to 99.4%. The probability of successful transitions at this level is $(1 - q) = 92.4\%$.

6.5 The effect of diversity on data systems

In the earlier parts of this chapter we have used SNR as the criterion by which to judge the effectiveness of a diversity system. This is an important parameter in analogue (particularly speech) transmissions since it is related to the fidelity with which the original modulating signal is reproduced at the system output. However the techniques of selection or combining diversity can equally be applied to the various data transmission formats discussed in Chapter 4 and in these systems fidelity is unimportant, provided the correct decision is made. In other words, in order to assess the effectiveness of diversity on data transmission systems we should determine the reduction in error rate which can be achieved from their use. We will restrict attention to binary FSK and PSK systems which produce results of the greatest interest as far as current mobile radio systems are concerned.

The form of the error probability expressions for FSK and PSK when the signals are subject to additive Gaussian noise have been given in Chapter 4 as

$$P_e(\gamma) = \tfrac{1}{2}\exp(-\alpha\gamma)\begin{cases}\alpha = \tfrac{1}{2} & \text{noncoherent FSK}\\ \alpha = 1 & \text{differentially coherent PSK}\end{cases} \tag{6.34}$$

$$P_e = \tfrac{1}{2}\operatorname{erf}c(\alpha\gamma)\begin{cases}\alpha = \tfrac{1}{2} & \text{coherent FSK}\\ \alpha = 1 & \text{ideal coherent PSK}\end{cases} \tag{6.35}$$

and we now need to examine the way in which these are modified by the use of various diversity systems which have the properties (in the presence of Rayleigh fading) discussed earlier in this chapter.

In Chapter 4 we saw how the expressions given in (6.34) and (6.35) are modified by the effects of fading. The technique used was to write down $P_e(\gamma)$ and integrate it over all possible values of γ, weighting the integral by the PDF of γ. For example, the error rate for non-coherent FSK could be expressed as

$$P_e = \frac{1}{2}\int_0^\infty \exp(-\gamma/2)p(\gamma)\mathrm{d}\gamma \tag{6.36}$$

We now use the same technique for diversity except that in this case, instead of using the expression for $p(\gamma)$ appropriate to Rayleigh fading, we use the expression appropriate to the CNR at the output of the diversity system. For a selection system the output CNR is given by eqn (6.5) as

$$p(\gamma_s) = \frac{M}{\gamma_0}[1 - \exp(-\gamma_s/\gamma_0)^{M-1}\exp(-\gamma_s/\gamma_0)$$

so the BER at the system output is the integral of P_e over all values of γ, weighted by this factor. As an example, consider a two-branch selection system with non-coherent FSK data.

$$P_{e,2} = \frac{1}{2}\int_0^\infty \exp(-\gamma_s/2)\frac{2}{\gamma_0}[1 - \exp(-\gamma_s/\gamma_0)]\exp(-\gamma_s/\gamma_0)\mathrm{d}\gamma_s$$

this is readily evaluated, yielding

$$P_{e,2} = \frac{4}{(2+\gamma_0)(4+\gamma_0)} \tag{6.37}$$

Note if $\gamma_0 \gg 1$ then $P_{e,2} = 4P_{e,1}^2 \quad \left[P_{e,1} = \frac{1}{2+\gamma_0}\right]$

For a maximal ratio combiner, the CNR at the output is given by eqn (6.9) as

$$p_M(\gamma_\mathrm{R}) = \frac{\gamma_\mathrm{R}^{M-1}\exp(-\gamma_\mathrm{R}/\gamma_0)}{\gamma_0^M(M-1)!}$$

which for a two-branch system reduces to

$$p_{M,2}(\gamma_\mathrm{R}) = \frac{\gamma_\mathrm{R}\exp(-\gamma_\mathrm{R}/\gamma_0)}{\gamma_0^2}$$

So, for non-coherent FSK transmissions

$$P_{e,2} = \frac{1}{2}\int_0^\infty \exp(-\gamma_\mathrm{R}/2)\frac{\gamma_\mathrm{R}}{\gamma_0^2}\exp(-\gamma_\mathrm{R}/\gamma_0)\mathrm{d}\gamma_\mathrm{R}$$

Again this is readily integrable, the result being

$$P_{e,2} = \frac{2}{(2+\gamma_0)^2} \tag{6.38}$$

$$= 2P_{e,1}^2 \quad \left[P_{e,1} = \frac{1}{2+\gamma_0}\right]$$

As a numerical example, consider a non-coherent FSK system with a BER of 1 in 10^3 in Rayleigh fading.

Using two-branch selection diversity, the BER is

$$4 \times (1 \times 10^{-3})^2 = 4 \times 10^{-6}$$

and with two-branch maximal ratio combining we get

$$2 \times (1 \times 10^{-3})^2 = 2 \times 10^{-6}$$

Turning now to a consideration of coherent detection systems, we can restrict attention to combining systems, since selection is meaningless in this context, and write that the averaged probability of error in general for these systems is

$$P_e = \frac{1}{2}\int_0^\infty \operatorname{erf}c(\alpha\gamma)p_\gamma(\gamma)\,\mathrm{d}\gamma$$

There is no convenient closed-form solution of this equation, but using the first term of eqn (6.9) to represent $p(\gamma_R)$ at low error rates, we can obtain a solution in the case of maximal ratio when all branches are similar. This solution is

$$P_e = \frac{1}{2(M-1)!}\int_0^\infty \operatorname{erf}c(\alpha\gamma_R)\frac{\gamma_R^{M-1}}{\gamma_0^M}\,\mathrm{d}\gamma_R$$

$$= \frac{1}{2\sqrt{\pi}}\frac{1}{(\alpha\gamma_0)^M}\frac{(M-\frac{1}{2})!}{M!} \tag{6.39}$$

where ! denotes the well-known gamma function.

An exact calculation using the full expression of $p(\gamma_R)$ gives the solid lines of Figure 6.12 for coherent FSK.

For ideal coherent PSK we would expect to obtain the same curves, but at 3 dB lower CNR, since $\alpha = 1$ for PSK compared with $\alpha = \frac{1}{2}$ for FSK. In a similar manner we expect DPSK, for which $\alpha = 1$, to achieve the same error rate at exactly 3 dB less CNR than non-coherent FSK, for which $\alpha = \frac{1}{2}$. To compare coherent and non-coherent systems we can compare DPSK with ideal PSK. It can be shown that, for the same error rate, PSK requires a smaller CNR on each branch, the relative factor being given by

$$\left[\frac{M!(-\frac{1}{2})!}{(M-\frac{1}{2})!}\right]^{1/M}$$

Thus at low error rates in Rayleigh fading conditions, DPSK is 3 dB worse than PSK, a result which was obtained earlier. However, the use of four-branch diversity reduces the difference to about 1 dB, which is the same as the difference under non-fading conditions.

As far as switched diversity is concerned, we can use the same mathematical technique to find the error rate, but care is needed since there is a discontinuity in the PDF at $\gamma_T = \gamma_s$, and different expressions apply above and below the threshold. Thus for switched diversity we can use eqn (6.33) to obtain the error rate as

$$P_e = \int_0^{\gamma_T} qP_e(\gamma)p_\gamma(\gamma)\,\mathrm{d}\gamma + \int_{\gamma_T}^\infty (1+q)P_e(\gamma)p_\gamma(\gamma)\,\mathrm{d}\gamma$$

$$= q\int_0^{\gamma_T} P_e(\gamma)p_\gamma(\gamma)\,\mathrm{d}\gamma + \int_{\gamma_T}^\infty P_e(\gamma)p_\gamma(\gamma)\,\mathrm{d}\gamma + \int_{\gamma_T}^\infty P_e(\gamma)p_\gamma(\gamma)\,\mathrm{d}\gamma$$

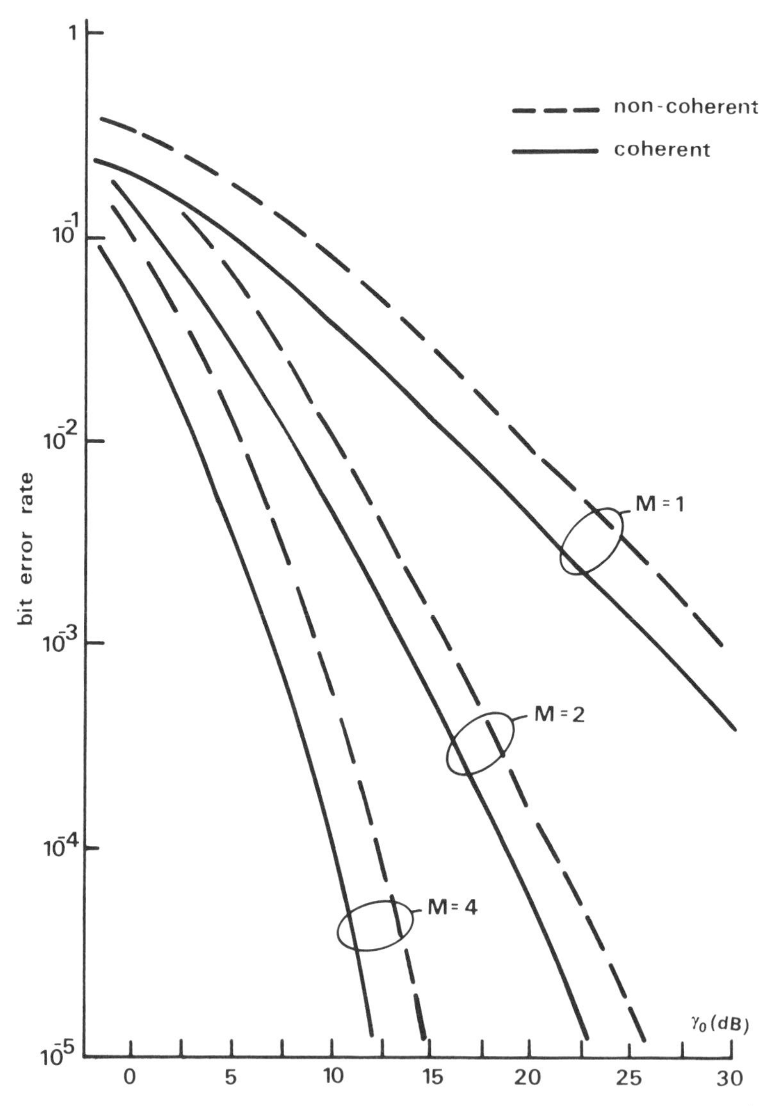

Figure 6.12 Bit error rate for FSK with maximal ratio diversity in the presence of Rayleigh fading.

The first and third terms can be combined to give

$$P_e = q \int_0^\infty P_e(\gamma) p_\gamma(\gamma)\,\mathrm{d}\gamma + \int_{\gamma_T}^\infty P_e(\gamma) p_\gamma(\gamma)\,\mathrm{d}\gamma \tag{6.40}$$

This is an expression in which both terms have the same integrand but the limits and weighting are different.

For non-coherent FSK, $P_e(\gamma)$ is given by eqn (6.34), so using this in eqn (6.40) we obtain

$$P_e = \frac{q}{2\gamma_0}\int_0^{\infty} \exp - \gamma\left[\frac{1}{2}+\frac{1}{\gamma_0}\right]\mathrm{d}\gamma + \frac{1}{2\gamma_0}\int_{\gamma_T}^{\infty} \exp - \gamma\left[\frac{1}{2}+\frac{1}{\gamma_0}\right]\mathrm{d}\gamma$$

$$= \frac{q}{2+\gamma_0} + \frac{1}{2+\gamma_0}\exp(-\gamma_T/\gamma_0)\exp(-\gamma_T/2)$$

But $\exp(-\gamma_T/\gamma_0) = 1 - q$ so

$$P_e = \frac{1}{2+\gamma_0}[q + (1-q)\exp(-\gamma_T/2)] \tag{6.41}$$

We note that if $\gamma_T = 0$, this reduces to $1/(2+\gamma_0)$, the expression for a Rayleigh-fading channel.

Similarly, for DPSK we obtain

$$P_e = \frac{1}{2(1+\gamma_0)}[q + (1-q)\exp(-\gamma_T)] \tag{6.42}$$

Finally in this section, it is worth noting that the ability of diversity systems to reduce the duration of fades implies that another very important advantage to be gained from the use of diversity is a significant reduction in the lengths of error bursts. Rayleigh fading tends to cause a burst of errors when the signal enters a deep fade, and since diversity tends to smooth out these deep fades it not only reduces the error rate, but also affects the error pattern by causing the errors to be distributed more evenly throughout the data stream. This in turn makes the errors easier to cope with, and if error-correcting codes are used to improve error rate, much shorter codes can be used in conjunction with diversity than would be necessary without it. However, it should be pointed out that, with a fixed transmitter power and channel bandwidth, diversity is a far more effective method of error reduction.

6.6 Practical diversity systems

Having established the various advantages that diversity reception has to offer, we now turn to a brief examination of the ways in which practical diversity systems can be realized. There are numerous possibilities [3, 6, 7], but when choosing a system for practical use it is necessary to make a comparison from the point of view of cost and effectiveness. Some of the possibilities realize their full potential only when integrated into the receiver design, and since in existing systems there is a large investment in receivers, it may well be that for them a low-cost system which can be designed in the form of an 'add-on' unit for conventional receivers is preferable to a unit with superior performance that requires an entirely new type of receiver.

Of the three basic schemes, equal-gain seems to be an optimum compromise between the complexity of having to provide branch weighting in maximal-ratio, and the smaller improvement yielded by selection. Also, equal-gain tends to come closer to maximal-ratio and depart from the performance of selection in situations very often encountered in the mobile radio environment, for example where there are correlated signal envelopes or one predominant wave. However, selection can perform better than the two combining systems where coherent noise is present, and this is sometimes the case in urban environments, polluted with man-made noise. Since selection may introduce its own switching noise, it is difficult to assess its practical overall superiority with respect to the other combining methods in this type of medium. Unfortunately, no practical comparative data between the various systems are readily available, and it does not seem that there is one 'ideal' system that will always outperform all others in the mobile radio environment. The remainder of this chapter is a brief survey of some techniques that have been used, or proposed, for mobile radio diversity systems. It is not intended as a comprehensive account–space restrictions would not permit such a venture.

However, before describing some of the practical techniques that can be used to implement diversity reception it is worth returning briefly to the question of predetection and postdetection systems. The distinction between the two was made at the beginning of section 6.2, but has not been apparent in the discussion above. Nevertheless, leaving aside selection and switched systems for the moment, there are very sound reasons, if a combiner system is to be used, to implement a predetection system. This is particularly so when analogue FM modulation is used.

In principle, it is irrelevant whether the signals are combined before or after demodulation when the demodulation process is linear, but of vital importance in any system where the detector has threshold properties (e.g. FM discriminators). This is because diversity methods using combining can produce an output SNR which is better than that of any of the inputs. If we have a number of branch signals, all of which are below the detector threshold, we should combine them before detection in order to produce a signal which is above the threshold. In this way we not only gain the diversity advantage, but also fully exploit the characteristics of the detector in further improving the SNR. This is obviously not the case when post-detection combining is used.

6.7 Predetection diversity

Diversity systems which coherently combine the input signals produce an output with better statistics than those which use selection or switching. They also have the potential to completely eliminate random FM, there are no switching transients, and in general as far as performance is concerned are superior, provided the four conditions mentioned in section 6.2 hold. The obvious disadvantage is cost, since they require more hardware; signal

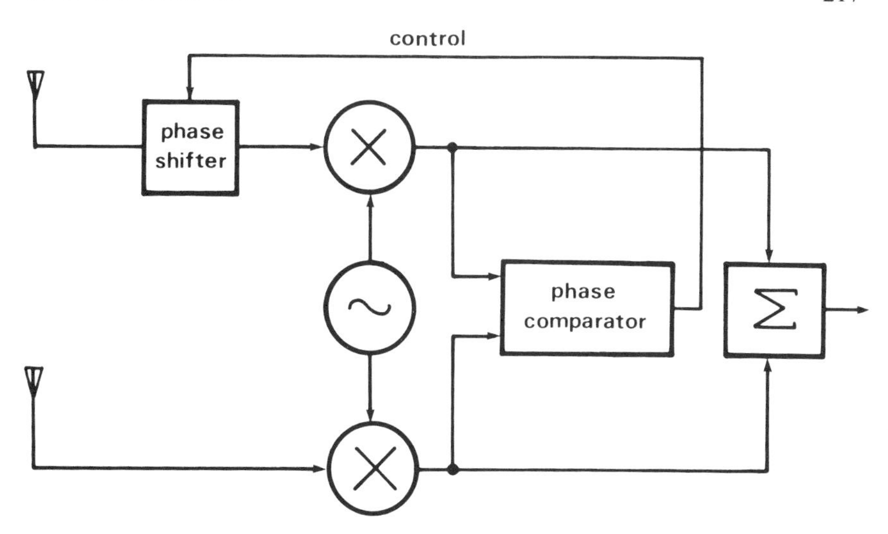

Figure 6.13 Principle of co-phasing, using a variable phase-shifter.

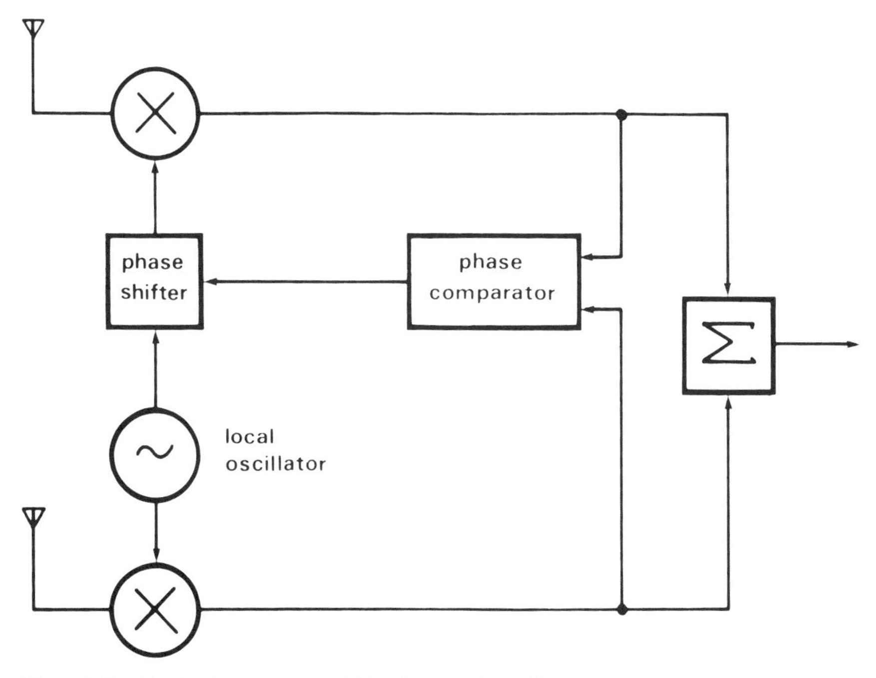

Figure 6.14 Co-phasing using a variable-phase local oscillator.

combining usually takes place at IF, so multiple front ends (RF amplifiers and mixers) are nearly always required.

The principle of one co-phasing method is illustrated in Figure 6.13, and several diversity systems have been based on variations of this theme. The signal from one antenna is passed through a phase shifter, and both it and the signal from the other antenna are down-converted using a common local oscillator. A phase comparator, operating at IF, senses any phase difference between the two signals and its output is fed back to control the phase shifter setting. An early version of this type of receiver [8] is shown in Figure 6.14. The need for a phase-shifter in the signal path is avoided by using the phase comparator output to control the phase of the approximate local oscillator, thus producing an equal-gain combiner. A more practical system is one in which the phase shifter is adjusted, as before, by a control voltage derived from the relative phase between the two RF signals (or between one of them and their sum) but all the processing is done at low frequencies [9]. The manner in which the control signal is derived is illustrated in Figure 6.15 which shows that if two RF signals are added together after one of them has been subjected to a small-deviation phase perturbation, then their sum contains an amplitude-modulation component at the perturbing frequency. This component can be detected by an AM receiver and filtered out from the information modulation. In practice, a low-frequency sinusoidal oscillator drives a small-deviation phase-perturbing network which is connected in series with the phase-shifter. The phase of the induced amplitude perturbation changes by 180° as the perturbed signal changes from lagging to leading the sum signal. When the RF signals are co-phased, only a small second harmonic is present, and a 'sense-of-correction' indication can therefore be obtained by low-frequency phase detection of the amplitude perturbations detected by the receiver, with the original perturbing oscillator. This procedure produces a positive or negative DC control signal that is used to determine the direction in which the phase-shifter must be adjusted in order to approach the optimum position [10]. The method is readily extended to M branches, provided $(M - 1)$ control circuits are used, each with its own perturbing frequency. For FM receivers, a dual technique has been proposed in which small-amplitude perturbations are applied to the RF signals [11]. The resulting phase perturbations in the sum signal can be detected by the receiver, and similar signal processing can be used to provide the control information. Various practical embodiments of the technique have been described for AM systems and a typical system is shown in Figure 6.16.

The major disadvantage in the practical systems is that, since it is not easy to build an analogue 360° phase-shifter for insertion in the RF path, it is usual to use a unit that is quantized into a number of steps. However, the association of a quantized phase-shifter with a 'sense-of-correction' detector leads to a situation in which the phase-shifters oscillate about their optimum position. This is due to the inability of such systems to detect the optimum position, the polarity of the control signal being reversed only after that position has been passed. This oscillation can be eliminated by using a true phase measurement scheme, and a compatible direct-reading method has been proposed [12]. In this method, the RF signals are single-sideband modulated without carrier

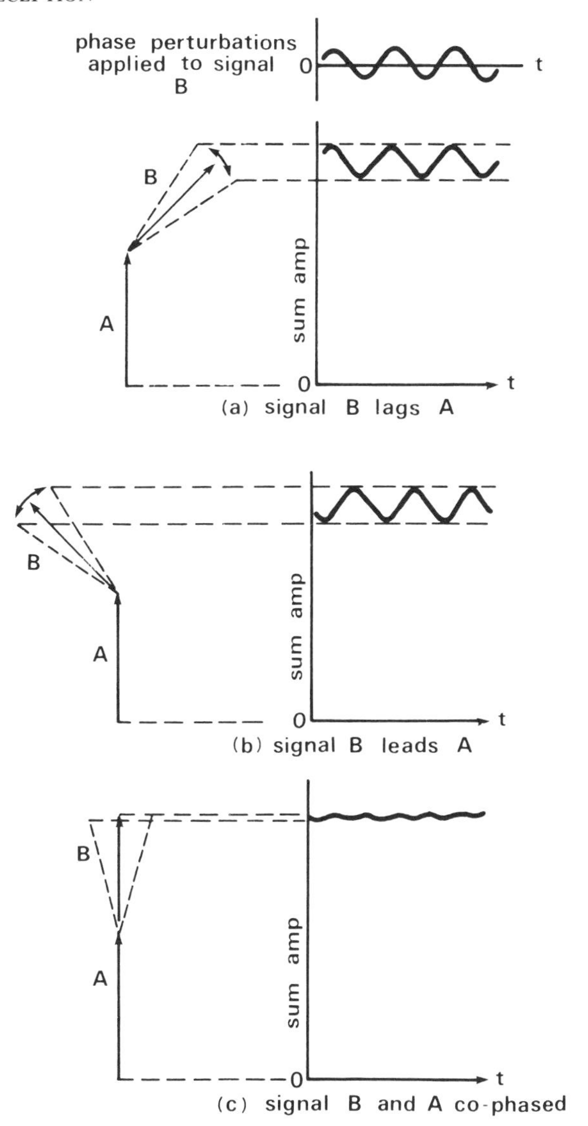

Figure 6.15 Method of generating an amplitude-modulation component by phase perturbations.

suppression by different low-frequency signals. The phases of the introduced sidebands are transposed on to the sum signal, and thus the phase of the detected modulation differs from that applied, by exactly the phase angle between the appropriate RF signal and the sum. It is thus possible to generate control signals which can be used to set the phase shifters.

Systems of this kind are mainly limited by the maximum rate of phase compensation attainable. This is determined by the number of input branches

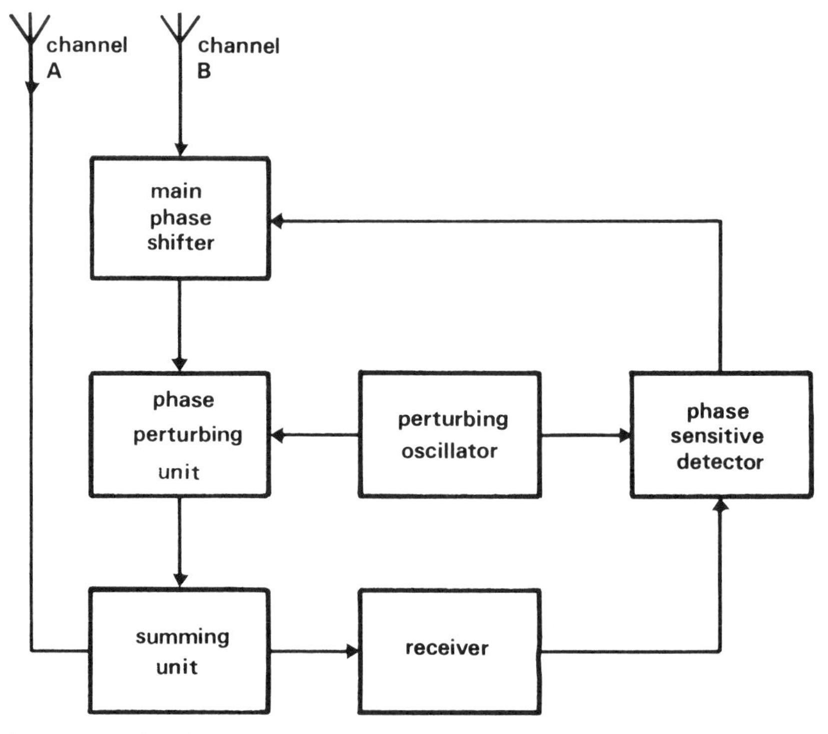

Figure 6.16 Predetection combiner using phase perturbations to derive control signals at low frequencies.

and the total processing bandwidth available for the perturbing signals, since these factors determine the response times of the filters used to isolate these signals after detection. The perturbing frequencies must be such that the control information passes through the receiver without causing interference to the wanted information sidebands, and sub-audio frequencies are commonly used. However, the increase in losses and the degradation in both noise figure and intermodulation performance caused by using units such as RF phase shifters and perturbation networks prior to the RF amplifiers, means that in normal circumstances the use of multiple front ends is to be preferred. This does not rule out the perturbation methods *per se* since the perturbation can be introduced either into the signal path after the RF amplifier, or added to the local oscillator outputs. However, in this case, if the perturbation contains amplitude modulation components, true multipliers must be used as mixers.

6.7.1 Phase-sweeping or 'mode averaging' diversity

A diversity system originally devised for HF communication links [13] has been considered for use in mobile radio systems [14]. The name 'mode averaging' originates from the initial use of the system to average over the

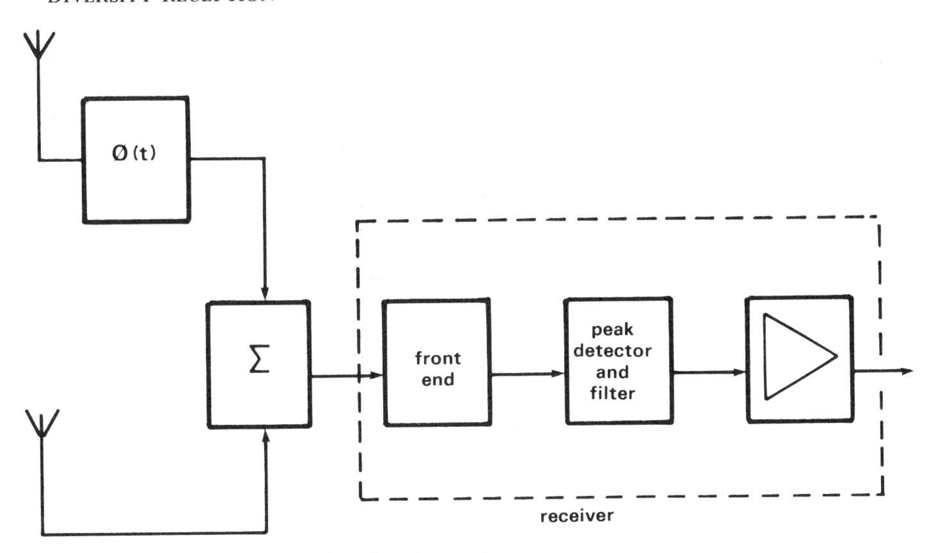

Figure 6.17 Basic phase-sweeping diversity receiver.

several modes of propagation in HF links, and there are various methods of implementation.

The principle of operation can be explained by considering that part of Figure 6.17 prior to the receiver. If the phase shifter is continously swept over the range 0–2π, then there are instants in time when the two signals come into phase and a peak occurs. If the rate at which the 2π phase excursion repeats is at least twice the highest modulation frequency expected in the received signal, then peaks occur at a rate to satisfy the sampling theorem. By using a peak detector and a suitable filter, demodulation of AM signals is possible giving the improvement expected from a diversity technique.

The way in which the phase-shifter sweeps out the 2π radians is of importance to avoid serious spectrum spreading and hence simplify signal retrieval while maintaining SNR. The best choice is to cause the phase to change linearly with time; this can be achieved by using a repetitive saw-tooth phase modulation, and is equivalent to frequency translating the whole of the spectrum passing through the phase-shifter by an amount $1/T$, where T is the repetition period. An experimental system as shown in Figure 6.17 has been evaluated, using a staircase approximation to give the required phase shift [14].

The phase-sweeping method becomes particularly attractive in association with a multiple-front-end receiver. Since the linear phase sweep is equivalent to a frequency shift, the desired effect can be achieved by connecting the antennas to front-ends which have local oscillators at different frequencies. To obtain good adjacent channel rejection, a suitably-tuned IF amplifier is used in each branch, and the signals are summed before demodulation and filtering [15].

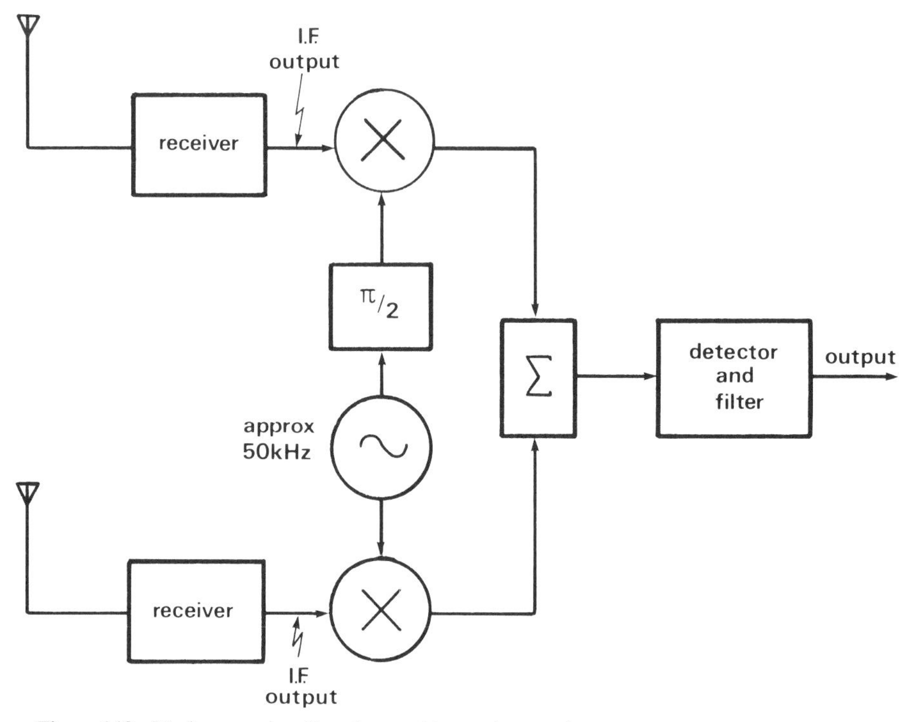

Figure 6.18 Mode averaging diversity combiner using quadrature oscillators.

An alternative method of implementation using two receiver front-ends is shown in Fig. 6.18. After amplification and down-conversion, the two IF signals are mixed with the outputs of quadrature oscillators and the mixer outputs are added and detected. The detector and low-pass filter produce a running average of the summed IF signal and this exhibits a much reduced fading depth.

6.8 Diversity systems using special receivers

A whole class of predetection combiners can be derived from a circuit sometimes known as the heterodyne phase-stripper. This circuit, shown in the block diagram of Figure 6.19, has the property that the phase of the output signal is independent of the phase of the input. To illustrate the action of the circuit, let us assume an input signal of the form

$$s_1(t) = R(t)\cos[\omega_c t + \phi(t) + \phi_0] + R_p \cos(\omega_p t + \phi_0) \tag{6.43}$$

in which the first term represents an amplitude and phase-modulated carrier, and the second is a pilot tone at a frequency sufficiently close to the carrier that it is valid to assume that the propagation medium affects the phase of both in the same way. The signal is multiplied by a local oscillator signal

$$s_2(t) = A\cos(\omega_0 t + \theta) \tag{6.44}$$

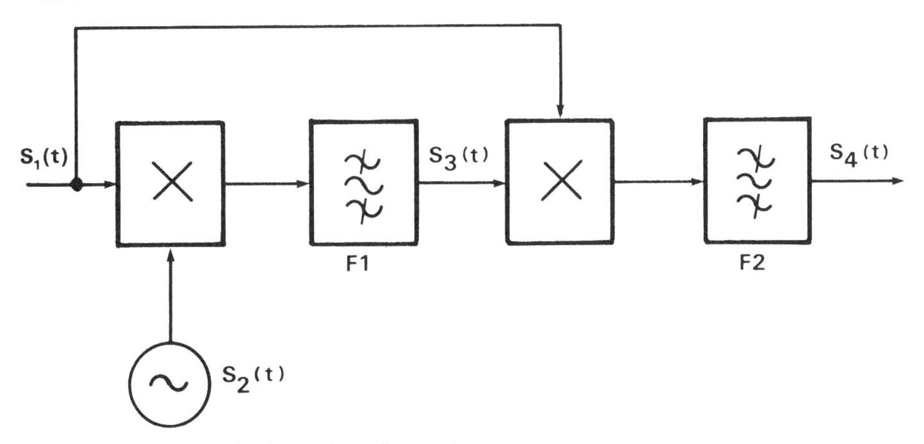

Figure 6.19 The double-heterodyne phase stripper.

at a frequency lower than that of the carrier, and the resultant signal is passed through a very narrow bandpass filter $F1$ which is centred at a frequency equal to the difference between the local oscillator and either the pilot-tone or the carrier. If the centre frequency corresponds to the latter case, $F1$ should be narrow enough to reject all modulation associated with the carrier. The output of $F1$ will then be of the form

$$s_3(t) = R_F \cos(\omega_F t - \theta + \phi_0) \tag{6.45}$$

with ω_F equal to either $(\omega_c - \omega_0)$ or $(\omega_p - \omega_0)$, and this is now multiplied by the fed-forward signal $s_1(t)$. Filter $F2$, which is wide enough to pass the modulation sidebands of the output, has a centre frequency given by the difference between the centre frequency of $F1$ and either the pilot-tone or the carrier. If the pilot-tone frequency is chosen to determine the centre frequencies of the filters, then the output of the circuit is

$$s_4(t) = k\{R(t)\cos[(\omega_c + \omega_0)t + \phi(t) + \theta] + R_p \cos(\omega_0 t + \theta)\} \tag{6.46}$$

and if the carrier frequency is chosen, the output is

$$s_4(t) = k\{R(t)\cos[\omega_0 t + \phi(t) + \theta] + R_p \cos[(\omega_p - \omega_c + \omega_0)t + \theta]\} \tag{6.47}$$

It can be seen that in both cases the static input phase ϕ_0 has been cancelled at the output, the phase of the local oscillator has been acquired, and the modulation has been preserved.

There are various alternative choices which can be made for the local oscillator and filter frequencies in order to obtain this cancellation of ϕ_0, the output being always at the local oscillator frequency. For example, if the frequency of the local oscillator is higher than that of the carrier and $\omega_F = \omega_0 - \omega_c$, then $F2$ is centred at $\omega_c + \omega_F$ to obtain cancellation, and the local oscillator frequency is regenerated at the output. However, in most practical embodiments of this technique it is usual to perform an initial frequency down-conversion, and then to use the circuit described with a local

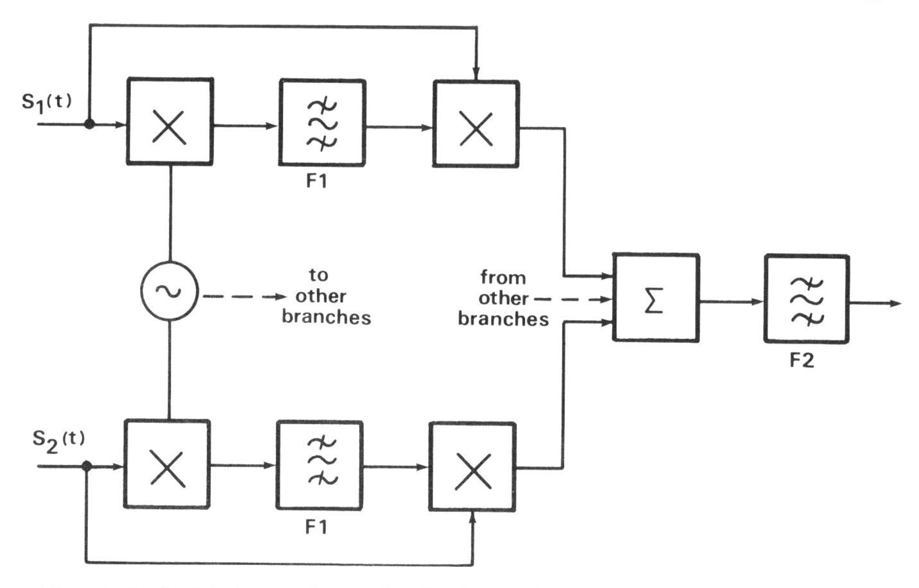

Figure 6.20 Predetection combiner using the phase-stripper principle.

oscillator at another lower frequency. In this way phase-stripping and conversion to a second intermediate frequency are both accomplished at the same time.

Since the phase of the output signal is that of the local oscillator, applying the same local oscillator to the various branches of a diversity reception system will automatically co-phase the various signals, provided that the 'static' phases vary slowly, or in other words that the fading rate is slow compared with the bandwidth of filter $F1$. In this case the signals can be added after the second multiplier, with filter $F2$ following the summing unit, and a predetection combiner results. Figure 6.20 shows this arrangement with only two diversity branches fully drawn.

Systems using this principle can be broadly classified into two categories: systems with and without pilot-tones. The following sections describe different methods of implementation.

6.8.1 Pilot-tone systems

The idea of the phase-stripper was originally applied to a diversity system that used a pilot-tone located in the middle of the baseband to compensate for phase changes associated with the information modulation [16]. The filters $F1$ in Figure 6.20 selected a frequency determined by the pilot-tone. However, if the pilot-tone is located outside the base-band, and can be recovered by $F1$, the local oscillator is not needed, since the recovered pilot-tone can be used in its place. In this case $F2$ can be a low-pass filter and its output is

$$s_4(t) = k\{R(t)\cos[(\omega_c - \omega_p)t + \phi(t)] + R_p\} \tag{6.48}$$

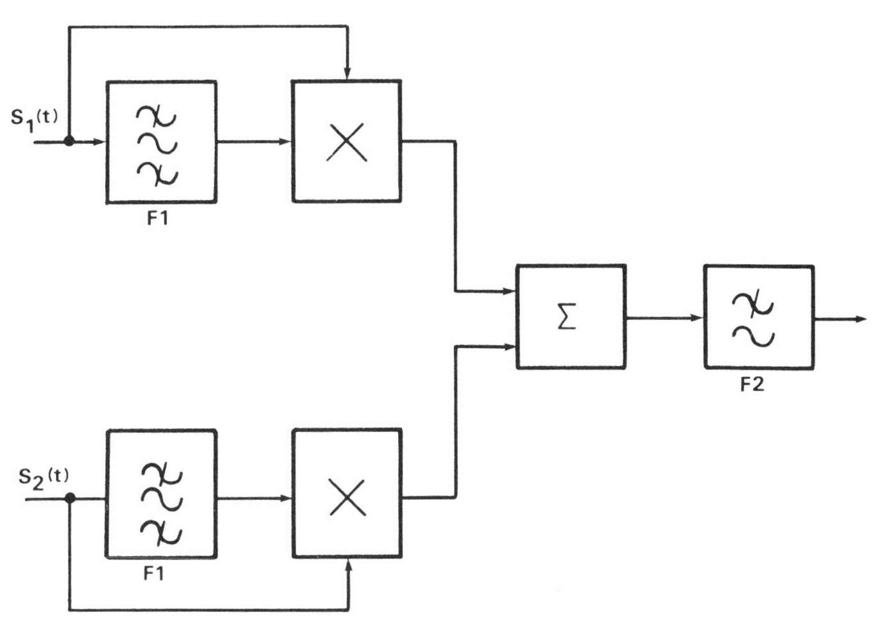

Figure 6.21 Phase-stripper combiner using a transmitted pilot tone in place of the local oscillator.

A block diagram of such a system [17] is shown in Figure 6.21.

In other embodiments of this principle, the filters $F1$ are eliminated altogether, and the remaining multipliers are replaced by either true squarers [18] or by devices whose characteristic exhibits a second-order non-linearity [17]. Filter $F2$ then selects the difference frequency $(\omega_0 - \omega_p)$ which is subsequently demodulated.

6.8.2 Systems without pilot-tone

It has been pointed out [19] that the principle of the phase-stripper can be applied to signals which contain no pilot-tone, provided a carrier is present in the received signal, and this is also apparent from the equations at the beginning of section 6.8. However, for the system to function correctly as a predetection combiner, the filters $F1$ must reject all the information sidebands, and care must be taken to isolate the local oscillator from the output of the circuit [20, 21]. It can be shown that in this case the circuit of Figure 6.20 is a maximal-ratio combiner, but if the second set of multipliers is replaced by mixers, it becomes an equal-gain combiner.

The fact that all the input signals are reduced to a common frequency and phase equal to those of the local oscillator points the way to an interesting solution [22, 23] which historically preceded those already mentioned. The combined output signal, suitably amplified, can be used as the local oscillator, forming a regenerative closed loop. Figure 6.22 shows such a system with

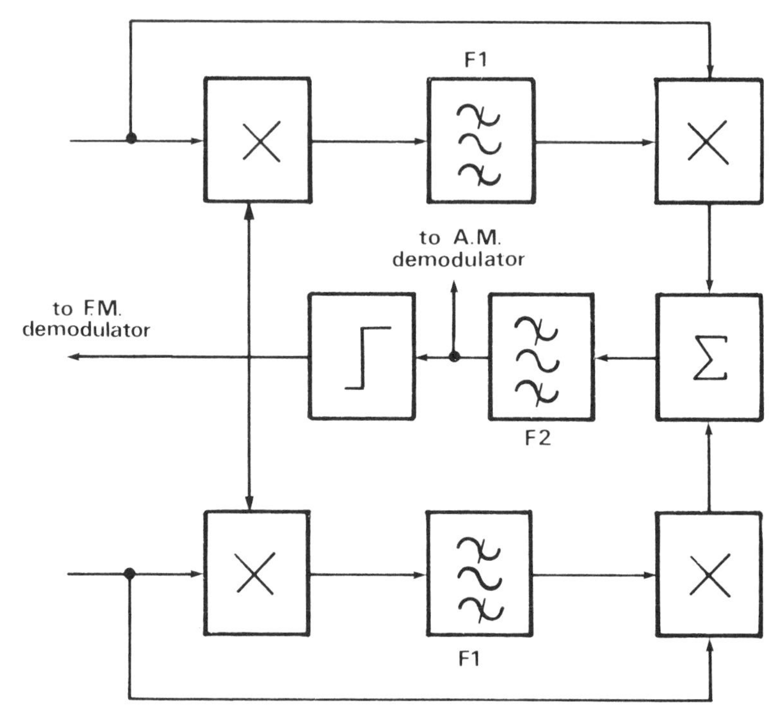

Figure 6.22 Closed-loop phase-stripping combiner with limiter (Granlund receiver).

outputs indicated for AM and FM demodulators. This solution has two main advantages over systems using a local oscillator. Firstly, good isolation between the local oscillator and the output is no longer necessary, and secondly, for FM signals the filters $F1$, although narrowband, do not need to be very sharp. The latter advantage results from the presence of modulation at both inputs to the first multipliers, and since $F1$ selects the difference frequency, modulation components in the range of interest will cancel. A theoretical analysis of this type of combiner has been published [24].

6.9 Switched diversity

Of all the diversity systems so far described, switched diversity is probably the simplest, and is directly compatible with existing standard receivers. The threshold voltage has to be obtained from a suitable point within the receiver, but this is normally a simple matter. In an experimental AM system [25], a suitable voltage was derived from the receiver AGC line; however, many modern FM receivers use integrated circuit IF amplifiers/demodulators, and these devices often have available a DC voltage proportional to the logarithm of the input signal strength (i.e. the signal strength in dB). The system shown in Figure 6.23 is of this type, and operates with a fixed threshold. Alternatively, if

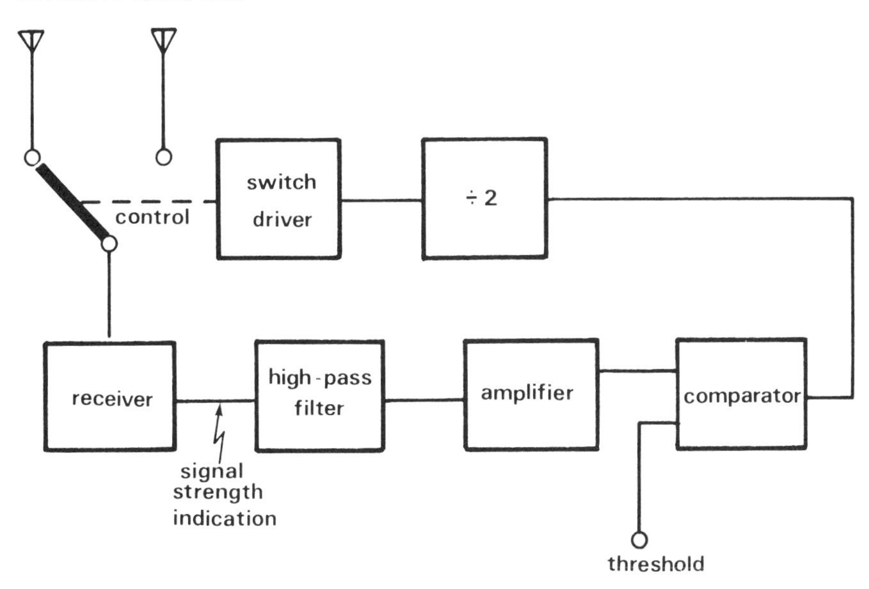

Figure 6.23 Switched diversity system.

the signal-strength indication is AC-coupled into the comparator, it is possible to set the reference voltage such that the threshold is placed a fixed number of decibels below the mean signal level (see Figure 6.9).

Two practical problems remain. Firstly, there is the need to design a high-quality antenna switch with a low insertion loss, high isolation of the unwanted input and good linearity. Switches using PIN diodes are usually satisfactory; an insertion loss of less than 0.5 dB and an isolation in excess of 60 dB are readily obtained. Secondly, there is the need to ensure that there is a minimum delay between the time the signal crosses the threshold and the time the switching takes place, otherwise there will be a degradation in performance. This problem has been analysed (ref. 3, Chapter 6), the general conclusion being that to retain significant diversity advantage the delay should not exceed about $0.2\tau_b$, where τ_b is the average duration of a fade below the threshold level.

The effectiveness of a practical predetection switched diversity system can be seen from Figures 6.24 and 6.25, which show cumulative probability distributions for several tests under mobile conditions at a carrier frequency of 100 MHz [25]. Figure 6.24 shows that the individual inputs from two antennas on a vehicle, although not identical, follow a close approximation to a Rayleigh distribution. The upper curves show the actual receiver input together with the theoretical curve, and a computed prediction of the output that would be obtained from an ideal selection-diversity system having these actual antenna voltages as its inputs. It can be seen that whereas true selection diversity would produce a very good output from these two inputs, the

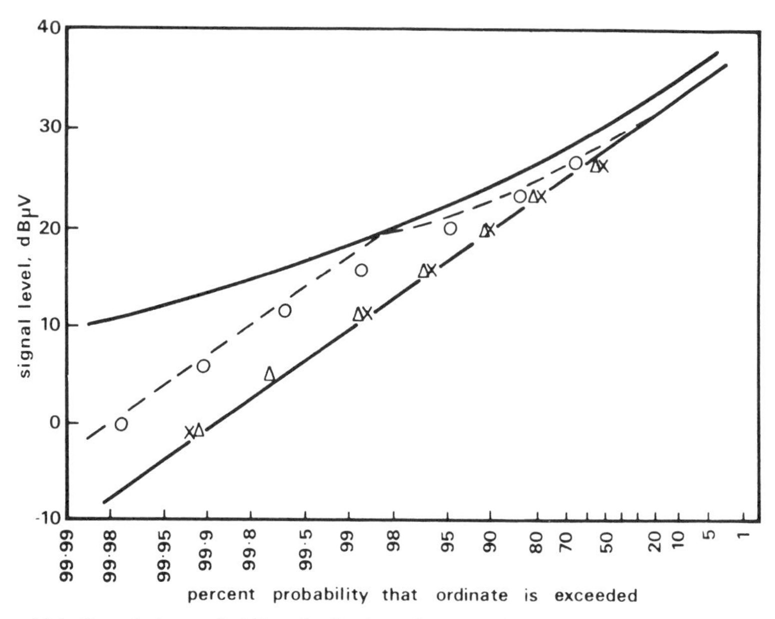

Figure 6.24 Cumulative probability distributions for a two-branch switched diversity system. × antenna 1; △ antenna 2; ○ measured system output; – computed output of an ideal selection system.

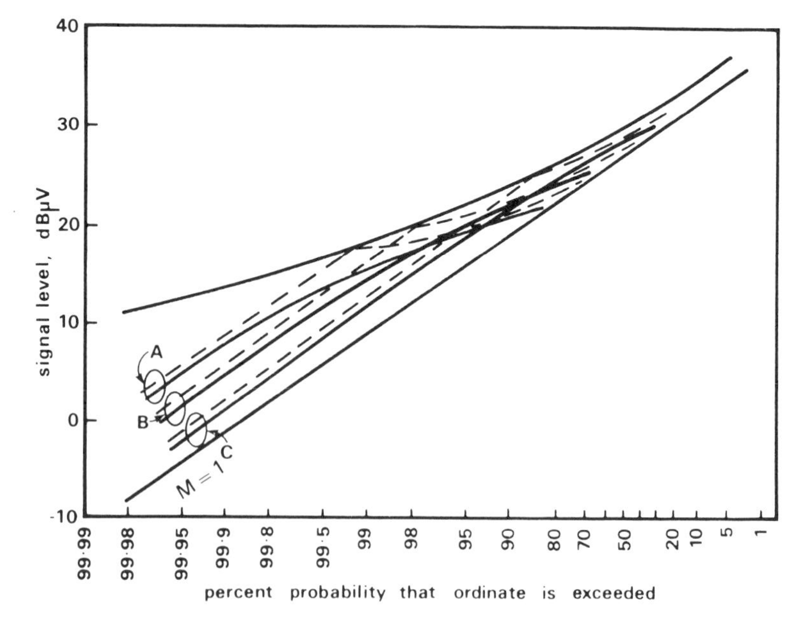

Figure 6.25 Theoretical and experimental cumulative probability distribution functions for a switched diversity system using various threshold settings. --- theoretical; ——experiment. $A = -18\,\text{dB}\,\mu\text{V}$, $B = -20\,\text{dB}\,\mu\text{V}$, $C = -22\,\text{dB}\,\mu\text{V}$.

performance of the switched system falls slightly below the theoretical prediction, particularly at levels close to the threshold, and for this test run the reliability at the threshold level is 94.5% compared with the theoretical value of 97.8%. Although this shortfall is partly attributable to the fact that the two inputs do not have identical mean values and do not follow an ideal Rayleigh distribution, a major influencing factor is the switching delay caused by time constants in the receiver. Ideally, the delay should be small compared with the average duration of a fade below the threshold level. Following the analysis in Chapter2 (eqn 2.12), and taking the example of $f = 100$ MHz, $v = 48$ km/h, $T = -10$ dB, the average fade duration is 26 ms. In the experimental system, the delay was approximately 5 ms, and there is clearly some advantage to be gained in a practical system by reducing this time.

Figure 6.25 shows some cumulative probability distributions of the receiver input, taken during various test runs using different threshold settings. These curves have all been reduced to the same mean value to make comparison easier. It can be seen that the theoretical and practical curves are closer at higher threshold settings. This is to be expected, since the average fade duration at higher thresholds is longer, and the switching delay becomes far less significant.

The switching performance of the system is summarized in Table 6.1, which gives typical figures at various threshold levels. Analysis of experimental data has shown that the average switching rate lies between 1.2 and 1.4 times the threshold level-crossing rates of the individual inputs. Again, following the analysis in Chapter 2 it is possible to predict the level crossing rate at any threshold. Taking the example of $f = 900$ MHz, $v = 48$ km/h, $T = -10$ dB, the level-crossing rate is 26.1 Hz, and hence the average switching rate to be expected in a practical system is between 31 and 37 Hz.

Another source of problems in predetection switched systems is that the input presented to the receiver may exhibit abrupt changes in both amplitude and phase, following each switching. This causes transient responses in the receiver which limit its performance and is another contributory cause to the difference between the theoretical and practical curves in Figure 6.25. In data communication systems these transients can cause errors and in general the BER improvements produced by switched diversity system fall somewhat below the theoretical predictions.

Table 6.1 Switching performance of 100 MHz diversity system

Length of test run m	Threshold setting dB below mean	Number of transitions Successful	Unsuccessful	% Successful Measured	Theoretical
150	5	46	14	76.6	78
150	7	48	7	87.2	85.5
150	10	40	4	90.9	92.4
150	12	26	2	92.8	95.2

The system shown in Figure 6.23 normally operates by switching to the alternative antenna whenever the signal falls below the threshold, irrespective of whether the new signal is acceptable or not, i.e. it uses the switch-and-stay strategy. Whereas this is the preferred mode of operation when the vehicle is moving reasonably quickly and fades occur frequently, it is clearly undesirable to allow this situation to prevail when the vehicle is moving very slowly, or is stationary in traffic. In the practical system shown in Figure 6.26, an additional circuit has been included to ensure that, if no fades have occurred for a few seconds, the system examines the other antenna to see whether its input is better. The basis of this circuit is a retriggerable monostable that is reset each time a switchover takes place. Provided the fades occur at a reasonable rate, the monostable is permanently in the triggered condition and so plays no part in the operation of the circuit. However, when no fades have occurred for a time equal to the period of this monostable, it produces an edge which causes the system to switch to the other antenna. If the input from this antenna is higher, the system remains connected to it. However if it is lower, there is a sudden drop in the signal level when the switch is made, which the system interprets as a fade, and returns the receiver to the original antenna. It therefore behaves according to the switch-and-examine strategy at slow fading rates, and the switch-and-stay strategy at higher fading rates.

6.10 Comparison

To conclude this section, it is of interest to compare briefly the various different diversity systems which have been described. Switched diversity is cheap, and can be designed in the form of an add-on unit for use with existing receivers. Although its performance is inferior to the combining methods in the presence of narrowband uncorrelated Gaussian noise, this may not be relevant in mobile radio, where the dominant interference is the (partially correlated) impulsive noise from vehicles.

The perturbation methods are also compatible with existing standard receivers, although they could be directly integrated into a future receiver design. The main limitation on these systems is the rate at which compensation is possible. This is set by the bandwidth available for the perturbing frequencies, limited basically to 300 Hz by spectrum overlapping, and difficulties in recovering the perturbing signals at the output of the receiver. It seems that their use may be restricted to frequencies up to the lower UHF mobile radio band.

A practical phase-sweeping system requires a receiver designed along conventional lines, but having more than one front end. Its performance is comparable with other combining methods.

The 'phase-stripping' combiner is basically the most elegant of those described. It requires a special design of receiver, but it does not have any

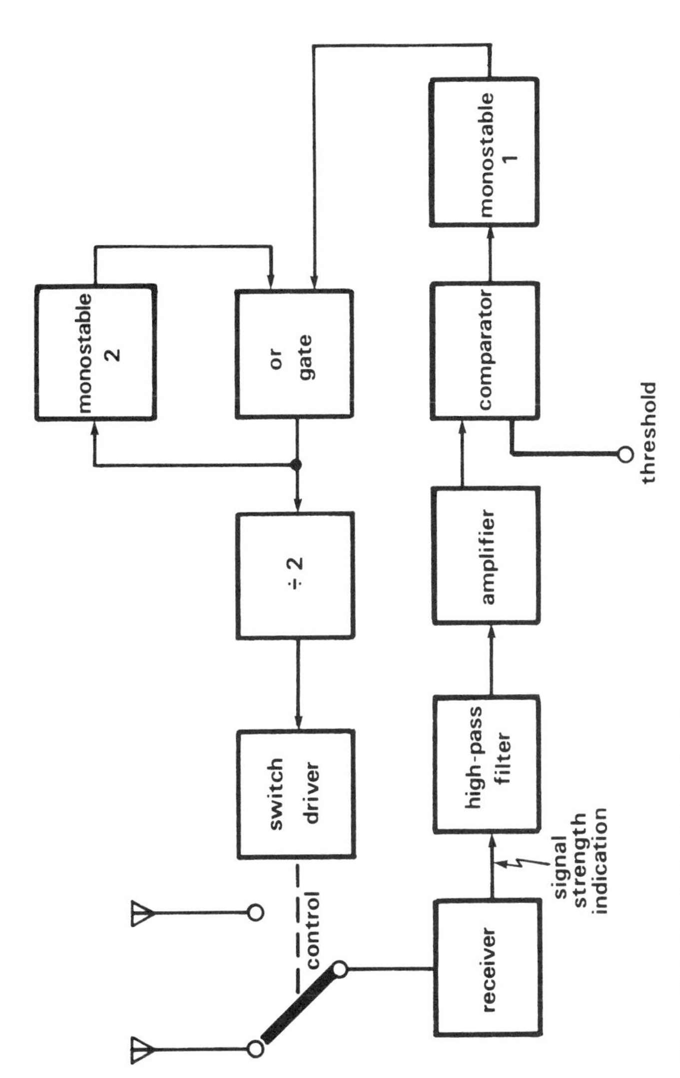

Figure 6.26 Practical switched-diversity system.

problems with regard to speed of compensation. Of the systems described, it is the only one which can completely cancel Doppler shift and random FM.

6.11 Postdetection diversity

Postdetection diversity is probably the most straightforward (if not the most economical) technique among the well-known diversity systems. The co-phasing function is no longer needed, since after demodulation all the baseband signals are in phase. The earliest diversity systems were of the postdetection type, where an operator manually selected the receiver that sounded best; this was, in effect, a form of selection diversity. In postdetection-combining diversity, the equal-gain method is the simplest.

Two or more separately received signals are added together to produce the combined output with equal gain in all the diversity branches. However, if this scheme is used in an angle-modulation system, the output SNR will be reduced drastically when the signal in one of the diversity branches falls below the threshold because the faded branch then contributes mainly noise to the combined output. In this case, maximal-ratio combining should be used to combine the signals, with each branch gain weighted according to the particular branch SNR. Postdetection maximal-ratio combiners thus require a gain-control stage following the detector. The required weighting factor for each branch can be obtained by using either a measure of the amplitude of the received signal envelope before detection or a measure of the out-of-band noise from the detector output. The first method provides an indication of the receiver input SNR only if the receiver noise is constant. The second method will provide a good indication of the receiver SNR even if the receiver input noise changes. An investigation of the performance of a postdetection maximal-ratio combining receiver for mobile radio has been made by Rustako [26], and a summary of the results is presented by Jakes [3].

In a system using angle modulation, the demodulated output signal level from the discriminator is a function of the frequency deviation only if the receiver input signal level is above threshold. The output noise level will vary inversely with the input down to the threshold, and then will increase nonlinearly below it. Brennan [1] has shown that it makes little difference to the performance of a postdetection combining receiver that utilizes angle modulation whether the weighting factors follow the output SNR exactly or if the receiver merely 'squelches', i.e. discards the output of that branch when the branch input falls below the threshold. This is because there is little gained in the combiner performance over the equal gain combiner by weighting the branch gain according to its SNR, if all the branch input signals are above threshold. Below threshold the noise increases rapidly, thereby reducing the output SNR by a significant amount; this means that the branch gain has to be reduced accordingly. Since the reduction in gain is so large, it makes little difference if the branch output is discarded altogether.

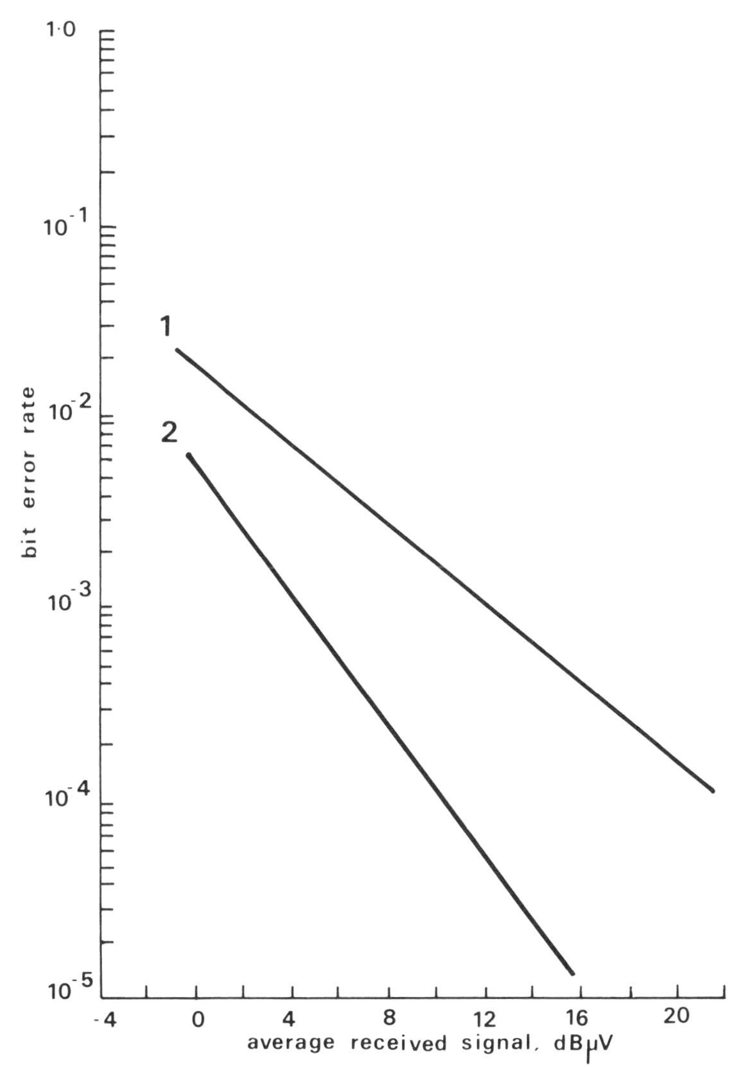

Figure 6.27 BER performance of an experimental 1200 bit/s direct-modulation FSK post detection combining system under Rayleigh-fading conditions. 1, single branch; 2, dual equal-gain combining.

A simple postdetection system designed to receive FSK data at 1200 bits/s directly modulated on to a carrier can be built by combining the outputs of two standard FM receivers [27]. Figure 6.27 shows the BER performance of such a system and demonstrates the improvement obtainable from the use of diversity. In the case described, the squelch circuits of the receivers can be used to make the system function as a hybrid between an equal-gain combiner and a selection diversity system. When the input signals to both receivers are above the squelch threshold the outputs are added equally; when one falls below the threshold the output from that branch is squelched and the system becomes a selection diversity receiver. Only in the unlikely event that both inputs are below threshold is there no output. To illustrate the point regarding the use of postdetection equal-gain combining with an angle-modulated signal, the BER measured with the squelch circuits inoperative [27] was found to be greater with diversity than in the case of a single receiver—this is due to the degradation in the average SNR of the summed signal.

6.11.1 Use of a modified phase-correction loop

In section 6.8 we discussed, at some length, the double heterodyne phase-correction loop which can be used in predetection diversity systems. The basic structure is shown in Figure 6.19, and it has been pointed out previously that $F1$ has to have fairly narrow bandpass characteristics.

A modified version of this system can be used in a postdetection combiner intended to receive binary digital FM signals which can be demodulated using either a conventional frequency discriminator, or differentially demodulated as a binary DPSK signal [28]. Both these non-coherent demodulation techniques are attractive for mobile radio because there is no requirement for the carrier-recovery function necessary in coherent demodulation schemes. We have seen earlier that the BER performance of non-coherent demodulators is slightly inferior to that of coherent ones, but the receiver structure is simplified and with discriminator detection a conventional FM receiver can be used for both voice and data transmissions. It is also possible to differentially demodulate a quaternary digital FM (QFM) signal as a quaternary DPSK signal.

In a conventional differential demodulator, the input signal and its delayed replica are multiplied to yield I- and Q-channel outputs. The conventional loop can therefore be modified to include the differential detection function, by using a delay line with a time delay τ as shown in Figure 6.28. Filter $F1$ is required only to reject the upper sideband component. If we assume as

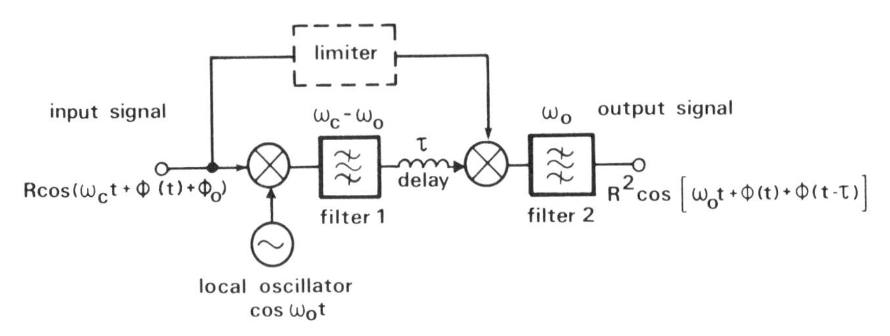

Figure 6.28 Phase correction loop modified by the inclusion of the differential detection function.

previously that the input signal is

$$s(t) = R\cos[\omega_c t + \phi(t) + \phi_0] \tag{6.49}$$

and the local oscillator is $\cos\omega_0 t$, then ignoring the gain constants of the multipliers, and assuming $(\omega_c - \omega_0)t = 2n\pi$ the output of any branch can be written as

$$\begin{aligned} s_0(t) &= R^2\cos[\omega_0 t + \phi(t) + \phi(t-\tau)] \\ &= R^2\cos[\phi(t) - \phi(t-\tau)]\cos\omega_0 t - R^2\sin[\phi(t) - \phi(t-\tau)]\sin\omega_0 t \end{aligned} \tag{6.50}$$

from which it can be seen that all branches are in phase and can be combined. If $\tau = T$ (the symbol period), then the I-channel output can be obtained by multiplication with the local oscillator output cos $\omega_0 t$ whilst the Q-channel requires multiplication with sin $\omega_0 t$. If $\tau \ll T$, then eqn (6.50) becomes

$$S_0(t) = R^2\cos\omega_0 t - \tau R^2\phi(t)\sin\omega_0 t \tag{6.51}$$

and the output due to frequency demodulation is available from the Q-channel.

The basic receiver structure using the modified phase correction loop is shown in Figure 6.29 which also illustrated how a limiter can be used in the feed-forward path. The basic structure is also applicable to frequency demodulation, and since the combining takes place after the frequency demodulation or differential demodulation operation, this is most correctly termed postdetection diversity. An analysis taking additive Gaussian noise into account shows that this postdetection system can achieve approximately the same performance as predetection systems using equal gain or maximal-ratio combining. Figure 6.30 shows the average bit error rate due to additive Gaussian noise in fading conditions. Random FM noise at the combiner output is a weighted sum of that from each branch and is substantially reduced.

6.11.2 Unified analysis of postdetection diversity

It is of interest to comment briefly on a unified analysis of postdetection diversity [29] in which the demodulated output of each branch is weighted by the vth power of the input signal envelope. Again, considering the possibility of differential or frequency demodulation, the optimum weighting factor has been shown to be $v = 2$. It can also be shown that weighting factors of $v = 1$ and $v = 2$ correspond, in the postdetection system, to predetection equal-gain and maximal-ratio combiners respectively, so a comparison can be made. Numerical calculations of bit error rate with minimum-shift-keying show that two-branch postdetection systems are only about 0.9 dB inferior to predetection combiners.

6.11.3 Postdetection selection and switched diversity

These two types of system can both be implemented in the postdetection

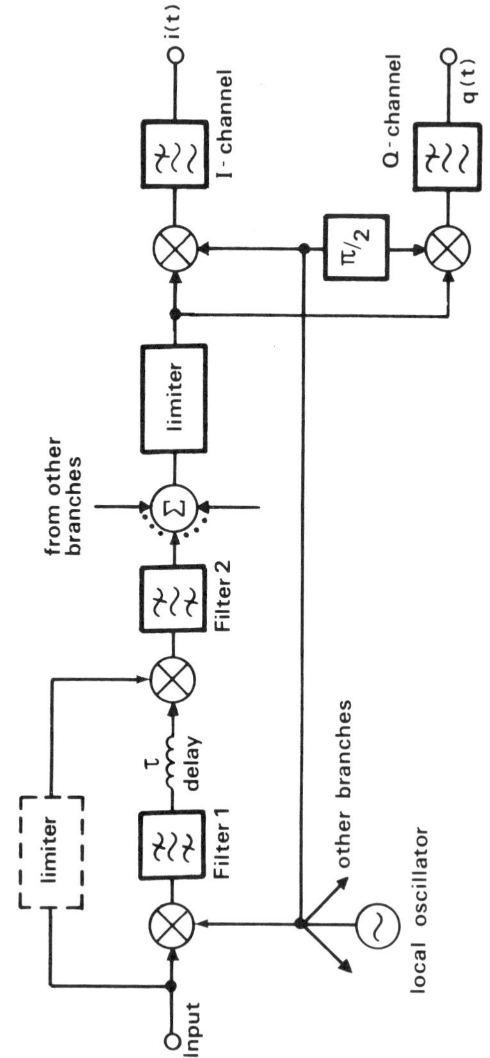

Figure 6.29 Basic receiver structure using the modified phase correction loop.

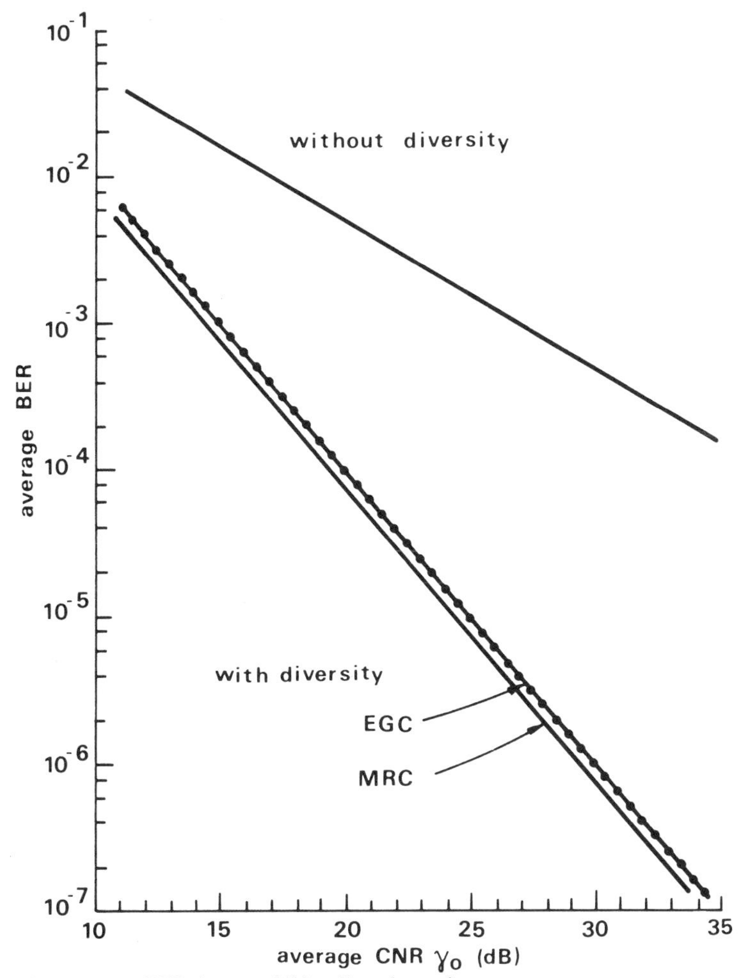

Figure 6.30 Average BER due to additive Gaussian noise.

format, with some advantages. As far as selection diversity is concerned, there are no amplitude transients, since the switchover takes place when the two signals are (nominally) of equal value. Abrupt changes in amplitude are still possible with switched diversity, but phase transients have no meaning in the postdetection context. As a result it is likely that in data communication systems the errors caused by the switching process will be much fewer with postdetection than with predetection systems. Figure 6.31 shows how a postdetection selection system can be built.

6.12 Time diversity

In our discussion of diversity techniques and systems, it has been made clear that in order to make diversity effective we need two or more samples of the

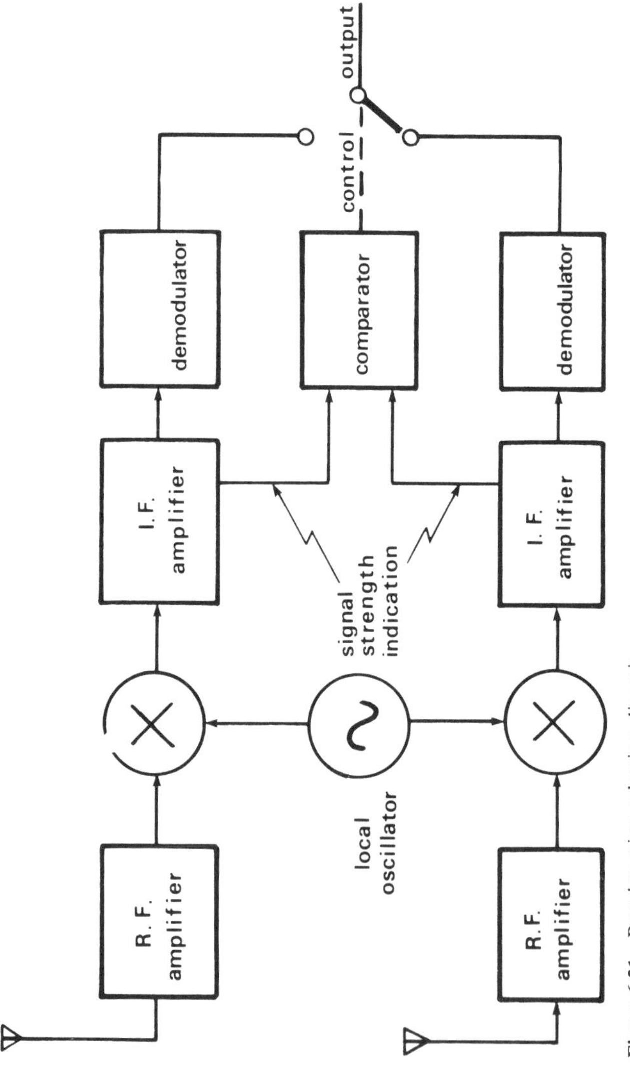

Figure 6.31 Postdetection selection diversity.

received signal which fade in a fairly uncorrelated manner. As an alternative to space diversity, we can obtain these independent samples from two or more transmissions sent over the mobile radio link at different times. Of course this cuts down the data throughput rate by an appropriate factor, but time diversity has several advantages, amongst which is that only a single antenna is used and there is no requirement for co-phasing or duplication of radio equipment. In principle it is simple to implement, although it is only applicable to the transmission of digital data, where the message can be stored and transmitted at suitable times.

The principal question to be answered in connection with time diversity is how far apart in time the two messages need to be in order to provide the necessary decorrelation. In practice, the time interval needs to be of the order of the reciprocal of the maximum baseband fade rate $2f_m$ i.e.

$$T > \frac{1}{2f_m} = \frac{\lambda}{2v} \tag{6.52}$$

For a vehicle speed of 48 km/h and a carrier frequency of 900 MHz the time separation required is 12.5 ms–the time separation increases as the fade rate decreases, becoming infinite when $v = 0$, i.e. when the vehicle is stationary. Theoretically the advantages are therefore lost, but at UHF the wavelength is so small that minor movements of people and objects ensure that the standing wave pattern is never truly stationary.

It is therefore worth examining the potential for time diversity in the mobile radio environment, and we take as an example the case when the same data is transmitted twice with a repetition period T. At the receiver a single antenna is used. If the original data stream is discontinuous, then it may be possible to use time diversity, by repeating words, without increasing the data rate. However, if the original data stream is continuous, the final data rate has to be doubled or tripled as appropriate. The way in which a data stream at $2n$ bits/s is generated from an original stream at n bits/s and a version delayed by a time T is shown in Figure 6.32. At the receiver the nth data element $a_n (n = \ldots -1, 0, 1, \ldots)$ is received twice, and the original and repeated data are demodulated from two samples of the fading signal received at different times. Hence the number of diversity branches is two, and this type of diversity is equivalent to a two-branch system with the signal envelopes $R(t)$ and $R(t-T)$.

One simple method of using the received data is to output the data element a_n which is associated with the larger signal envelope. In this case the system is directly analogous to selection diversity, with the resultant signal envelope after selection represented as

$$R_0(t) = \max\{R(t), R(t-T)\}$$

as shown in Figure 6.33.

Analysis has shown that the average fade duration and level-crossing rates

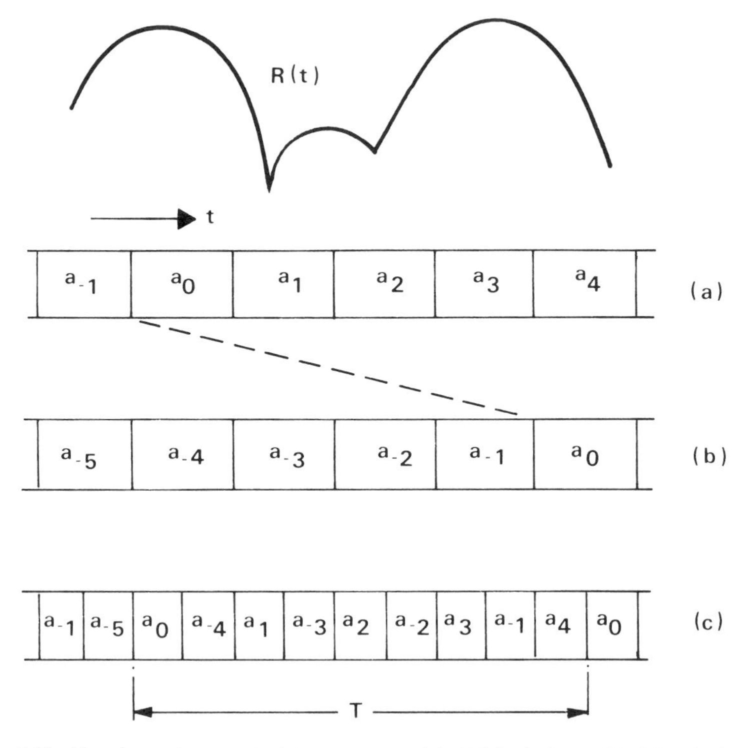

Figure 6.32 Signal envelope and data sequence. (*a*) Original data; (*b*) delayed data; (*c*) transmitted data.

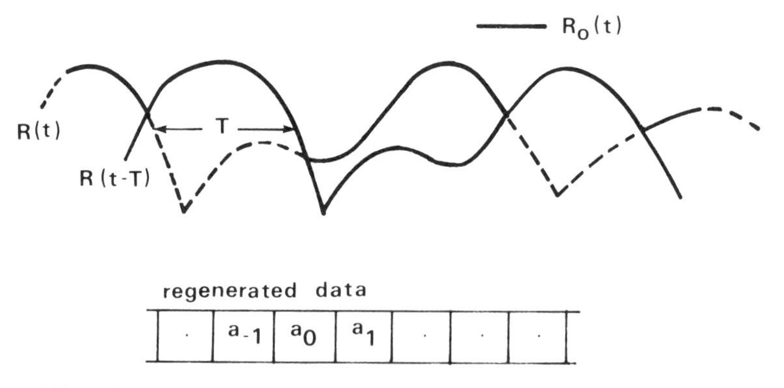

Figure 6.33 Relationship between $R(t)$, $R(t-T)$, $R_o(t)$ and the regenerated data.

are substantially reduced by the use of time-diversity provided certain criteria are met [30]. Some diversity advantage is still obtainable at a normalized repetition period $f_m T = 0.2$ which corresponds to $T = 5$ ms at a carrier frequency of 900 MHz and a vehicle speed of 48 km/h, and time diversity is thereby effective in reducing the rate at which error bursts occur. Although the average fade duration is halved for large repetition periods, the value for $f_m T = 0.1 \sim 0.3$ is longer than in the no-diversity case. In other words, time diversity reception creates an average burst error length longer than for the case of no diversity in this range of $f_m T$. To obtain some diversity advantage $f_m T$ should exceed about 0.5.

An alternative method of employing time diversity which avoids the necessity to monitor the signal strength associated with the reception of each data symbol is to transmit the sequence not twice but three times and to form an output by a majority decision (on a symbol-by-symbol basis) on the three versions received. This is simpler, but now eats seriously into the data throughput rate.

6.13 Discussion and conclusions

Switched diversity is potentially an economical diversity method, since it is simple in concept and can be very easily added on to the system currently in use. Laboratory experiments have produced encouraging results, and, with careful design, switched diversity could be an effective way of improving the performance of mobile radio data systems at very low cost.

Predetection combining can increase the SNR before detection, which in an FM system produces an effect similar to threshold extension, and in the presence of uncorrelated Gaussian noise its performance is very good. Systems intended for use only with angle-modulated signals can be designed using a very simple loop filter with a corresponding reduction in cost.

Postdetection combining may not be economical, since it involves duplication of the predetection parts of the receiver. However, it is effective in reducing BER and can be implemented using existing techniques.

All the diversity schemes produce a substantial improvement in signal quality or a reduction in the BER over that obtainable from a single receiver. Direct comparison of the results with the theoretical predictions is difficult for several reasons. First, the theory gives the BER as a function of γ, the CNR, and this is difficult to measure in practice. Secondly, field-trial results are corrupted by ignition noise that is not taken into account by the theory, and the detection process in the practical receiver is somewhat often different from that assumed in the theory. Nevertheless, the improvements obtained by using diversity as opposed to a single receiver are substantial and of the same order as theory predicts and in practice it is the improvement obtained, which is the factor of greatest importance. Alternative techniques such as error-correcting codes have the disadvantage that they eat very seriously into the information bandwidth and hence reduce the effective data rate. With the exception of time diversity most diversity techniques do not do this (or need not do it), and typically a two-branch diversity system produces an improvement at the 99% reliability level comparable with that obtainable from a 12-dB increase in transmitter power. It removes the vast majority of signal dropouts, making speech more readable, and produces a reduction in error rate of more than one order of magnitude. There are also other advantages, such as reduction of random FM; with some kinds of predetection combiner the random FM can be completely eliminated, and this can be an important consideration at higher carrier frequencies.

As far as implementation is concerned, although predetection combining can outperform other systems in the presence of uncorrelated Gaussian noise, this may not be representative of the conditions prevailing in mobile radio, where there are many noise sources such as the ignition systems of other vehicles in close proximity to the radio installation. The total inputs (signal plus noise) to the various branches may then be sufficiently correlated to impair the performance of combining systems, and selection diversity (or in practice switched diversity) becomes the optimum technique in these circumstances.

References

1. Brennan, D.G. Linear diversity combining techniques. *Proc. IRE* **47** (1959) 1075–1102.
2. Kahn, L.R. Ratio squarer. *Proc. IRE* **42** (Nov. 1954) 1704.
3. Jakes, W.C. (ed.) *Microwave Mobile Communications*. Wiley, New York (1974).
4. Lee, W.C.-Y. Mobile radio performance for a two-branch equal-gain combining receiver with correlated signals at the land site. *IEEE Trans.* **VT-27** (Nov. 1978) 239–243.
5. Adachi, F. Feeney, M.T. and Parsons, J.D. Effects of correlated fading on level crossing rates and average fade durations with predetection diversity reception. *IEE Proc.* **135 F** (1) (Feb. 1988) 11–17.
6. Parsons, J.D., Henze, M., Ratliff, P.A. and Withers, M.J. Diversity techniques for mobile radio reception. *Radio and Electronic Engineer* **45** (7) (July 1975) 357–367.
7. Lee, W.C.-Y. *Mobile Communications Engineering*. McGraw-Hill, New York (1982).
8. Adams, R.T. and Mindes, B.M. Evaluation of IF and baseband diversity combining receivers. *IRE Trans on Communications* **C3-6**, (June 1958) 8–13.
9. Withers, M.J., Davies, D.E.N. and Apperley, R.H. Self-focussing aerial arrays for airborne communication, in Blackband, W.T. (ed.), *Signal Processing Arrays*, Technivision Services, Workingham, Berks. (1968) 319.
10. Parsons, J.D. and Ratliff, P.A. Diversity reception for mobile radio. *Radio and Electronic Engineer* **43** (5) (May 1973) 317–324.
11. Parsons, J.D. and Ratliff, P.A. Self-phasing aerial array for FM communication links. *Electron. Lett.* **7** (13) (July 1971) 380–381.
12. Henze, M. and Parsons, J.D. Experimental dual-diversity single-receiver predection combiner for UHF mobile radio. *IEE Proc.* (*Electronic Circuits and Systems*) **1** (1) (Sept. 1976) 2–10.
13. Villard, O.G. (Jr)., Lomasney, J.M. and Kawachika, N.M. A mode-averaging diversity combiner. *IEEE Trans. on Antennas and Propagation* **AP-20** (4) (July 1972) 463–469.
14. Withers, M.J. A diversity technique for reducing fast-fading. *Proc. Conf. on Radio Receivers and Associated Systems, IERE Conf. Proc.* **24**, (1972).
15. Kawachika, N.M. and Villard, O.G. (Jr.) Computer simulation of h.f. frequency-selective fading and performance of the mode-averaging diversity combiner. *Radio Science* **8** (3) (March 1973) 203–212.
16. Bello, P. and Nelin, B. Predetection combining with selectively fading channels. *IRE Trans. on Communications Systems* **CS-10** (3) (March 1962) 32–42.
17. Withers, M.J., Davies, D.E.N., Wright, A.H. and Apperley, R.H. A self-focusing receiving array. *Proc. IEE* **112** (9) (Sept. 1965) 1683–1688.
18. Gilbert, E.N. Mobile radio diversity reception. *Bell Syst. Tech. J.* **48** (Sept. 1969) 2473–2492.
19. Cutler, C.C., Kompfner, R. and Tillotson, L.C. A self-steering array repeater. *Bell. Syst. Tech. J.* **42** (5) (Sept. 1963) 2013–2031.
20. Black, D.M., Kopel, P.S. and Novy, R.G. An experimental u.h.f. dual-diversity receiver using a

predetection combining system. *IEEE Trans on Vehicular Communications* **VC-15** (2) (Oct. 1966) 41–47.

21. Boyhan, J.W. A new forward-acting predetection combiner *IEEE Trans. on Communication Technology* **COM-15** (5) (Oct. 1967) 689–694.
22. Granlund, J. Topics in the design of antennas for scatter. *Tech. Rep* **135**, MIT Lincoln Lab., Lexington, Mass. (Nov. 1956) 105–113.
23. Adams, R.T. and Smith, H.L. Reply to 'Digital transmission capabilities of a transportable troposcatter system'. *IEEE Trans. on Communication Technology* **COM-16** (4) (April 1968) 345–348 (correspondence).
24. Halpern, S.W. The theory of operation of an equal-gain predetection regenerative diversity combiner with Rayleigh fading channels. *IEEE Trans on Communications Technology* **COM-22** (8) (Aug. 1974) 1099–1106.
25. Parsons, J.D. Experimental switched-diversity system for VHF-AM mobile radio. *Proc. IEE* **122** (8) (Aug. 1975) 780–784.
26. Rustako, A.J. Performance of a two-branch post-detection combining VHF mobile radio space-diversity receiver. MS thesis, Newark College of Engineering (1969).
27. Parsons, J.D. and Pongsupaht, A. Error rate reduction in VHF mobile radio data systems using specific diversity reception techniques. *IEE Proc. Part F*, **12** (6) (Dec. 1980) 475–482.
28. Adachi, F. and Parsons, J.D. Postdetection diversity using a modified phase correction loop for digital land mobile radio. *IEE Proc. Part F*, **134** (1) (Feb. 1987) 27–34.
29. Adachi, F. and Parsons, J.D. Unified analysis of postdetection diversity for binary digital FM mobile radio. *IEEE Trans. VT* (to be published).
30. Adachi, F., Feeney, M.T. and Parsons, J.D. Level crossing rate and average fade duration for time-diversity reception in Rayleigh-fading conditions. *IEE Proc. Part F*, (to be published).

7 Using the radio channel in cellular radio networks

7.1 The radio channel as a system component

It is apparent from the foregoing chapters that the characteristics of the radio path which exists between the transmitter and the receiver are complex and heavily dependent on the nature of the physical environment in which the equipment operates. In addition, the influence of man-made noise on the performance of the receiver end of the radio connection greatly influences the extent to which the receiver is able to recover the original transmitted information.

It has also been seen that there are procedures, such as the implementation of diversity techniques, by which the effects of the multipath transmission mechanism can be minimized in order to make the best of a difficult situation and provide the system and equipment designer with optimized radio links between the various elements of the radio network. The objectives of the remaining chapters are now to examine how the system designer can manipulate the other parameters at his disposal to achieve performance targets defined by service requirements and user needs. From the overview of Chapter 1 it is clear that these requirements and needs are extremely diverse, and an immediate problem arises in finding means to categorize them in a way which identifies common factors so that major system design strategies can be perceived.

In global terms, principal needs are clear. Systems must:

(i) Accommodate a large and expanding user population in a (more or less) fixed allocation of spectrum
(ii) Provide for speech telephony and data transmission, sometimes in separate dedicated networks and sometimes in the same networks
(iii) Be compatible as far as possible with the line networks (PSTN or ISDN).

The system designer will also have in mind that user terminals need to be manufacturable at modest cost, and that not every theoretically possible technique can be implemented in available technology—some signal processing strategies, for instance, may have to wait for substantial advances in VLSI methods, and so on. Plainly, system design involves compromises, and to illustrate how these have been optimized in two very different systems, the remainder of this book is devoted to considering the major features of existing analogue cellular radio systems on the one hand and the evolving pan-European digital cellular radio network on the other.

7.2 Wideband versus narrowband

These terms have significance according to the applications to which they refer, and it is necessary therefore at this stage to decide what 'narrowband' and 'wideband' mean in the context of the Rayleigh channel. Recalling the definition of coherence bandwidth in Chapter 2, this may be used as a reference. A 'narrowband' system is then one which operates in a channel substantially narrower than the coherence bandwidth, whereas a wideband system occupies channels greater than, often by many times, the coherence bandwidth. In the former case, the effects of multipath propagation operate on all the components of the signal together so that the received signal, while varying in amplitude and accompanied by added noise and interference, remains relatively undistorted in comparison with the transmitted original. In the latter case, however, the effects of selective fading across the transmitted signal bandwidth result in gross distortions of the received signal as far as some features are concerned—power spectral density for instance. On the face of it, it would appear that wideband systems have little to offer in the context of high-capacity applications, and that the narrower the channel which can be utilized, the greater the capacity should be. But this is a simplistic view; wideband systems offer a much greater opportunity for the application of sophisticated modulation, coding and signal processing techniques which permit a receiver to distinguish features of the wanted signal in the presence of relatively high levels of interference even when, on a spectrum analyser, wanted and interfering signals look very similar. This is an extremely significant factor and one which will recur throughout the discussions of cellular radio technology. As a result, the development of cellular radio systems has seen an evolutionary process take place in which the issues of available technology and equipment cost have dominated system design in the introduction of first-generation services, and the emergence of VLSI with greatly enhanced performance and low power consumption has paved the way for the introduction of wideband digital systems as the second-generation standards.

7.3 Cellular radio fundamentals

7.3.1 Why 'cellular'?

In scanning the literature describing cellular radio systems in their entirety, the reader is prompted to conclude that cellular radio represents a completely new approach to service provision. Although as far as the complete system is concerned, this is largely true, in considering the radio segment, the emergence of cellular technology has been more a process of logical development than revolutionary innovation. The traditional problem faced by mobile radio system designers has been how to balance the apparently conflicting requirements of area coverage and user capacity. These requirements conflict because, if a base station is to provide service to mobiles over a wide area, it must have high power and be situated on the highest point available in the required coverage area. But following this strategy means that the channels allocated to the transmitting site cannot be re-used for another service for a very considerable distance. Since the spectrum available is fixed, and the traditional technologies using AM and FM have dictated division of the spectrum into channel segments on an FDM basis, only a small total user population per channel can be achieved by the high-power hilltop site approach. Recognizing this, the regulatory authorities have, since the early introduction of mobile radio on a substantial scale, restricted base station transmitting power in order to improve frequency re-use opportunities, and this has obliged system designers to follow other strategies to achieve area coverage. These involve developing an infrastructure of fixed radio or line links to connect a number of base stations via a central control point, so that the service transmissions can be radiated simultaneously from each to locate a wanted mobile. Then, if appropriate, a selection procedure can be utilized to use only the nearest base to the mobile to sustain subsequent message exchanges. Clearly, to create national networks by this technique is costly and complicated and, in any case, an obvious alternative is available. A national network of immense capacity already exists in the PSTN. This could be used as the basic infrastructure to connect base stations together—if there are many connection points between base stations and the fixed network, then individual base stations need cover only small areas and enormously greater frequency re-use becomes possible. The potential of this strategy was appreciated more than two decades ago, but its implementation required two major developments. Firstly, much higher radio frequencies than those common in PMR would have to be used so that radio coverage from an individual base could be defined and restricted to a known area or 'cell'. Frequency re-use could then be maximized. This had implications in both transceiver technology and the allocation of appropriate spectrum. Secondly, means would have to be found to detect and monitor the location of mobiles relative to base stations, so that calls to mobiles could be directed to the appropriate base station, and mobiles could gain access to the network by addressing the base which offered the best radio path. This required a new generation of electronic exchange or 'switch' and, more significantly, low-cost processing power at base stations and in each mobile terminal to handle the system 'housekeeping'.

The early 1970s saw the emergence of the required technology, the World

Administrative Radio Conference (WARC) in 1979 approved allocations of frequency spectrum in the 800/900 MHz bands, and the scene was set for the introduction of cellular radio on a commerical basis.

7.3.2 Frequency re-use strategies

7.3.2.1 The cell 'cluster'. Frequency re-use is therefore the fundamental target of cellular radio systems, and implementing this to maximum effect has been the subject of much study since the inception of the cellular principle. For planning purposes, the area to be provided with radio coverage must be divided into cells in a regular fashion, and this implies representation of each cell by a regular polygon chosen so that the polygons will form sets (or

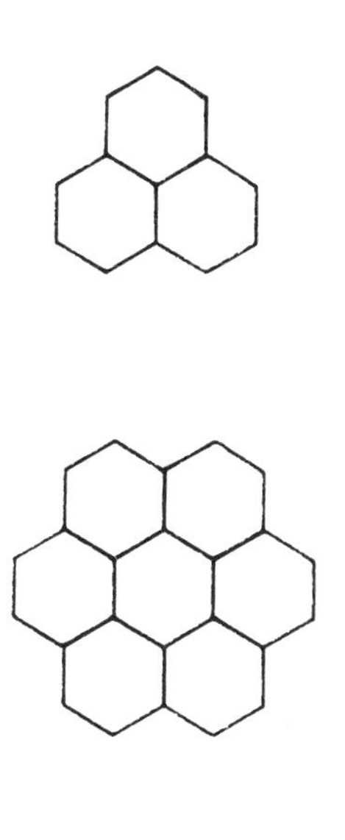

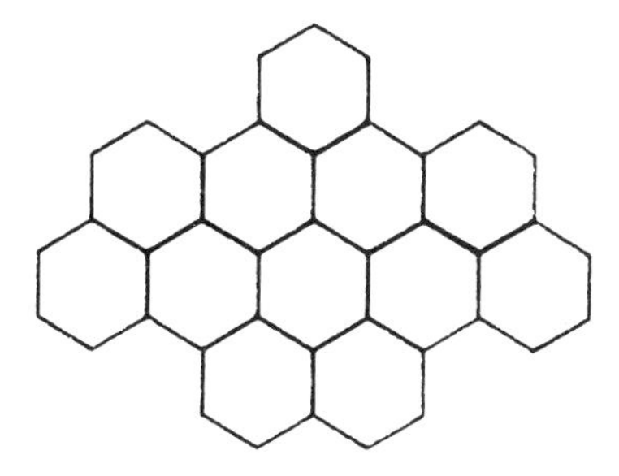

Figure 7.1 3-, 7- and 12-cell clusters.

'clusters') which will fit together—'tessellate'—without leaving gaps. The hexagon is an ideal choice, approximating to a circle, but offering a wide range of cluster sizes according to the number relationship

$$C = i^2 + ij + j^2 \tag{7.1}$$

where i and j are positive integers or zero, and $i \geqslant j$. Any value of C given by this relationship will produce clusters which tessellate. The further selection of a specific cluster size, C_L, from the set C is determined by considerations of how the available spectrum will be utilized. For the present, a narrowband channel scenario will be assumed, with an allocated block of spectrum B_T Hz wide available for division into channels B_C Hz wide, giving a total of $B_T/B_C = N_C$ channels in total. If the N_C channels are divided equally among all the cells in the cluster, then the first obvious trade-offs are seen from Figure 7.1. The larger the cluster size, the fewer channels are available in any individual cell to support the required traffic. On the other hand, if the cluster size is large, then the ratio of the distance which separates a cell from the equivalent member of any other touching cluster to the cell radius is large.

This relationship is an important one in cellular coverage planning, and may be related to cluster size as follows.

In Figure 7.2, seven cell clusters are shown, with the area of each cluster itself represented by a hexagon. The cluster hexagon has radius R_L, each individual cell has radius R_C and the distance between cells in adjacent clusters which use the same frequency is D. This last quantity is often referred to as the *repeat distance* or *re-use distance*. It is readily seen that the ratio of the areas of the

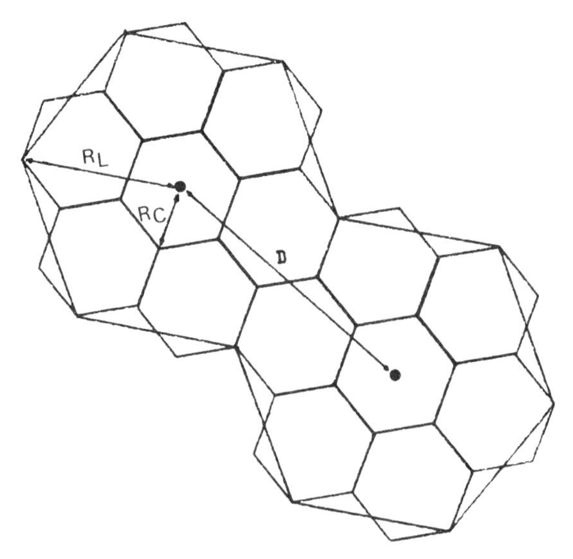

Figure 7.2 Calculating the D/R ratio.

cluster and an individual cell is equal to the cluster size, i.e. 7, and, since the area of a hexagon is proportional to the square of its radius, $R_L/R_C = \sqrt{7}$. Straightforward trigonometry reveals that the distance between the centre of the large hexagons is given by $D = R_L \cdot \sqrt{3}$, giving $D/R_C = \sqrt{3} \cdot \sqrt{7}$. Similar arguments can be used to show that this relationship generalizes to a cluster of C_L cells to give $D/R_C = \sqrt{3C_L}$.

So as to maximize frequency re-use, it is desirable to bring cells using the same frequencies as close together as possible, which would suggest aiming at a small cluster size (e.g. $C = 3$ or 4, which are both solutions of eqn 7.1); but there are other constraints which derive from the operating characteristics of the transmitter and receiver equipment which influence the choice of C_L.

7.3.2.2 Adjacent channel and co-channel considerations. In a narrowband system realization, it is important to consider the relationship between the choice of cluster size and the way in which the receivers in the system operate in the radio environment created by the frequency re-use strategy. Receivers on a certain channel can tolerate only a certain level of signal present at the input on an adjacent channel without degradation in performance (the adjacent-channel performance). Furthermore, in systems using angle modulation the receiver will discriminate in favor of a strong signal and suppress weaker interfering signals on the same frequency to a certain extent (the co-channel performance). The level of interference relative to a wanted signal which the receiver will tolerate, or the *protection ratio*, is however, also a function of the deviation ratio of the modulation. The greater this is, the better will the receiver discriminate against interference. In the cellular frequency re-use strategy, a receiver in any cell operating on any channel is subjected to co-channel interference from the equivalent cells in surrounding clusters, and to adjacent channel interference from utilization of the adjacent channels within its own cluster. Again, a compromise must be sought among conflicting optima for individual parameters. Ideally, a wide-angle deviation gives better protection against co-channel interference, but if this is increased beyond a well-defined limit, it will result in sideband energy extending outside the allocated B_C bandwidth, with consequent implications that affect receiver response to similar emission on adjacent channels.

In the system developed for implementation in the UK, the Total Access Communications System (TACS), the compromise has involved setting the deviation ratio in the FM analogue speech channels to give an emitted spectrum which exceeds the allotted B_C (25 kHz) by a modest amount but, at the same time, requires that a frequency re-use pattern be adopted which avoids the use of adjacent channels in any pair of adjacent cells. This clearly precludes use of $C_L = 3$ or $C_L = 4$, which were proposed earlier as optimal from the frequency re-use standpoint, since in a three-cell cluster each cell is adjacent to the other two, and in the four-cell case, two cells are adjacent to the other three.

The next value of C_L in the set C is 7, and this permits the adjacent channel requirement to be more closely approached, since a member of the outer ring of cells is adjacent to only three of the remaining six cells. However, the centre cell is adjacent to all the others. The next value of $C_L(12)$ does permit the adjacent channel requirement to be met completely, but at the expense of an increased D/R_C ratio and reduction in the number of channels per cell.

However, it has been found that co-channel rather than adjacent channel interference dominates the receiver-determined aspects of the choice for C_L, and, in consequence, the seven-cell cluster has been generally regarded as the optimum compromise.

7.3.2.3 Other cell illumination possibilities. Throughout the foregoing, a tacit assumption has been made that the transmitters illuminating the cells are placed at the cell centres and radiate uniformly in all directions. It was on this basis that the seven-cell arrangement was accepted as the realistic minimum configuration. However, whilst mobiles must of necessity have omnidirectional antennas because their orientation relative to base stations is random, some advantages are obtainable from deploying directional transmitting and receiving antennas at base stations. To appreciate the reasons for this requires investigation into how system capacity may be increased in areas of high user demand relative to adjacent areas with fewer users. The basic cellular principle requires that, since the number of available channels is fixed, capacity increase is achieved by reducing the size of cells in areas of high demand and implementing more concentrated frequency re-use. With cell-centre illumination, this implies creating new small cells within the overall cluster patterns as in Figure 7.3, but there are clearly constraints on how frequencies are re-used within the small cells, since this re-use must not infringe the rules determining allocation of frequencies for the large cell pattern. In addition, there are two further considerations.

(i) If a large number of small cells is created, then each will require its own base-station facilities and this means, apart from increasing the amount of base-station transceiver equipment deployed, that suitable transmitting sites must be found to accommodate them.

(ii) the radio propagation characteristics in very small cells differ somewhat from those in large cells (as may be seen from Chapter 2). The inverse fourth-power law relationship between signal strength and distance is approximated most closely at some distance from a transmitter rather than near to it, and in turn this means that in small cells the carrier-to-interference ratio deteriorates. To keep the carrier-to-interference ratio constant as cell size is scaled down would require some proportional increase in transmit frequency, an option which is not available to the system designer.

The use of directional base-station antennas provides the additional degree of freedom necessary to overcome these drawbacks. In Figure 7.4, a conven-

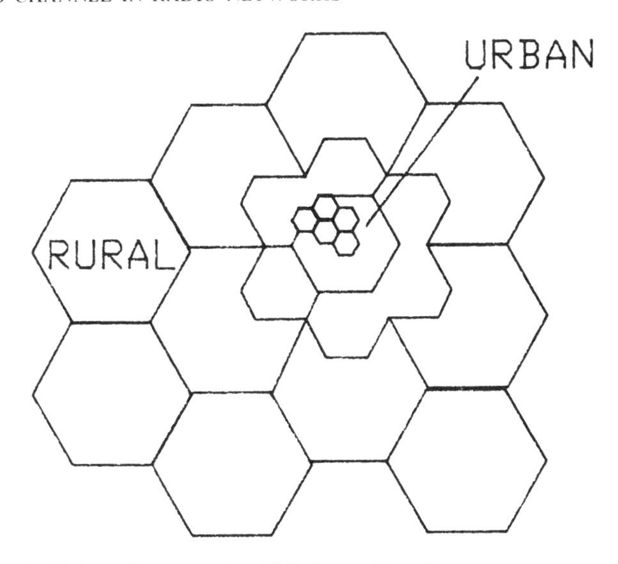

Figure 7.3 Subdividing cells to cater for high demand in urban areas.

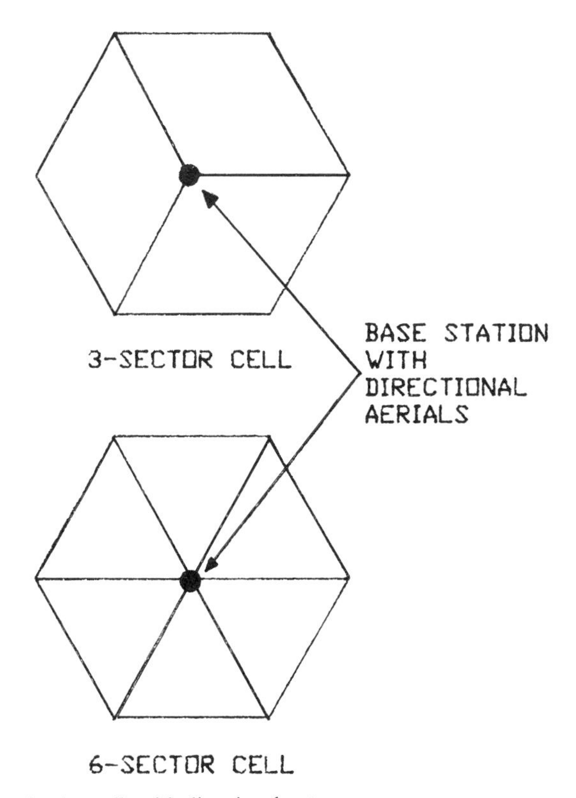

Figure 7.4 Illuminating cells with directional antennas.

tional hexagonal cell is shown 'sectored' (or sectorized in the USA) into three or six. The three-sector arrangement increases the effective number of cells in a seven-cell cluster from 7 to 21, but since 21 is also a solution of eqn. (7.1), the advantage obtained is principally that of achieving 21-cell re-use capability from seven base station sites. However, in the six-sector case, a four-cell cluster arrangement can be sectored into 24 effective cells, which has the further benefit of increasing the number of channels available in each cell before sectoring. The reduction in basic cluster size without an accompanying degradation in carrier-to-interference ratio at mobiles is achieved by choosing the frequency re-use pattern such that the front lobe of any one base station transmitter antenna illuminates only the back lobe of its co-channel counterpart. This strategy has been used by one of the UK service providers (Cellnet) to satisfy heavy user demand in London.

It should be borne in mind that the strategies outlined above are by no means the only possibilities available, and much effort has been put into further alternatives by cellular radio system designers in USA, Scandinavia, Europe and Japan. The foregoing does, however, demonstrate how strategies for maximizing user capacity rely on structuring of frequency re-use patterns to establish a carrier-to-interference ratio at a mobile anywhere within the cell clusters which will enable it to sustain the required output quality, in the case of speech telephony, or the required target bit error rate if the message traffic is data.

Equally, it can now be seen how the use of digital techniques, which offer satisfactory performance at greatly reduced effective carrier-to-interference ratios, can permit further reduction in cluster size with attendant increase in user capacity.

References

1. Young, W.R. Advanced Mobile Phone Service: Introduction Background and Objectives. *B.S.T.J.*, **58** (1) (January 1979) 1–14.
2. MacDonald, V.H. The Cellular Concept, *B.S.T.J.*, **58** (1) (January 1979) 15–41.
3. Huff, D.L. The Development System, *B.S.T.J.*, **58** (1) (January 1979) 249–69.
4. Appleby, M.S. and Garrett, J. The Cellnet Cellular Radio Network, *British Telecom Engineering*, **4** (July 1985) 62–69.

8 Analogue cellular radio systems

8.1 Channel structures

In the last chapter, the principles of channel allocation and frequency re-use were set out in general terms. It is now possible to examine in more detail the system requirements of the cellular radio network and to see how these are met in the available radio channels; that is, to assess the implications of the channel characteristics in structuring the different messages which sustain voice circuits to mobiles and hand-portables.

To illustrate this, reference will be made principally to the UK cellular radio system, the Total Access Communications System (TACS), which was inaugurated as a operating service in 1985. TACS is, however, a close relative of the earlier Advanced Mobile Phone System (AMPS), developed over many years before this in the USA. Other systems which were developed in parallel in Scandinavia (the Nordic Mobile Telephone System, NMT), in Germany (Netz C and D), and in Japan (the Nippon Advanced Mobile Telephone System, NAMTS) also have many similar features, and comparisons of various operational aspects are made towards the end of this chapter. It will be seen from these, however, that the systems have many features in common, and therefore, in discussing TACS/AMPS, the principal aspects of all the major analogue cellular radio systems will be touched on.

The basic traffic carried in cellular networks falls into two categories: voice (or data), which is output to the system users, and control signals, which are invisible to the users but essential to sustain the message channels as the mobile terminals move around in the cellular radio environment, passing from one cell to another and from one cluster to another. The mobile comprises a single transmitter and a single receiver so that, although these are both frequency-agile and can change their operating frequencies according to instructions from the network and also according to their own in-built programs, once a conversation is in progress any housekeeping data exchanges between mobile and bases must take place on the voice channel. On the other hand, when the mobile terminal is not in use for conversation, data exchanges are still necessary so that the network is updated about the location of the mobile in the radio cells and can have continuous knowledge of its operating status.

A further consideration is a measure of the imbalance between the quality of radio transmissions on the base-to-mobile path (the *forward* path) relative to the mobile-to-base path (the *reverse* path). This is due to big differences in output transmit power between bases and mobiles, particularly in the case of hand-portables, and to the variable orientation of antennas at mobiles. In addition, the radio environment at cell base-station sites is different to and more interference-prone than that in which individual mobiles operate remote from base. This means that it may be necessary to use different transmission strategies, particularly for data, on forward and reverse paths. The conclusion must therefore be that a total of four different signalling paths need to be identified separately, as illustrated in Figure 8.1. The Forward Control Channel (FOCC) and Reverse Control Channel (RECC) are used to maintain contact between mobiles and bases when no subscriber conversations are in progress, and also for setting up calls; the Forward Voice Channel (FVC) and Reverse Voice Channel (RVC) maintain data exchanges when calls are in progress.

A further consideration is that data on the FVC and RVC clearly must be formatted to use the standard voice channel, but can the FOCC and RECC be assigned to channels which have different characteristics from voice channels? In fact, even though these control channels in TACS are a dedicated subset of the total, they nevertheless have to be configured for compatibility with the standard voice channel. The reason for this is that in a busy system using many voice channels per cell, the volume of data exchange exceeds the capacity

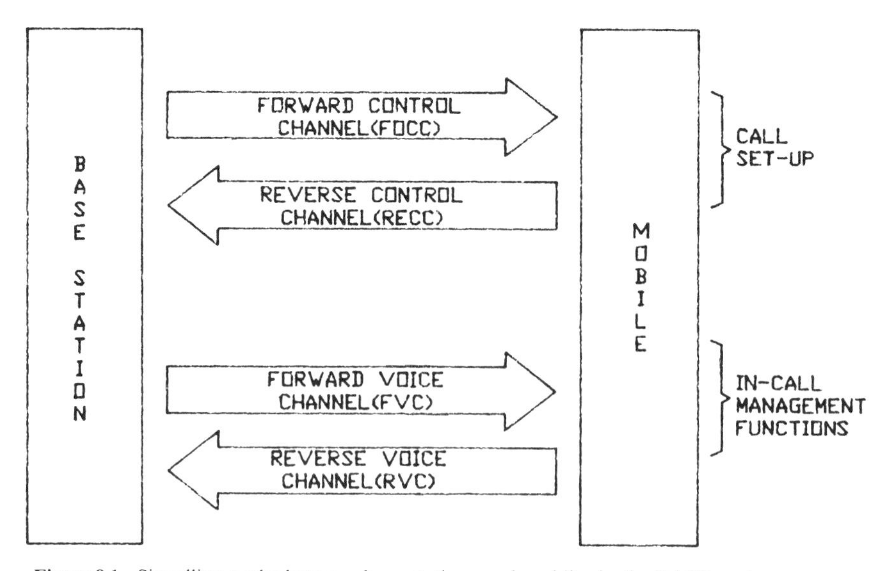

Figure 8.1 Signalling paths between base stations and mobiles in the TACS systems.

of the dedicated control channels, and so to avoid preventing users from accessing vacant voice channels because the data channels are saturated, the system needs the capability to reassign some voice channels as data channels to optimize total network throughput. The situation is shown in Figure 8.2.

The overall picture which emerges therefore is of a system essentially configured in the radio segment to carry high-quality voice traffic with four different formats of data signals constrained to live within the voice-channel structure. The FVC and RVC data messages must be formatted in short bursts which can be inserted from time to time into the voice path, but for such short times that the users are unaware of their existence. The FOCC needs to be transmitted continuously, so that whenever a mobile in a cell switches on or enters the cell from outside the cellular coverage area, it can find out from the FOCC what channels are allocated for voice traffic in the cell, whether access is inhibited for making calls because the voice channels are all busy or the RECC is in use, and so on. The RECC, however, needs to be activated only occasionally to update the network with any new information about when the mobile moves from one cell into another, wants to initiate a call, and so on. More details of the data message structure will be considered later, but it is appropriate now to look at the specification of the radio transmissions in both directions.

8.2 Specifications for the radio equipment

Details of the overall specifications, including message sequences, etc are set out in [1].

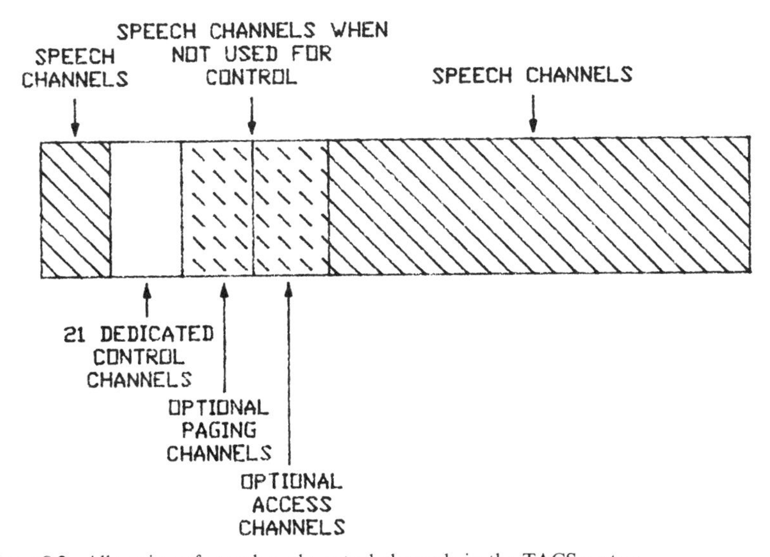

Figure 8.2 Allocation of speech and control channels in the TACS system.

8.2.1 RF power levels

Base station transmit levels must be chosen to give the coverage dictated by the size of cell which each transmitter serves, and since these differ in radius from 1 km in densely populated areas, to 15 km in rural environments, a wide range of transmit levels is appropriate. A maximum of 100W erp per channel is, however, set as an upper limit for the largest cells.

Mobiles have different output levels according not only to their type, but also according to demands of the radio path in use. Base stations instruct mobiles to decrease or increase their output power levels to maintain an acceptable received signal at the base receiver. This minimizes the total amount of emitted RF power from the community of mobiles in each cell, and by so doing minimizes mutual interference, particularly where high-power mobiles coexist with low-power hand-portables.

Four classes of mobile are available (Table 8.1). Mobile station power levels can take values between the maxima defined in Table 8.1, and a common minimum of −22 dBW according to the mobile attenuation code (MAC) transmitted by the base station, as shown in Table 8.2.

8.2.2 Modulation

Voice. As indicated in the last chapter, the voice modulation characteristics are chosen to achieve the best compromise between bandwidth occupancy and

Table 8.1 Mobile station power levels

Nominal erp		Output power from TX port (assuming 1.5 dB dipole antenna gain)
Class 1	10 dBW (10 W)	8.5 dBW (7.00 W)
Class 2	6 dBW (4W)	4.5 dBW (2.8 W)
Class 3	1.6 dBW (1.6 W)	0.5 dBW (1.1 W)
Class 4	−2 dBW (0.6 W)	−3.5 dBW (0.45 W)

Table 8.2 Mobile attenuation codes

Mobile station power level (PL)	Mobile attenuation code (MAC)	Nominal erp (dBW) Mobile 1	station 2	power 3	class 4
0	000	10	6	2	−2
1	001	2	2	2	−2
2	010	−2	−2	−2	−2
3	011	−6	−6	−6	−6
4	100	−10	−10	−10	−10
5	101	−14	−14	−14	−14
6	110	−18	−18	−18	−18
7	111	−22	−22	−22	−22

capability to reject co-channel interference. This means frequency modulation with maximum permissible deviation ratio. In conventional PMR equipment, the peak deviation permissible to confine transmitted energy to a 25 kHz channel is 5 kHz, but in TACS this is increased to 9.5 kHz to improve carrier-to-interference performance. The implications of this are shown in Figure 8.3, and indicate why no allocation of adjacent channels in the same cell is realistically possible and why allocations of adjacent channels in adjacent cells are also minimized.

The pre-emphasis characteristic is +6 dB/octave between 300 Hz and 3 kHz.

Data. Frequency shift keying (FSK) is used for the signalling activities between bases and mobiles in both directions. The deviation used is less than for speech, since the baseband data signal has different spectral characteristics from speech. The basic data rate is 8 kbits/s and the deviation 6.4 kHz, a binary 1 producing $(f_c + 6.4)$ kHz and 0 $(f_c - 6.4)$ kHz, where f_c is the carrier frequency. Additionally, the baseband data streams in both directions are further encoded such that each non-return-to-zero binary 1 becomes a 0-to-1

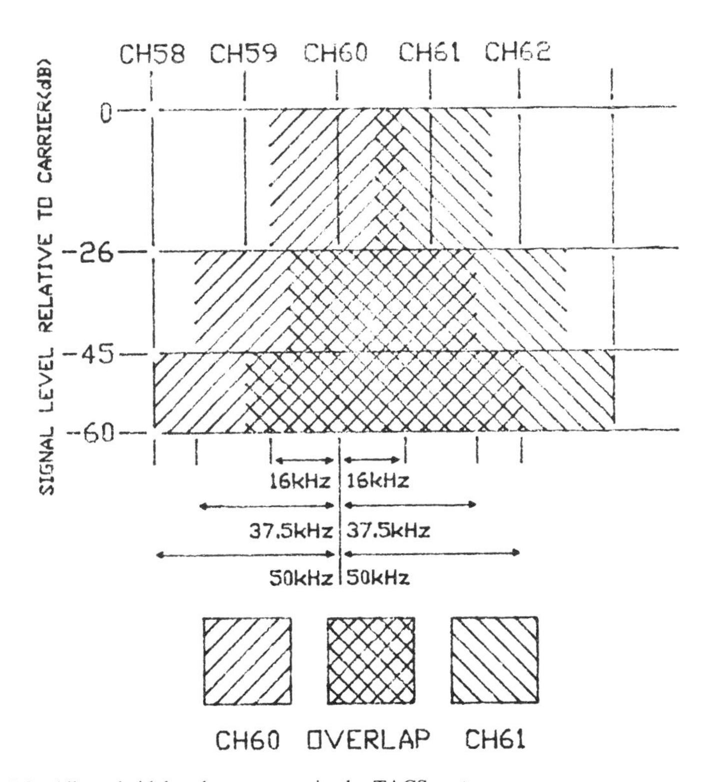

Figure 8.3 Allowed sideband occupancy in the TACS system.

transition and each NRZ zero becomes a one-to-zero transition. This is Manchester encoding, and effectively increases the data rate to 16 kbits/s, a strategy which assists the receiving end in recovering the basic data clock.

8.2.3 Spectrum and channel designation

The spectrum assigned to the TACS system is part of a wider allocation which includes provision for the new pan-European digital system which will begin to be implemented in 1991 (the subject of the next chapter). When the TACS service was initiated, operating licences were awarded to two consortia and the available spectrum divided equally between them. In addition, other prior users of the allocated spectrum had to have time to relocate so that in 1985, when service commenced, the allocations were as shown in Figure 8.4. This

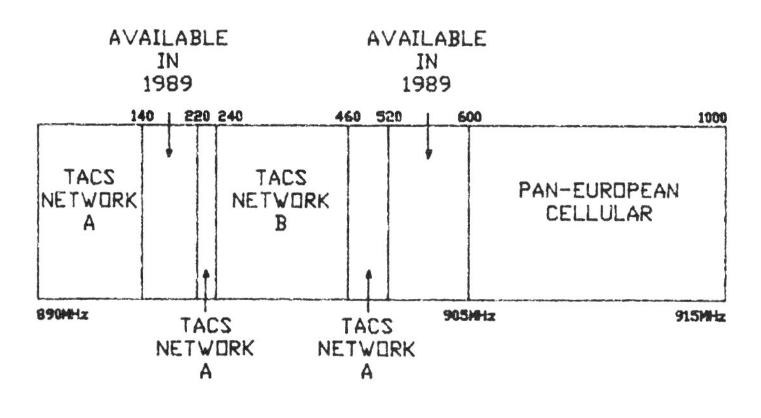

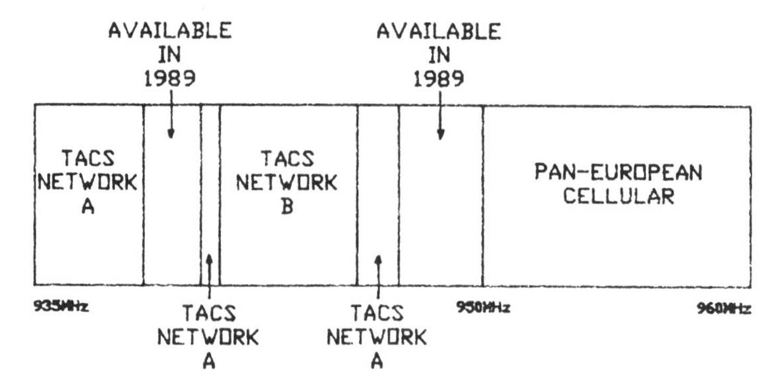

Figure 8.4 The 1985 spectrum allocated in the UK to cellular radio services.

gave each operator 220 channels, with extension to 300 channels each scheduled for 1989.

Two frequency allocations are required to support duplex telephony, and the frequencies shown in Figure 8.4 correspond to the mobile assignments; base stations have their transmit and receive bands transposed. When any transmit frequency is selected in either direction, the corresponding receive channel is automatically chosen to be at 45 MHz separation.

Subsequently, in order to satisfy a massive demand in the Central London area, further frequencies outside the original allocation have been made available for use 'within a 6-mile radius of Charing Cross'. This has given both system operators a further 200 channels, and a provision of another 120 channels should this become a necessity. This extends the frequency range of Extended TACS (or E-TACS) to that shown in Figure 8.5. This is clearly not without some attendant practical difficulties, since the lower limit of the base transmit band is now only 12 MHz from the upper limit of the mobile transmit band, and this makes demands on filter design.

Figures 8.4 and 8.5 indicate the extremities of the allocated frequency bands; channel centre frequencies are offset by 12.5 kHz so that channel 1 in the original 1985 assignment corresponded to a carrier frequency of 935.0125 MHz at the mobile station receiver, and 890.0125 MHz at the base station receiver.

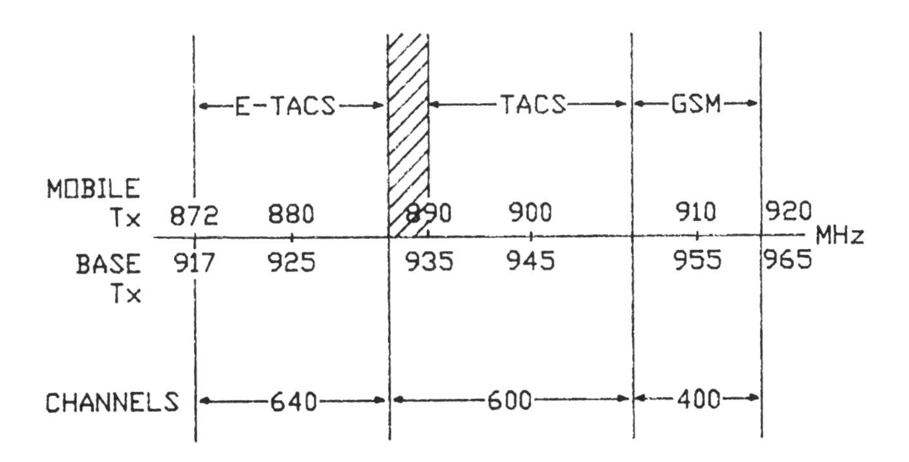

Figure 8.5 Spectrum extension in Central London—the E-TACS allocation.

8.3 Network control activity

8.3.1 Dedicated control channels

Recalling Figure 8.2, it is seen that the 21 channels from the available set are dedicated to the system management and call set-up functions of the FOCC and RECC [2, 3]. In terms of function, the control channels are divided into three categories: dedicated control channels (DCCs), paging channels and access channels. This does not mean to say that these functions are always separated on to channels at different frequencies within the control channel set. In areas of low user density, it would be unreasonably costly in equipment terms to provide three control channels to support modest voice traffic, so it is usual in such circumstances to multiplex all three control functions on to a single data carrier. However, the functions remain distinct and may be considered separately.

The forward control channel (FOCC) comprises the DCC (and the other two channels in the multiplex case), and is the basic co-ordinating channel for the network. It is transmitted continuously, because (i) on switch-on, all mobiles scan the FOCC channel(s) to find the identity numbers of paging channels, and (ii) mobiles rely on monitoring the level of the received FOCC to determine when they are approaching the edge of a cell coverage area and need to be passed on to an adjacent cell base station.

The paging channels, as their name suggests, are used to alert particular mobiles to an incoming call, but they also carry more general network information specific to the locality—channel numbers of the access channels in use, access methods to be used, traffic area identities and so on.

Mobiles use the access channels to contact the base stations to acknowledge paging messages, update the network with their locations by registering with the base stations offering the best radio paths, and to initiate outgoing calls.

All three types of control channel carry sequences of data blocks called 'overhead messages' which contain a multiplicity of status information fields and instructions.

The contents of overhead messages depend on the type of channel which carries them, but when DCC, paging and access channels are multiplexed together, the overhead messages contain all the information in one continuous bit stream.

Before looking further at the signalling formats, it is appropriate to consider the way in which data messages are sent over the speech-compatible channels.

8.3.2 Protecting data messages

Two methods are employed to protect the data against co-channel interference, a necessary procedure since data messages have none of the inherent redundancy of speech and consequently are very vulnerable to corruption. The first is to repeat each transmitted data word several times and take a majority decision on the data word content on a bit-by-bit basis. The second is to use an error-detecting and correcting code associated with each data word. The objective is to obtain a 95% accuracy in data word transmission on the majority decision basis, and to use the error-correcting code to improve this to 99.9%. The code used is a BCH (Bose–Chaudhuri–Hocquenghen) systematic linear block code, which appends a 12-bit parity word to each data block in all the messages transmitted. The number of repeats used is not, however, the same for all data links (FOCC, RECC, FVC and RVC). In the case of base-to-mobile transmissions in the FVC, 11 repeats are used, the remaining links using 5 repeats.

The situation is summarized in Table 8.3.

Table 8.3 Protection of data words

Signalling channel	Data bits	Parity bits	Repeats
Base-to-mobile (forward) control channel	28	12	5
Mobile-to-base (reverse) control channel	36	12	5
Base-to-mobile (forward) voice channel	28	12	11
Mobile-to-base (reverse) voice channel	36	12	5

Further details of the generator polynomials for the codes, all of which have distance 5, are obtainable from [1]. Readers unfamiliar with coding techniques will find them described in [4]. The codes in question have an error correcting capability of one bit and can detect up to four bit errors, so that data word reception is executed as follows. When a message is received, a bit-by-bit majority decision is carried out on the repeated blocks; this corrects many of the errors introduced in transmission at this stage. If a single bit error remains, the parity word is then used to correct it, but if the parity check detects more than one bit in error, then the message cannot be corrected and is therefore rejected. In the case of the Forward Voice Channel, 11 repeats are used to make sure that 5 are received in the mobile.

8.3.3 Signalling formats

With the foregoing background, it is now possible to appreciate the signalling frame structure in the four data link paths, and these are set out in Figures 8.6 and 8.7, showing the control channel and voice channel data formats respectively.

The FOCC is structurally the most complex and contains three distinct sets of information:

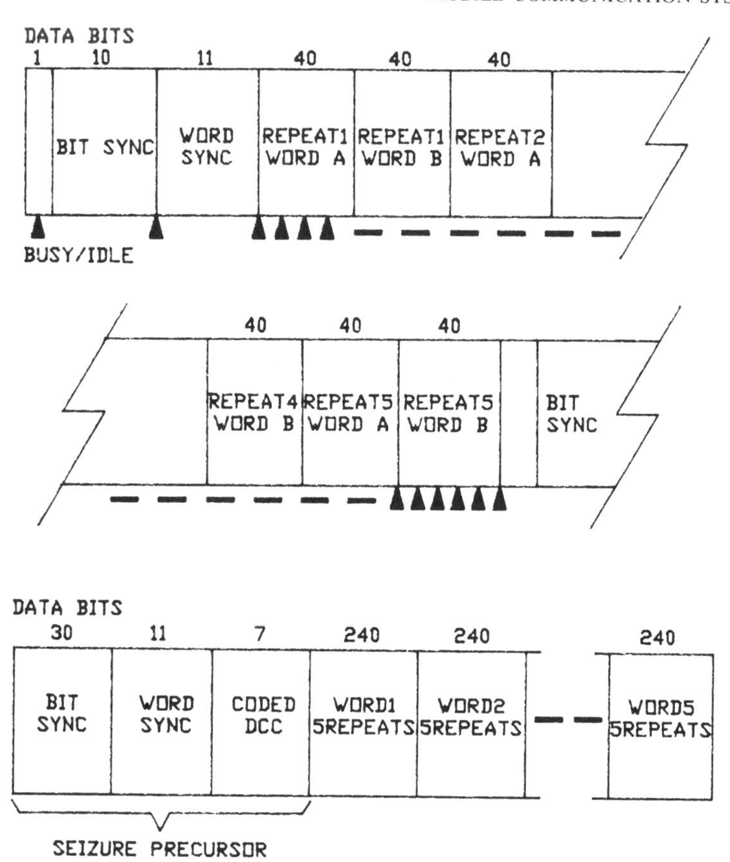

Figure 8.6 Forward and reverse control channel message formats.

(i) Busy/idle information, which indicates the status of the associated RECC (whether any mobile is in the process of updating location or status, or is initiating a call)
(ii) Information data word A, which is addressed to even-numbered mobiles
(iii) Information data word B, which is addressed to odd-numbered mobiles.

The busy/idle bits are always located in the same positions in the frame for ease of extraction and are additional to the information bits within the words. The complete frame is 463 bits long and is repeated continuously.

The data words themselves are complex packages comprising three further categories of message:

(i) Mobile station control messages
(ii) Overhead messages
(iii) Control filler messages.

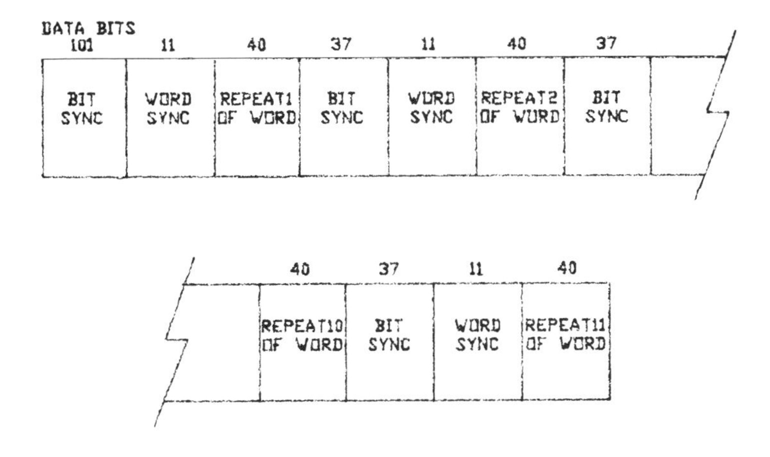

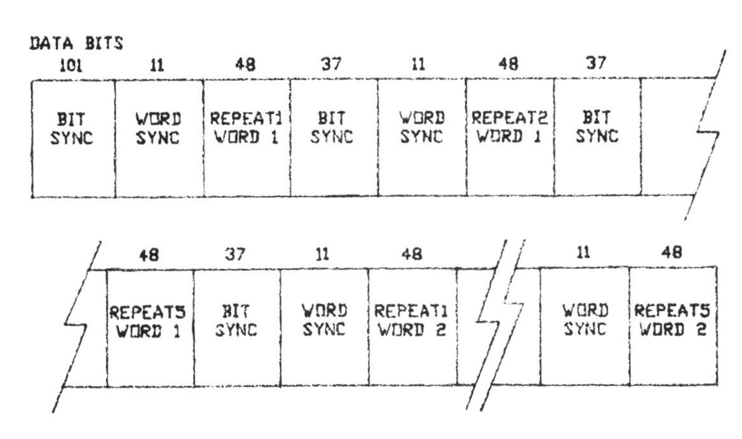

Figure 8.7 Forward and reverse voice channel message formats.

The control filler messages are necessary to make sure that, even when there is no active data to transfer, the FOCC remains continuous for the benefit of mobiles using its measured level.

The format of the overhead message which carries system parameter information is shown in Figure 8.8. The symbols have the following meanings:

T1T2	Type field—set to '11', indicating an overhead word
D'C'C'	Digital colour code field
AID1	First part of the traffic identification field
RSVD	Reserved for future use; all bits must be set as indicated
NAWC	Number of additional words coming field. In word 1, this field is set to one fewer than the total number of words in the overhead message train.
OHD	Overhead message type field. The OHD field of word 1 is set to '110', indicating the first word of the system parameter overhead

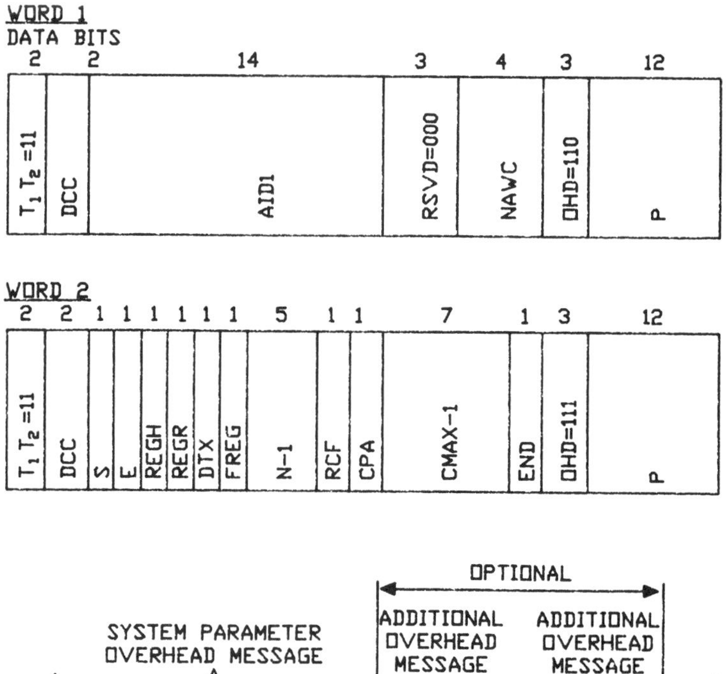

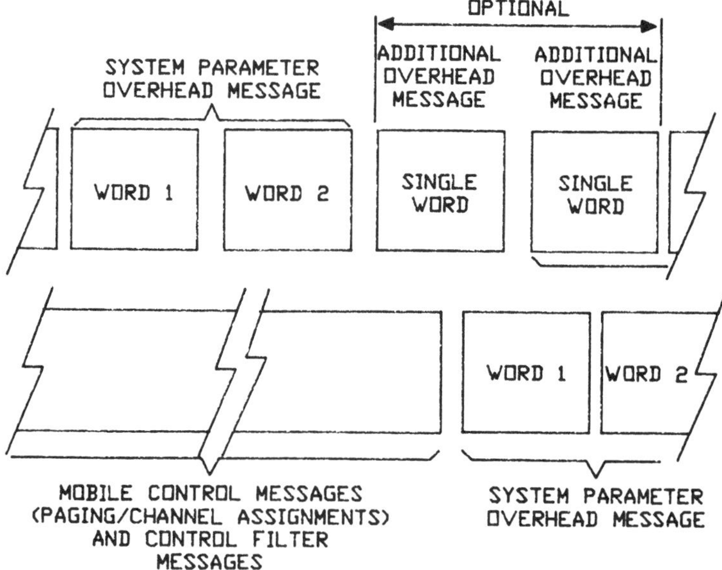

Figure 8.8 System parameter overhead message and structure of FOCC message stream.

	message. The OHD field of word 2 is set to '111', indicating the second word of the system parameter overhead message.
P	Parity field
S	Serial number field
E	Extended address field
REGH	Registration field for mobile stations operating on their preferred system

REGR	Registration field for mobile stations not operating on their preferred system
DTX	Discontinuous transmission field
FREG	Forced registration field
N-1	N is the number of paging channels in the system
RCF	Read control filler field
CPA	Combined paging/access field
CMAX-1	CMAX is the number of access channels
END	End identification field. Set to '1' to indicate the last word of the overhead message train; set to '0' if not last word

Further data can be sent to mobiles if the Read Control Filler (RCF) field is set to indicate that more data should be recovered from the control filler message appearing on the control channel. The system parameter overhead message is sent regularly, and if any additional overhead messages are required they can be appended to the system parameter overhead message, as shown in Figure 8.8, before multiplexing into the format of Figure 8.6.

8.3.4 Supervision

Several of the data field symbols listed above require further explanation in the context of the overall network operation. One of these, the Digital Colour Code (D′C′C′), together with two other signals represent the last components of the signalling infrastructure for TACS. These other two signals are the Supervisory Audio Tone (SAT) and the Signalling Tone (ST) and their roles are as follows.

Once a voice channel has been established, there is clearly a need to monitor the progress of the call as the mobile moves in the Rayleigh fading environment which influences the signal level in its own radio path to base, and also influences the local level of interfering emissions on the same channel from adjacent cell clusters. Left to its own devices, if the mobile experienced co-channel interference temporarily at a greater level than the wanted signal, this interference would capture the receiver and an unwanted conversation would intrude, causing annoyance and jeopardizing privacy of voice channels. The SAT provides the necessary safeguard against this.

The SAT is not in fact a single tone, but three tones spaced close together at around 6 KHz. The SAT colour code fields (SCC) appear in the mobile control message words on the FOCC as two bits, as shown in Table 8.4.

Table 8.4 SAT colour code

Bit pattern	SAT frequency
00	5970 Hz
01	6000 Hz
10	6030 Hz
11	(not a channel designation)

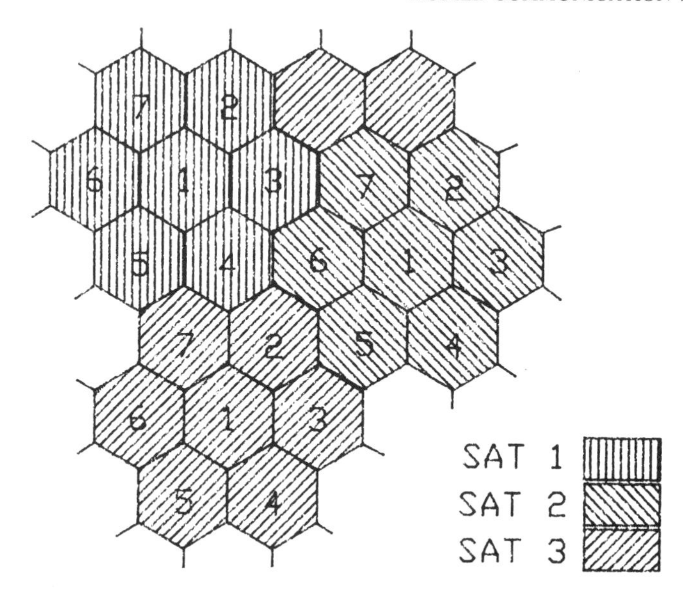

Figure 8.9 Supervising audio tune (SAT) assignment to cell clusters.

The SAT operates as follows. In setting up the voice channel, the base informs the mobile which SAT is appropriate to the complete cluster of cells in which the call is being handled. The mobile then retransmits (or 'loops back') the defined SAT to confirm that the correct connection is made, and continues to transmit this SAT throughout the call. The other two SAT frequencies are assigned to adjacent cell clusters, as in Figure 8.9, and it is now apparent that if a temporarily strong co-channel interferer is sensed by the mobile or base, originating in an adjacent cluster, it will carry the wrong SAT and the audio output will be muted. The only time the SAT is itself muted during a call is when wideband data are being transmitted from the mobile on the RECC.

The Digital Colour Code serves a similar purpose on the FOCC and RECC, but is digital in form in order to be compatible with the rest of the message stream. The equivalent loop back in the control channels is achieved by the mobile retransmitting a coded version of the D'C'C', as in Table 8.5. The final signalling tone (ST) is an 8-KHz tone transmitted by the mobile when it is being run or clearing down a call, to indicate that the receiver is '*on-hook*'.

8.4 System operation

8.4.1 Principal functions

We are now in a position to see how the various signalling functions work together to execute the management of the activities which are required in the operational system. It will be recalled that the primary network tasks are,

Table 8.5 Digital colour code

BIT SYNC	WORD SYNC	CODED DCC*	FIRST WORD REPEATED 5 TIMES	SECOND WORD REPEATED 5 TIMES	THIRD WORD REPEATED 5 TIMES
30	11	7	240	240	240

seizure precursor

BIT SYNC = 1010...1010
WORD SYNC = 11100010010

[*Digital colour code]
Reverse control channel message stream (mobile-to-land)

firstly, to keep track of where all the active mobiles in the system are at any one time, so that calls can be directed to them, and secondly, to manage the process of hand-off by which a mobile engaged in a voice conversation is passed from one base station to another as it passes across cell boundaries. For their part, the mobiles are responsible for making sure that, even when they are not handling voice traffic on their voice channels, they are updating the network about their whereabouts within the cell clusters.

Although the PSTN handles all the normal dialling and routing activities and carries all the final voice traffic, there is clearly a second layer of data base and network management required to support the functions just mentioned. This layer is provided by the Mobile Switching Centres (MSC) which each handle the activities of a number of cell clusters. Figure 8.10 shows MSCs connected to the base stations on the various cells in its traffic area. The MSCs are responsible for keeping the record of the location of all the active mobiles and, so that calls can be routed from the network to any cell, the MSCs must also be connected together as well as to the PSTN. The MSCs are in fact computer-controlled telephone exchanges, specifically designed to support a continuously changing set of subscribers. This record-keeping activity is only possible provided the mobile can continually update the MSCs with their location by a process known as 'registration'. This is the process referred to earlier in which, when not making a call, the mobiles continually monitor one of the common signalling channels and listen to the codes generated by the base stations which identify the traffic area. If the mobile starts to detect errors in the received data stream, it concludes that the signal strength has dropped below a usable level, and rescans all the signalling channels to search for a higher signal level corresponding to a different base station. If it acquires a new, stronger channel and detects a new traffic area, it calls the base station which is generating it, identifies itself, and the new base in turn passes on the new arrivals identification to the MSC, which updates its data base appropriately.

In 'in-call hand-off', a mobile engaged in a call has its signal strength monitored by the base station handling the call, and if the signal strength drops below a certain threshold, then the base alerts the MSC that the mobile

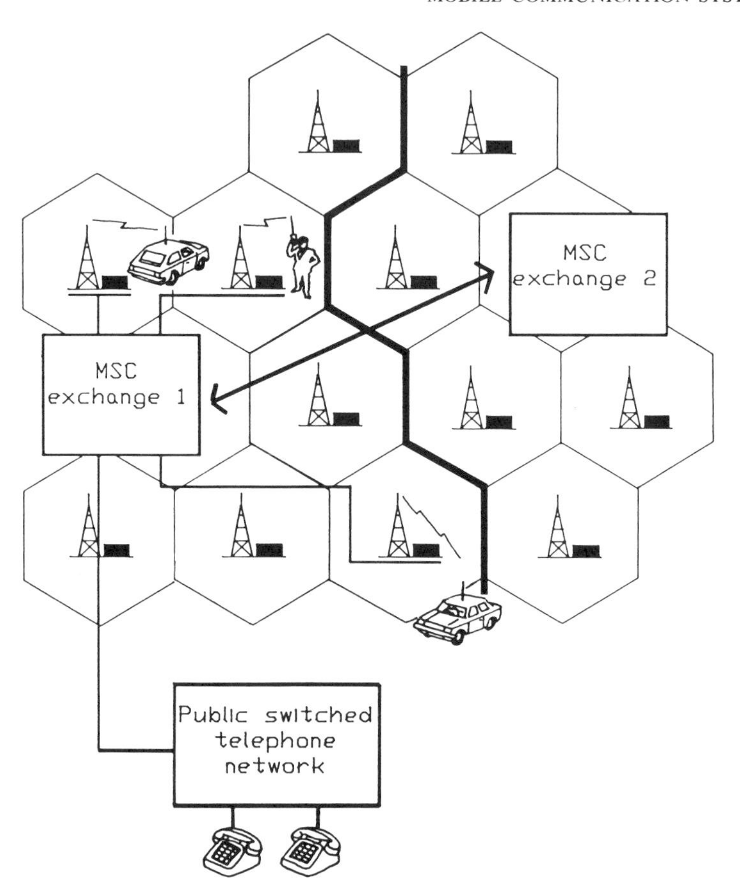

Figure 8.10 Mobile switching centres.

in question is probably nearing the edge of the cell and may require to be handed off to another base. The MSC then commands the scanning receivers in surrounding cells to monitor the call in question and choose the best cell to transfer the call to. Once the optimum cell to hand-off to has been identified, the MSC alerts the new base station of the need for a channel and when this has been selected, the original base instructs the mobile to transfer to the new channel.

One further factor which must be taken into account in the UK is the existence of two complete networks operated by separate organizations—Racal-Vodafone and Cellnet. Each has half the available spectrum, but both systems have the capability of handling mobiles from the other network if service from the other network is not available. Users, however, subscribe to only one of the networks (the 'primary' network) and attempt to use this wherever they can, reverting to the 'secondary' network if they cannot. The secondary network operator subsequently charges use back to the primary provider, who passes the charge on to the customer.

8.4.2 Mobile scanning

This last factor is responsible for the format of the scanning procedure which mobiles adopt at switch-on. All mobiles are programmed with the channel numbers of the DCCs in both networks, but those of the primary network are marked. At switch-on, the mobile scans the 21 DCCs of the primary network, looking for the channels with the strongest and second-strongest signals. It first takes up the strongest channel and looks for overhead messages. If it is successful in reading these, it will find information about the channel numbers of the paging channels; if unsuccessful, it will try the second strongest DCC and try again. If this strategy fails, it is likely that the mobile is out of range of its primary service provider's network, so it repeats the whole procedure with the DCCs of the secondary network.

When the mobile has obtained information about the paging channels it scans these, again looking for the strongest and second-strongest signals, taking up the strongest initially and attempting to read the overhead messages. These contain information about the traffic area in which it is operating and a number of additional network parameters; they also stimulate a registration message on the RECC. As before, failure to read the strongest channel causes the mobile to try the second-strongest, and if this too fails then the whole scanning procedure is restarted from the beginning. The entire procedure, if the second-strongest choice in each case finally proves successful, can take about 17 seconds, but if first tries are successful, this falls to between 5 and 10 seconds.

Once settled on a paging channel, the mobile enters an 'idle' state, and continues to monitor the messages being transmitted. If the mobile moves away from the base station to which it is listening, then it detects a fall in signal level, and when messages cease to be readable, the mobile re-enters the scanning routine looking for a more suitable base station.

8.4.3 Registration

Two types of registration are available in the TACS system: 'forced' and 'periodic'. Forced registration requires that whenever the mobile performs a re-scan operation and finds a new paging channel, it compares the traffic area identity with the one it was receiving before the re-scan and, if a change is detected, it re-registers with the new base station, updating its memory with the new identity at the same time.

Periodic registration allows the mobile to store the last four traffic areas it has visited in memory, together with a numeric indicator for each. It then counts every reception of a traffic area indication in the memory, and when this totals the number of counts corresponding to the numeric indicator, it transmits a registration message and re-starts the count. If the mobile receives a new traffic area identifier, it adds this to the store (deleting the oldest), and registers with the new traffic area. This procedure gives the network considerable flexibility in controlling registration information.

8.4.4 Call origination

All registrations and call initiations require access, and the mobile obtains this by first scanning the network's access channels, whose numbers it will have obtained from the overhead messages on the paging channels. As with other scanning operations, strongest and second-strongest channels are noted, and monitored to obtain access instructions from the overhead messages on the access channels. Once the mobile has received the parameters of the access channel, it checks the status of the busy/idle bit stream from the base station. If this indicates 'idle'—that is, no other mobile is attempting an access—then the mobile waits for a randomized delay time, starts to send its message and continues to monitor the busy/idle bit stream, waiting for this to set to 'busy', indicating that the access channel has been captured. To make sure, however, that the mobile has captured the access channel and that the busy status is not the result of some other mobile obtaining access, the time is measured between starting to send the message and the busy/idle stream setting to 'busy'. If this interval is correct, the mobile continues to send its message. If the interval is too short or too long, the mobile concludes that some other mobile has captured the channel, aborts the message and starts again.

After sending its complete access message, the mobile remains on the access channel awaiting instructions regarding which voice channel to go to and which SAT it should use. Registration accesses receive only a confirmation message which returns the mobile to 'idle'.

8.4.5 Call receipt

A call originating in the PSTN is passed by the MSC to the base in the known location of the wanted mobile. A paging call is then transmitted on all the paging channels in the appropriate traffic area (the network does not know which paging channel the mobile has settled on). When the mobile receives the paging message, it accesses the base station in the usual way, but indicates that the access is a response to a paging message. The base then allocates a speech channel and SAT to which the mobile responds, and then transmits an 'alert' message causing the mobile to ring and to transmit the continuous 8 kHz signalling tone. When the user answers, the tone stops and the call proceeds.

8.4.6 Hand-off

This is the MSC-controlled process described above, under 'principal functions'. Having located a best cell to handle the mobile, whose falling received-signal strength at the base station sustaining its call has initiated the field-strength measurement in surrounding cells, the MSC requests the best cell base station to allocate a voice channel. The base notifies its choice to the MSC, which then commands the original base to instruct the mobile to change channel. A short signalling burst is sent to the mobile on the voice channel, giving the new channel number, and the mobile re-tunes. The perceptible part of this process to the user, the actual re-tuning of the mobile, takes about 400 ms, during which the receiver is quiet.

8.4.7 Call termination

If the mobile user terminates a call, the mobile signals the termination when the user replaces the handset, by sending a long burst (1.8 s) of 8 kHz signalling tone to the base station, and then re-enters the control-channel scanning procedure.

If a PSTN caller clears down first, a release message is sent to the mobile which acknowledges by sending a burst of 8 kHz signalling tone before resuming the control-channel scanning procedure.

8.5 Some system comparisons

Returning to an earlier comment that all the analogue cellular systems implemented in various countries have many features in common, it is of interest to compare major system parameters in relation to TACS. The principal design considerations centre on two issues: control strategies, and designs for defined carrier/interference operation. The control channel commitment of the TACS/AMPS system at 21 channels is the highest, with NAMTS using only one. The second aspect, of carrier-to-interference ratio, is closely related to the choice of FM deviation ratio, which was discussed in detail in the TACS/AMPS case. In the NMT and NAMTS systems, sideband energy is constrained to occupy only the 25 kHz allocated channel, while in TAC/AMPS the modulated carrier occupies around 32 kHz. The former strategy permits use of adjacent channels in the same cell (if need be), while the latter precludes this. However, in clusters of seven cells or more, this limitation of TACS/AMPS is not significant, and the system has the considerable advantage of better carrier/interference performance.

Comparisons between cellular systems can be interpreted in several ways—in particular the calls per cell per busy hour is sometimes referred to as a measure of 'efficiency'. But this is probably misleading, since in implementing a system in areas of high demand, a number of complete cell clusters will be deployed, and all the 300 available channels put into service in each. This means that 300 base station transceivers will be required per cluster whichever system is chosen, and, to cater for the same level of demand, the same number of complete clusters would need to be set up. Therefore, the total number of calls per cluster per busy hour is probably a better guide to relative performance. It is seen that on this basis, there is very little to choose between all the systems cited. Of course, the systems which use larger numbers of cells per cluster need, on the face of it, more base station sites, but since the 9- and 12-cell frequency re-use patterns involve cell sectoring, this factor is not particularly significant.

The history of system introduction in Europe tends to confirm the general conclusion that the differences among the competing systems relate to second-order issues, and countries not committed to developing their own national systems based on their own telecommunications industries have faced a difficult choice.

Since, although the systems are similar in general principles, they are incompatible, it is easy to appreciate why pressure to define a common standard for Europe had become very great indeed by the mid-1980s, with several major manufacturers each developing equipment based on digital techniques. The final chapter of this book looks at the progress of these systems to field trials and the subsequent definition of the pan-European digital system.

References

1. United Kingdom total access communications system, mobile station—land station, compatability specification, **Issue 3** (March 1987) British Telecom Research Laboratories.
2. ibid. section 3.7.
3. ibid. section 2.7.
4. Lin, S. and Costello, D.J., *Error Control Coding: Fundamentals and Applications* Prentice-Hall (1983).

9 Digital cellular radio systems

9.1 Digital versus analogue for second-generation cellular systems

In the last chapter, the use of established analogue modulation techniques was described in the context of high-capacity cellular radio telephone applications. Conclusions regarding existing systems operation and future developments may be drawn, and these centre on the issue of capacity. Already, first-generation systems are showing signs of capacity saturation in major urban centres, even with a relatively modest total user population. Clearly, major capacity increase will be called for to meet future demand, and second-generation systems must address this challenge. What are the options available? To extend existing system operations, more spectrum would have to be made available. Alternatively, new approaches to system design must be found to utilize spectrum more efficiently.

Two aspects of existing systems define their capacity limits: the bandwidth occupied by each voice channel, and the minimum carrier-to-interference ratio in which mobile and portable equipments can operate. In looking at future systems, further factors must be considered. For a fixed allocation of spectrum, massive increase in capacity implies corresponding reductions in cell size, and this means that signalling activity will be increased as more rapid hand-offs occur and as the cell base stations are required to handle more access requests and registrations from the community of mobiles and portables in each cell. Improved efficiency of access and hand-off must be sought in next-generation systems.

Returning therefore to the basic choice between analogue and digital approaches to a capacity increase of at least an order of magnitude, it might appear that much of this increase could be obtained by implementing a narrowband single-sideband modulation strategy—as was noted in Chapter 4, amplitude companded SSB systems have been operated experimentally at 2.5 kHz per channel bandwidth on a 3 or 4 kHz channel centre spacing. In addition, TTIB and FFSR offer means for overcoming most of the difficulties of operating SSB systems in the Rayleigh fading radio environment. However, this argument ignores the issue of required C/I for satisfactory performance. Taking this into account, the inability of linear modulation schemes to offer discrimination against co-channel interferers greatly reduces the gain obtainable from reduced channel occupancy by each telephone conversation. Lee [1] and Prabhu and Steele [2] have separately estimated that although gains are available from implementation of SSB techniques, these are not great enough to justify their introduction in the cellular environment. Clearly, the technology must develop in a different direction.

In looking at digital techniques in general terms, it is sensible to formulate criteria by which the available techniques can be assessed. Several authors have made suggestions in this direction—Maloberti's [3] are a good summary.

The digital system must have six major attributes:

(i) High subjective voice quality
(ii) Low infrastructure cost
(iii) Low mobile station cost
(iv) High radiofrequency spectrum efficiency
(v) Good capability to support hand-held stations
(vi) Ability to support new services.

These lead to three major criteria as a basis for choice:

(i) The cost of the system (mobile equipment and infrastructure)
(ii) Spectrum utilization
(iii) The risk factor of the particular solution.

What then are the technical possibilities available from the digital approach and how can these be exploited? Of primary interest is the capability of signal processing strategies to extract digital information from received signals in the presence of high levels of noise and co-channel interference. If this aspect is the dominant criterion in system design, the known capability of spread spectrum systems operating in virtually zero signal-to-noise ratio conditions would point towards a wideband digital approach to the cellular system problem, and indeed, this has been explored in a system, CD900, which will be described later. However, when it is intended that the new system shall have high capacity but low terminal cost, then implementation of a very advanced and sophisticated technology presents some difficulty, and alternatives must be considered.

At one extreme, a single channel per carrier (SCPC) approach might be attractive, but this does not offer the opportunity to exploit a major attraction of the wideband approach. This arises because, as has been noted at the beginning of Chapter 7, the coherence bandwidth in the Rayleigh channel at 900 MHz is narrow (around 100 kHz)—if transmissions are narrow relative to this, then fading will affect all the components of the complete signal similarly, so that the only way to avoid breaks in transmission due to multipath fading is to exploit diversity techniques. However, if the transmitted signal is wider than the coherence bandwidth then only parts of it will suffer fading at any one time as the receiver moves about in the multipath environment. The wideband approach, therefore, offers the opportunity to use the multipath propagation process to advantage. But clearly, the channel does not need to be wideband in the spread spectrum sense, it only needs to be wider than the coherence bandwidth.

Three possible basic strategies, therefore, can be perceived:

(i) Single channel per carrier (usually simply referred to as FDMA).
(ii) Wideband—each carrier uses all or a substantial fraction of the available spectrum allocation (wideband TDMA).
(iii) Intermediate—each carrier carries a modulation rate which results in bandwidth occupancy of more than the coherence bandwidth, but many such carriers need to be used in a FDM mode to occupy the available spectrum. This is generally now referred to as 'narrowband TDMA' or 'FDM-TDMA'.

9.2 Choice of basic system architecture

It has to be said at this point that a choice of one of the above approaches in preference to the others is not a simple one and, in fact, is likely to be a function of many factors other than the fundamental one of user capacity. The prevailing spectrum availability and regulatory environment, the maturity of essential enabling technologies, and so on will have major influence on choice. This situation emerged very clearly in the preparations made by European cellular system manufacturers for the decision-making at the end of 1986 which determined the basic system architecture for the intended pan-European digital radio service to be introduced in 1991. All three approaches were offered, including one mixed system (the Philips MATS-D) which sought to exploit the dissimilar problems of up and down links from mobiles to bases by operating single channel per carrier on the uplink and TDM on the downlink.

The next step is clearly to set out the advantages and disadvantages of each, and then to see how these various aspects are exploited or overcome in particular systems.

9.2.1 Single channel per carrier, FDMA

Advantages:

(i) Individual channels operate well within the coherence bandwidth, so that equalization of the channel will not materially improve performance, and capacity increase is achieved by reducing the information bit rate and finding efficient channel codes.

(ii) Technological advances required for implementation are relatively modest, and a system can be configured so that subsequent improvements in terms of speech coder bit rate reduction could be readily incorporated.
(iii) The system can be flexible, as are the existing cellular schemes in handling large rural cells and small urban cells.

Disadvantages:

(i) Since the system architecture, based on FDMA, does not differ markedly from the analogue equivalent, the improvement available in capacity relies on operation at a reduced C/I ratio. But the narrowband digital approach gives only limited advantages in this regard so that only modest capacity improvements could be expected from a given spectrum allocation.
(ii) Narrowband technology involves narrowband filters, and because this is not realizable in VLSI, this may set a high cost 'floor' for terminals even under volume production conditions.
(iii) The maximum bit-rate per channel is fixed and small, inhibiting the flexibility in bit-rate capability which may be a requirement for computer file access in some applications in the future.

9.2.2 Time division multiple access

Advantages—wide and narrow-band:

(i) Offers the capability of overcoming Rayleigh fading problems by appropriate channel equalization.
(ii) Has flexibility in bit rate, not only for multiples of basic single channel rate but also submultiples for low bit rate broadcast-type traffic.
(iii) Potentially integrable in VLSI without narrowband filters, giving a low cost floor in volume production.
(iv) Offers opportunity for frame-by-frame monitoring of signal strength/bit error rates to enable either mobiles or bases to intitiate hand-off and to execute this rapidly.

Disadvantages and wideband versus narrowband:

(i) For mobiles and particularly hand-portables, TDMA on the uplink demands relatively high peak power in transmit mode, which shortens battery life between recharges.
(ii) Exploiting the advantages of matched filtering and correlation detection to the full demands a substantial amount of signal processing which, again, in current technologies is demanding in power consumption and introduces delay in the signal speech path.

Comparing wideband and narrowband TDMA approaches, it is seen that the disadvantages are more pronounced in the wideband case than in the narrow, deriving from the need for considerable advances in signal processing techniques and enabling technologies before effective implementation can take place. The issue of high portable peak transmit power is particularly significant in the rural situation of large cells which, in the wideband case, would limit effective cell radius as far as hand-portables are concerned to a maximum appropriate to a dense user-demand urban area.

The potential of wideband TDMA systems is clearly great, however, and it has to be concluded that all three basic approaches to digital cellular system design have their place in future service provision with single channel per carrier systems nearest to the marketplace, narrowband TDMA offering a way forward for second-generation systems to take over from established analogue systems as they become saturated, and wideband technology contributing to a possible third-generation system.

9.3 Essential techniques for digital implementation

So far, the arguments in favour of a digital approach to cellular system design have been set out, but regardless of the choice of wideband TDMA, single channel per carrier or narrowband TDMA, a number of aspects of subsystem design are essential to realization of the systems. These relate to speech coding, channel coding, and channel equalization. These will be considered next, before turning attention to the further problems of channel frequency hopping and spectrum spreading which represent the remaining 'tools' needed to realize operating systems.

9.3.1 Speech coding

It has already been noted that reduction in channel bandwidth is an important factor in improving spectrum utilization in analogue cellular systems, and while the improvements in user capacity possible with digital systems relate more to operation in reduced C/I environments, the need to constrain the bandwidth occupied by the baseband signal prior to modulation is nevertheless of importance if the C/I benefits are not to be sacrificed. In the digital context, bandwidth translates into bit rate, and the nature of the problem is readily perceived by reference to established experience in the line telecommunications industry. For many years, normal speech traffic in much of the PSTN has been carried as digital signals using Pulse Code Modulation. As indicated in Chapter 4, there are two aspects to this process: the analogue signal must be sampled at a rate at least twice as high as the highest frequency components in the analogue signal, and the speech signal must be processed by companding to minimize the number of quantized amplitude levels into which each sample is encoded. Simply following this procedure has resulted in a standard required for the preservation of 'toll quality' speech of 64 kbits/s. Again recalling Chapter 4, any digital modulation method which may be used to transmit this bit rate on a carrier will occupy a very much wider bandwidth than that occupied by an analogue FM carrier. Techniques must be sought, therefore, to preserve the essential elements of speech while reducing the bit rate of the baseband digital signal which describes it.

A number of methods have been developed for realizing this, and research is continuing to secure further bit rate reductions relative to those rates which are currently accepted (16 kbits/s for full toll quality, 9.6 kbits/s and 4.8 kbits/s for reduced-quality services). Speech coding is now a highly developed and specialized science, and here it is possible to do no more than give an outline of the principal techniques which are now in use.

Sub-band coders. This technique relies on dividing up the baseband speech signal into a number of sub-bands and then coarsely encoding the amplitude of the signal contained within each sub-band.

There are a variety of strategies for performing the sub-band energy encoding, and there is clearly a choice of how many sub-bands to use. Figure 9.1 shows a typical sub-band coder implementation in general terms—the baseband speech from 0-4 kHz is divided into eight sub-bands of 500 Hz each. The uppermost sub-band, from 3.5 to 4 kHz, is neglected, while the others are allocated code bits according to a semi-adaptive bit allocation algorithm. In the receiver, high-frequency regeneration can be used to improve the subjective quality of the speech.

Predictive coding. This strategy relies on a different approach. Here the encoding and decoding process is set up by modelling the vocal tract in the

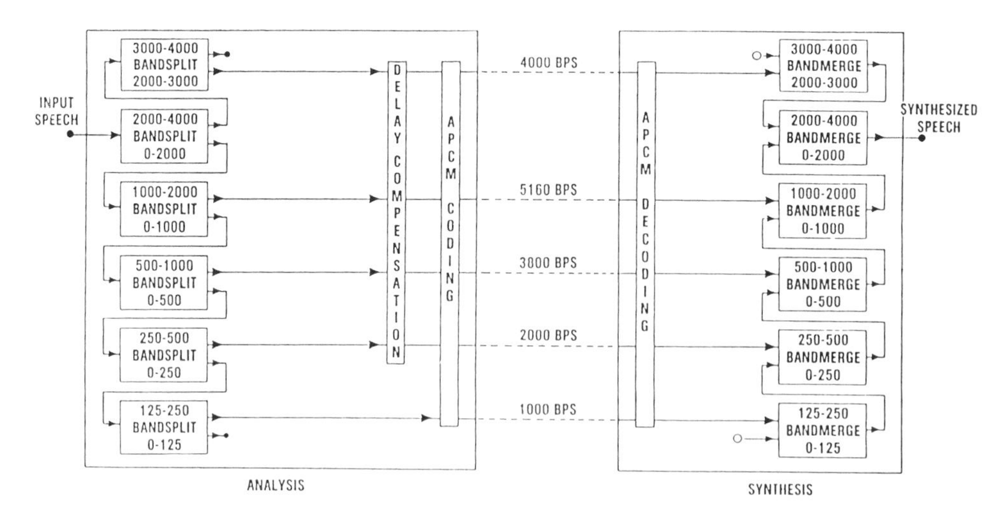

Figure 9.1 Sub-band speech coder.

speaker by a linear predictive filter, a process accomplished by analysing the speaker's utterances in blocks of samples corresponding to a few tens of milliseconds and then transmitting data about the characteristics of the filter and information about how the filter should be excited. The process is referred to as 'Linear Predictive Coding' (LPC). There are a number of variations distinguished by the way in which the excitation signal is transmitted along with the filter coefficients. Clearly the excitation signal can be allowed only a narrow analogue bandwidth (corresponding to a restricted bit rate), and an obvious technique to implement is one somewhat akin to delta modulation in which a residual signal is transmitted after comparing the prediction results with the original.

Residual excited linear predictors (*RELPs*). These use similar principles. Again there are several types, of which the simplest is the baseband RELP. In this type of coder, the input speech is fed through an inverse filter using the derived prediction coefficients which are regularly updated to follow the changes in the speech signal. The inverse filtered speech is then compared with the original, and a residual signal obtained. The prediction coefficients are calculated so as to minimize the residual signal which is then transmitted in some form together with the prediction coefficients. In the baseband RELP coder, the residual signal is low-pass filtered and then sampled at a low rate; some artificial reconstruction of lost high-frequency components in the receiver is then required.

Alternative strategies rely on using more sophisticated coding techniques for transmitting the prediction residual signal. In Multi-pulse Excitation (MPE-LPC), the prediction residual is approximated by a sequence containing a reduced number of samples. These are then encoded in amplitude and time of occurrence, and transmitted with the predication coefficients. In the decoder, the samples are used to excite the prediction filter, but additional signal processing involving spectral weighting functions is required as well. This, together with the need to implement iterative optimization procedures to set up the pulse positions and amplitudes, makes implementation of this type of coder complex.

An alternative with reduced complexity restricts the positions and amplitudes of the excitation pulses to a defined set corresponding to a regular a grid This strategy, Regular Pulse Excitation (RPE-LPC) may be seen as a compromise between baseband and multipulse approaches in terms of performance and complexity, but with potential for further development.

The emergence of increasingly powerful digital signal processing chips in parallel with fundamental research into speech coding techniques continually offers new opportunities to increase the sophistication of coding strategies without incurring unacceptable penalties in processing delay. Recently, increasingly impressive results have been obtained with both sub-band and linear predictive coders as a result of implementing more powerful processing algorithms. In sub-band coders, adaptive quantization of the individual sub-band signals offers and attractive enhancement; in LPC techniques, additional facilities in long-term prediction and in accelerating the processes of solving the linear equation set which determines the prediction coefficients, are making substantial progress. Code-excited linear prediction (CELP), a particularly effective technique of recent origin, uses a stored code book of speech elements which, in conjunction with linear predictive coding, gives a robust coder with good transmission bit rate economy. It is to be anticipated that the commitment to research and development currently associated with speech coding work will continue to yield steady and significant improvements.

9.3.2 Channel coding

Channel coding has been touched on elsewhere in this book, previously in the context of data transmission, so it is appropriate to consider (i) the implications of carrying real-time speech as well as data over the channel, and (ii) the consequences as far as coding is concerned of following wideband or narrowband strategies in cellular system architecture.

The real-time speech requirement means that the much-used protocol for data packets known as ARQ, or automatic retransmission request, cannot be applied—an ARQ system needs only to utilize a sufficient amount of coding to enable the receiver to detect errors in a packet and initiate a retransmission request if errors are found. Real-time speech does not allow time for this procedure. The channel characteristics are significant because, in a narrow channel, the effects of multipath transmission are to produce rapid deep fades during which bursts of errors can be expected. In the wideband channel, on the other hand, a much greater degree of statistical independence among errors is found. These considerations have much influence over the coding strategies employed to protect digitized voice transmissions. A further factor relates to the type of speech coder in use—speech coders are generally considered satisfactory only if they will tolerate an incidence of bit errors of roughly 1% or less, but some data bits in coder output words are often more susceptible than others to errors (the coder performance is degraded more by errors in some parts of the word than others). These sensitive bits need more robust protection by coding than others.

The essential process of coding are readily appreciated. The output bit stream from the source (speech coder) is divided up into blocks of predetermined length and the additional bits are added, but the sequence of these redundant bits is derived systematically from the information bits. The complete code word is then longer than the original, and since the original source bit rate is required to sustain the message traffic in real time, the effect of the coding overhead is to require a higher bit rate in the channel than that produced by the source.

The redundant bits in the code word can be used simply to detect errors in the information block, to correct for errors uniformly distributed throughout the code word, and to correct for burst errors. Generally, the more that is demanded of a code, the greater the coding overhead becomes and, as with any signal processing function, the complexity needed to generate the code and the delay involved in decoding also become greater. Clearly, the objective in code design and in choosing an appropriate code for given requirements lies in minimizing the coding overhead while still achieving an acceptably low probability of incorrectly decoding the wanted information bits. This strategy is assisted by treating those bits within a message which need strong protection separately from those which are less critical.

Two strategies are widely referred to for structuring the code words. The most commonly used is the linear block code, and in particular, the Bose–Chaudhuri–Hocquenghem (BCH) variant. An alternative technique is offered by convolutional coding, of which the best known type is the Reed–Solomon code. Insofar as generalizations are possible, Reed–Solomon codes perform well in narrowband channels where burst errors predominate, and BCH codes work well where uniformally distributed statistically-independent errors are encountered.

One further strategy, widely used where burst errors occur, involves bit interleaving. When a succession of code words is being transmitted, if the information bits from the source are dispersed among a number of code words in series, then a burst error which affects a large part of a complete code word affects only a small part of the original source output. In other words, bit interleaving has the effect of dispersing burst errors over a number of information blocks. There are, of course, limits to how far this procedure can be taken, since in the decoding process, the decoder must wait for all the code words over which the original information bits have been dispersed before it can recover the message, and processing delay is an inevitable consequence.

9.3.3 Channel equalization

Reference was made earlier to the potential in wideband systems for channel equalization as a means of combating the effects of multipath transmission. The function of the equalizer is readily understood from considering the effect of transmitting an impulse into the channel and observing the received signal at a remote point in a multipath environment as a function of time. The received signal is dispersed in time, as illustrated in Figure 2.1 which shows energy arriving at different times according to the different paths traversed from transmitter to receiver. The equalizer's function in conceptual terms is therefore to take the various time-dispersed components, weight them according to the known characteristics of the channel, and sum them together inserting appropriate delays among the components to achieve, as nearly as possible, a replica of the original transmitted signal. The problems encountered in the mobile radio channel, however, are accentuated by two factors: (i) the channel is not static, so that as the mobile moves through the multipath environment, the equalizer must continually adapt to the changing channel characteristics which vary at the fast fading rate; and (ii) since signal components arrive at the mobile from all directions relative to the fixed base station, independent of the motion of the mobile, Doppler spread arises as well, giving a channel response as shown in Figure 2.21.

Despite these difficulties, two techniques have proved successful in providing adaptive equalization for the Rayleigh channel. The first of these is shown in Figure 9.2 and is the Decision Feedback Equalizer (DFE). The upper section of this—the forward path—by itself is a linear transversal equalizer which is a standard device in which the values of the received signal are linearly weighted by the equalizer coefficients C_n and summed. The algorithm used to update the coefficients C_n is either 'zero forcing', which forces the combined signal and equalizer impulse response to zero at all but one of the T-spaced instants, or 'least mean square', which minimizes the mean square error at the equalizer output. In the DFE of Figure 9.2, the feedback cancels interference from symbols which have already been detected.

The alternative to channel equalization is implementation of the Viterbi algorithm which performs Maximum Likelihood Sequence Estimation on the basis of knowing how the channel distorts received signals after demodulation. The Viterbi receiver generates a signal set which includes the channel distortion by measuring the channel impulse response, and then compares the actual received signal with the known possibilities on a bit-by-bit basis.

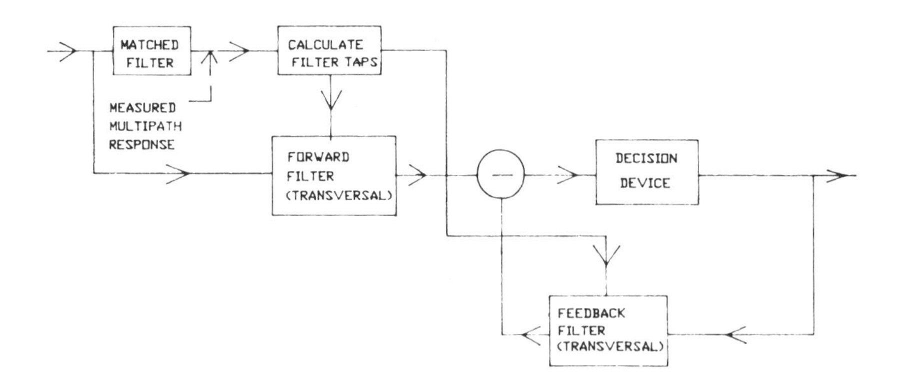

Figure 9.2 Decision Feedback Equalizer.

Both methods require that the receiver be given frequent opportunities to reassess the status of the channel, and this means that a training sequence of known symbols must be transmitted regularly to provide the basis for updating. The major factor in determining which strategy to deploy for a practical system is the kind of delay spread anticipated. The Viterbi algorithm (VA) has much to recommend it in performance terms, but becomes disproportionately complex to implement relative to DFE if delay spreads in excess of around 10 μs are encountered. At the time of writing, the debate on the relative merits of DFE and VA in the context of particular system architectures is by no means resolved.

9.3.4 Frequency hopping

It is apparent from the above, that a number of factors dictate the format of digital signals in any of the possible system configurations outlined in section 9.2. Coding and channel equalizer training require that transmitted bits be grouped into blocks or frames of fixed duration. Bit interleaving can be extended over a number of frames so that if a frame is lost due to fading or co-channel interference, messages can still be exchanged satisfactorily. If signals are formatted into fixed-length frames, then the possibility exists, in single channel per carrier and narrowband TDMA, to randomize both co-channel interference and the effects of multipath propagation by frequency hopping. Figure 9.3 illustrates the effects of applying the frequency hopping strategy to an FDM-TDMA system.

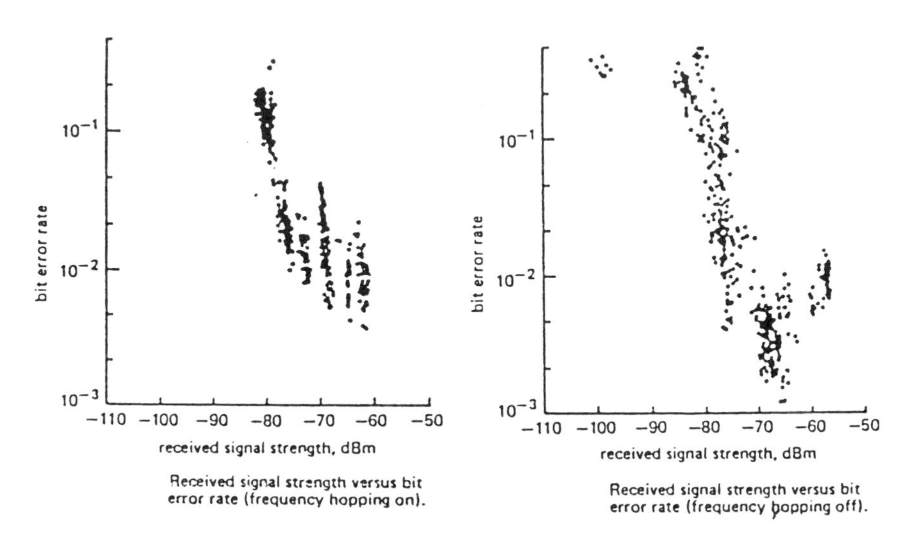

Figure 9.3 Effects of frequency hopping on bit error rates: frequency hopping on (left) and frequency hopping off (right).

9.4 Example systems

Provision of a single, unified cellular radio system for all countries in Europe has provided a unique opportunity and challenge to the European telecommunications industry. By the end of 1986, major companies had largely completed the development of possible systems competing for the pan-European standard, and these differed greatly in basic architecture and complexity. Trials in early 1987 revealed that, in broad terms, all the contenders were viable and it was concluded that by combining the best features of all, an optimum system could be defined. In the context of this overview of digital cellular possibilities, two of the competing systems are outlined before considering the final European choice.

9.4.1 An asymmetrical system—MATS-D

Before looking at the MATS-D system architecture, it is helpful to examine one final aspect of system design which has been referred to only in passing previously. This is the use of code division multiplexing (CDM) as a means of separating simultaneous users of the system radio segment. The essence of CDM is the use of an additional dimension in signal space which enables transceivers which occupy the same bandwidth (in contrast to FDM systems) and exist at the same time (in contrast to TDM systems) still to be separable. This is done by spreading the wanted transmission bandwidth, multiplying the wanted digital signal by a spreading code at a substantially higher bit rate. (The resulting bits are usefully referred to as 'chips' to distinguish them from message bits.) If the spreading code has the spectral properties of a pseudo-random binary sequence, then the result of the multiplication is a noise-like signal, but when the spread spectrum signal is processed through a correlator, which has knowledge of the sequence used to perform the spreading in the transmitter, then the wanted signal is recovered. If many sequences can be identified which have low cross-correlation characteristics then an equivalent number of spread spectrum transmissions can co-exist but remain orthogonal. The general theory of Direct Sequence Spread Spectrum systems is beyond the scope of the present text—and indeed no current or proposed cellular system for early implementation uses this technique—but it has the attraction of achieving, in principle, two important objectives. Spreading can be used to generate a transmission of many times the coherence bandwidth to optimize the effectiveness of channel equalization; spread spectrum systems will also function in very bad signal-to-noise or signal-to-interference environments, which is an important factor in frequency re-use strategies. Details of spread spectrum methods are to be found in references [4]–[5].

Returning to the MATS-D system, the asymmetry referred to relates to the separate optimization of up- and downlinks from mobile-to-base station and vice versa. Taking the downlink first, here the traffic to all mobiles is co-ordinated from the central base station interface with the PSTN. A time division multiplex of signals is clearly attractive here from a hardware standpoint, in that it is unnecessary to couple individual channels through a complex frequency-selective multicoupler as is required in FDM systems. Although delays and Doppler shift can differ among the community of mobiles served by the base, the individual mobiles can still extract wanted message traffic without problems of synchronization. As far as mobiles are concerned, the problem of co-channel interference is very important, so that a wideband spread spectrum signal from base will give the mobiles the optimum opportunity to differentiate wanted from interfering base station transmissions. A further level of multiplexing is achieved by transmitting in phase-quadrature so that the final wideband transmission has a complex structure as shown in Figure 9.4. The result is a multiplex of 64 traffic channels on a single carrier supporting 1.248 Mchips/s. The system is based on 16 kbit/s speech, and there are 4 channels time-multiplexed per code layer. Eight of these code layers are superimposed after the code spreading to form a CDM signal. This carries 32 speech channels; two of these signals are modulated in quadrature, which results in a 64-channel RF signal.

Each of the 16 CDM channels (8 in each plane of phase) has a total capacity of 78.0 kbits/s. By using an alphabet of four data symbols in every CDM channel, where two bits are coded into one symbol, the symbol rate is 39 ksymbols/s. The four symbols are represented by two different antipodal spreading codes; with spreading of the symbol rate of 39 ksymbols/s by 32, the chip rate of 1.248 Mchips/s results.

The uplink, by contrast, presents very different problems. To use TDM on the uplink, a mobile must be able to deliver signal bursts at the correct time to the base station. Since mobiles are all at different distances from the base, this means they must have means for advancing or retarding their messages by the correct amount. In addition, whereas mobiles are subject to unfavourable signal-to-interference situations, base stations are often better placed and are less in need of the spread spectrum advantages offered to mobiles. Transmit power considerations are also important, since burst mode TDMA requires a high ratio of peak-to-average transmit power from the mobiles and, for hand-portables in particular, high peak transmit powers are difficult to generate consistent with conserving battery power. MATS-D, therefore, uses single channel per carrier demand-assigned access on the uplink.

9.4.2 A wideband TDMA system—CD900

This system develops a different strategy from MATS-D—a basic premise is taken that full advantage from the wideband approach can be obtained only with a wider transmission bandwidth than the 1.5 MHz or so of MATS-D, and by preserving a symmetrical configuration on up- and downlinks. The wide bandwidth occupancy in CD900 is achieved by targeting a high bit-rate TDMA scheme with a large number of channels (time slots per frame) and supplementation by spectrum spreading. To achieve high user capacity, a concentrated frequency re-use strategy has been devised, and this is illustrated

code	CDM	:	8
time	TDM	:	4
phase	quadrature	:	2

8 * 4 * 2 = 64 channels per carrier

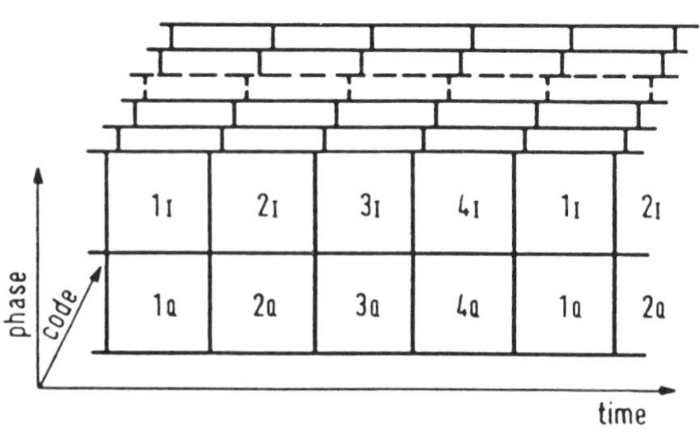

Figure 9.4 MATS-D wideband transmission multipex.

in Figure 9.5. Each base station illuminates three cells, but sequentially in time; with the TDMA structure shown, 20 channels are allocated to each cell on each carrier. The allocation of 45 MHz in the 900 MHz band would then be occupied with carriers spaced at 6-MHz intervals.

The attractions of concentrated frequency re-use (effectively based on a 3-cell cluster) and the capability of the system to use a high chip rate of 3.99 Mchips/s gross, made this scheme a strong contender for pan-European acceptance. However, the technological problems of implementing such an extremely sophisticated system presented the planners with the prospect of a long delay to introduction of a pan-European system, and it was decided to accept a less ambitious approach.

9.4.3 The GSM system—narrowband TDMA

Groupe Spéciale Mobile (GSM) of CEPT were clearly faced with a difficult choice, perceiving features of merit in all the systems offered. The advantages of TDMA as far as base station design is concerned are evident in permitting many channels to be supported by a single base station transceiver, but the number of channels per carrier, and consequently the bit rate, should not be too high to create difficulties for mobiles in having time to scan a number of base station transmissions to determine hand-off requirements. Additionally, the number of channels per carrier should not be too high if peak power output requirements from hand-portables are to be met and the equalizer complexity/performance compromise within reach of short- to medium-term DSP

Table 9.1 Parameters for radio interface chosen by GSM

TDMA 8 channels/carrier
270 kbit/s per carrier
GMSK modulation 200-kHz carrier spacing
Frequency hopping
Adaptive equalization
13 kbit/s speech coding using RPE/LPC
Forward error correction using multi-rate punctured convoluted coding and giving overall channel rate of 22.8 kbit/s

developments is to be maintained. The final choice was for a symmetrical system with the parameters set out in Table 9.1.

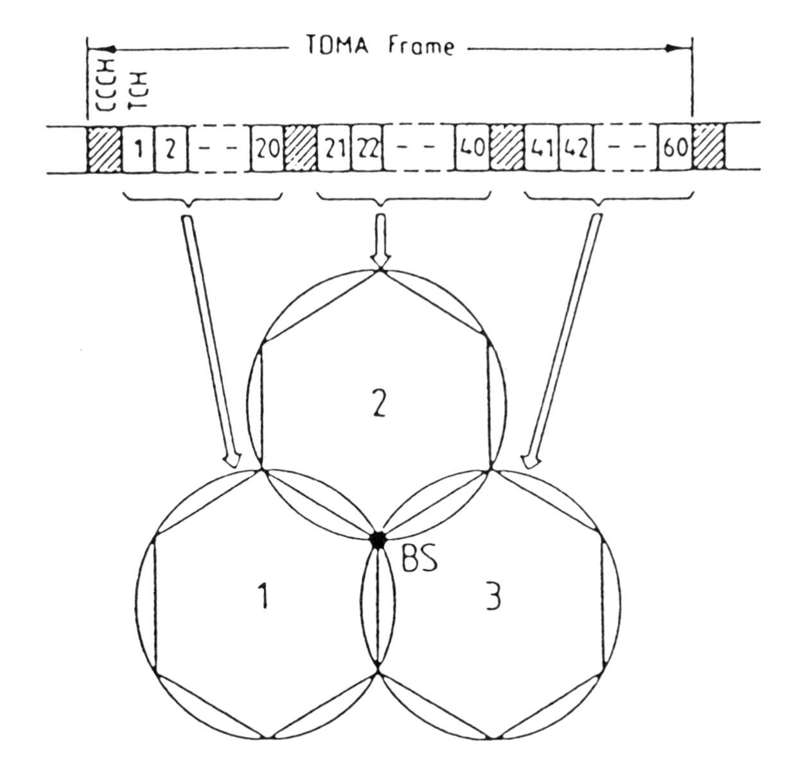

Figure 9.5 CD 100 frequency re-use strategy.

9.5 Postscript

The brief outlines above of digital cellular radio developments stimulated by the prospects for a pan-European system can do little more than set the scene for the future. The gulf in technological terms between the traditional stand-alone private mobile radio schemes using AM or FM and the GSM cellular scheme scheduled to enter service in 1991 is immense by any standards. But already planning is under way for generations of systems beyond GSM.

The introduction of cellular systems in Europe, North America and Japan has brought awareness of the possibilities of radio-connected communications terminals not only to the closed communities of specialist users served by PMR, but also to the public at large. The public expectation is now boundless, and the pressure to create the 'personal communications' environment in which every adult carries a universal personal terminal to meet the bulk of his or her communications needs is already being felt in the international telecommunications planning community. Progress is certain to be rapid and sustained, with technologies currently discarded as too ambitious being regularly re-appraised and developed. The challenges and the opportunities for mobile radio engineers have never been greater.

References

1. Lee, W.C.Y. PacTel mobile companies, comparing FM and SSB, *Communications* (November 1985) 98–111.
2. Prabhu, V.K. and Steele, R. Frequency-Hopped single-sideband modulation for mobile radio. *Bell Syst. Tech. J.* **61** (7) (September 1982) 1389–1411.
3. Maloberti, A. Choice criteria for the radio-subsystem of a digital mobile system in the 900 MHz band. *2nd Nordic Seminar on Digital Land Mobile Radio Communications* (Stockholm 14–16 October 1986).
4. Cooper, G.R. and McGillen, C.D. *Modern Communications and Spread Spectrum*, McGraw-Hill (1986).
5. Ziemer, R.E. and Peterson, R.L. *Digital Communications and Spread Spectrum Systems*, Macmillan Publishing Company, New York (1985).

Index